WELDING

PRINCIPLES AND APPLICATIONS

NINTH EDITION

Study Guide/Lab Manual

Larry Jeffus

Australia • Brazil • Mexico • Singapore • United Kingdom • United States

Study Guide/Lab Manual to accompany Welding: Principles and Applications, 9th Edition

Larry Jeffus

SVP, Higher Education & Skills Product: Erin Joyner

VP, Product Management: Mike Schenk

Product Director: Mathew Seeley

Senior Product Manager: Katie McGuire

Product Assistant: Kimberly Klotz

Director, Learning Design: Rebecca von Gillern

Senior Manager, Learning Design: Leigh Hefferon

Senior Learning Designer: Mary Clyne

Marketing Director: Sean Chamberland

Marketing Manager: Andrew Ouimet

Director, Content Creation: Juliet Steiner

Manager, Content Creation: Alexis Ferraro

Senior Content Manager: Sharon Chambliss

Digital Delivery Lead: Amanda Ryan

Production Service/Composition: SPi Global

Art Director: Jack Pendleton

Cover Designer: Erin Griffin

Cover image(s): Hypertherm, Inc.

For product information and technology assistance, contact us at **Cengage Customer & Sales Support, 1-800-354-9706** or **support.cengage.com**.

For permission to use material from this text or product, submit all requests online at **www.cengage.com/permissions**.

Library of Congress Control Number: 2020906223

Book Only ISBN: 978-0-357-37769-7

Cengage

200 Pier 4 Boulevard

Boston, MA 02210

USA

Cengage is a leading provider of customized learning solutions with employees residing in nearly 40 different countries and sales in more than 125 countries around the world. Find your local representative at **www.cengage.com**.

Cengage products are represented in Canada by Nelson Education, Ltd.

To learn more about Cengage platforms and services, register or access your online learning solution, or purchase materials for your course, visit **www. cengage.com**.

Notice to the Reader

TABLE OF CONTENTS

Chapter 1   Introduction to Welding   1

Chapter 2   Safety in Welding   15

Chapter 3   Shielded Metal Arc Equipment, Setup, and Operation   27

Chapter 4   Shielded Metal Arc Welding of Plate   37

Chapter 5   Shielded Metal Arc Welding of Pipe   73

Chapter 6   Shielded Metal Arc Welding AWS SENSE Certification   101

Chapter 7   Flame Cutting   127

Chapter 8   Plasma Arc Cutting   163

Chapter 9   Related Cutting Processes   183

Chapter 10 Gas Metal Arc Welding Equipment, Setup, and Operation   201

Chapter 11 Gas Metal Arc Welding   211

Chapter 12 Flux Cored Arc Welding Equipment, Setup, and Operation   261

Chapter 13 Flux Cored Arc Welding   269

Chapter 14 Gas Metal Arc and Flux Cored Arc Welding of Pipe   315

Chapter 15 Gas Metal Arc and Flux Cored Arc Welding AWS SENSE Certification   333

Chapter 16 Gas Tungsten Arc Welding Equipment, Setup, Operation, and Filler Metals   375

Chapter 17 Gas Tungsten Arc Welding of Plate   389

Chapter 18 Gas Tungsten Arc Welding of Pipe   429

Chapter 19 Gas Tungsten Arc Welding AWS SENSE Plate and Pipe Certification   449

Chapter 20 Shop Math and Weld Cost   487

Chapter 21 Reading Technical Drawings   503

Chapter 22 Welding Joint Design and Welding Symbols   515

Chapter 23 Fabricating Techniques and Practices   529

Chapter 24 Welding Codes and Standards   551

Chapter 25 Testing and Inspection   561

Chapter 26 Welding Metallurgy   571

Chapter 27 Weldability of Metals   593

Chapter 28 Filler Metal Selection   605

Chapter 29 Welding Automation and Robotics   615

Chapter 30 Other Welding Processes   621

Chapter 31 Oxyacetylene Welding   629

Chapter 32 Brazing, Braze Welding, and Soldering   685

INTRODUCTION

The need for skilled welders has grown rapidly over the past few years and is expected to continue growing as many of the current welders retire. Individuals that master the required welding skills and technical knowledge place themselves in an ideal position to get a high paying welding job. There are many types of welding jobs, for example: you might want to work in an assembly or fabrication shop where you go to work at the same place every day. You might want to work on construction jobs where you can have the opportunity to travel locally or nationally from one job site to another. Once you have developed the skills, you could open your own welding shop. There is a wide range of opportunities for welders.

Mastering welding and cutting skills requires that you dedicate the time to practice. Welding is much like other skilled activities such as athletics or playing a musical instrument—the more you practice the better you become. Practicing welding helps you build muscle memory, which allows you to manipulate the torch, electrode, or gun automatically so that every weld you make is consistent. The welding and cutting lab practices in *Welding: Principles and Applications*, Ninth Edition are organized so that you will be building more advanced skills as you progress through the lab activities.

In addition to developing your welding and cutting skills, you must comprehend the science and technology of welding. Not only do you need to be able to make a weld or cut, you must understand what can affect welds or cuts. That knowledge will allow you to make any changes in the machine settings or your technique to ensure a satisfactory weld each time. Also you must be able to work in the lab safely, read mechanical drawings, know how to fabricate and assemble weldments, anticipate the effects that the weld or cut will have on the metal, solve basic math problems as they relate to welding, and explain why one welding process would be preferred over another.

This *Study Guide/Lab Manual* is designed to focus your studies and assist in organizing your learning of welding principles and applications. Once the reading assignments in the text are completed, the instructor will ask you to answer the questions in the appropriate section in this guide. Do not guess the answers to the questions. If you do not know the answer, leave the space blank. You can then identify the topics that you need to review again until you understand the information. Ask your instructor for help if you continue to have difficulty with any of the welding principles. You will be a better welder if you combine welding skills with a firm understanding of the how and why of welding.

Introduction to Welding

■ PRACTICE 1-1

Name _____ Date _____

Class _____ Instructor _____ Grade _____

OBJECTIVE: After completing this practice, you should be able to identify items that are manufactured using various welding processes.

EQUIPMENT AND MATERIALS NEEDED FOR THIS PRACTICE

Paper and pencil.

INSTRUCTIONS

Answer the following questions by looking around your school, home, and community to identify items manufactured with the following welding process.

1. List four items that are manufactured using the oxyfuel welding, brazing, or cutting process.

         Process   Item

   a. _____ , _____

   b. _____ , _____

   c. _____ , _____

   d. _____ , _____

2. List four items that are manufactured using the shielded metal arc welding process.

   a. _____

   b. _____

   c. _____

   d. _____

3. List four items that are manufactured using the gas metal arc welding process.

   a. _____

   b. _____

   c. _____

   d. _____

4. List four items that are manufactured using the flux cored arc welding process.

   a. _____

   b. _____

   c. _____

   d. _____

5. List four items that are manufactured using the gas tungsten arc welding process.

   a. _____

   b. _____

   c. _____

   d. _____

INSTRUCTOR'S COMMENTS _____

_____

_____

■ PRACTICE 1-2

Name _____ Date _____

Class _____ Instructor _____ Grade _____

OBJECTIVE: After completing this practice, you should be able to identify those jobs available in your local newspaper's employment section that require welding skills.

EQUIPMENT AND MATERIALS

Paper and pencil.

INSTRUCTIONS

Using your local newspaper, make a list of the jobs listed in the help wanted section that might require welding skills.

a. _____

b. _____

c. _____

d. _____

e. _____

f. _____

g. _____

h. _____

i. _____

j. _____

INSTRUCTOR'S COMMENTS _____

_____

_____

CHAPTER 1: QUIZ 1

Name _____   Date _____

Class _____   Instructor _____   Grade _____

INSTRUCTIONS
Carefully read Chapter 1 in the textbook and answer each question.

MATCHING
In the space provided to the left of Column A, write the letter from Column B that best answers or completes the statement in Column A.

Column A	Column B
_____ 1. Some of the names used to refer to _____ include gas welding and torch welding.	a. automatic
_____ 2. Shielded metal arc welding (SMAW) is often called _____ welding, rod welding, or just welding.	b. Certified Welding Inspector (CWI)
_____ 3. Welding is defined as "a joining process that produces _____ of materials by heating them to the welding temperature, with or without the application of pressure or by the application of pressure alone, and with or without the use of filler metal."	c. solid electrode wire
_____ 4. Very little in our modern world is not produced using some type of _____ process.	d. type of equipment
_____ 5. Welding processes differ greatly in the manner in which heat, pressure, or both heat and pressure are applied and in the _____ used.	e. continuously fed
_____ 6. In OF welding and TB, a high-temperature flame is produced at the torch tip by burning _____ and a fuel gas.	f. semiautomatic
_____ 7. Shielded metal arc welding (SMAW) uses a _____ stick electrode that conducts the welding current from the electrode holder to the work, and as the arc melts the end of the electrode away, it becomes part of the weld metal.	g. gas
_____ 8. Gas tungsten arc welding (GTAW) uses a _____ electrode made of tungsten.	h. integrity
_____ 9. Gas metal arc welding (GMAW) uses a _____ that is continuously fed from a spool, through the welding cable assembly, and out through the gun.	i. welding
_____ 10. Flux cored arc welding (FCAW) uses a flux core electrode wire that is _____ from a spool, through the welding cable assembly, and out through the gun.	j. manual
_____ 11. There are a number of thermal cutting processes such as oxyfuel cutting (OFC) and _____.	k. stick

_____ 12. Oxyfuel gas cutting uses the high-temperature flame to heat the _____ of a piece of steel to a point where a forceful stream of oxygen flowing out a center hole in the tip causes the hot steel to burn away, leaving a gap or cut.

l. oxyacetylene welding (OAW)

_____ 13. Plasma arc cutting (PAC) uses a stiff, highly ionized, extremely hot column of _____ to almost instantly vaporize the metal being cut.

m. welder fitters

_____ 14. In a _____ operation, the welder is required to manipulate the entire process.

n. oxygen

_____ 15. In a _____ operation, filler metal is added automatically, and all other manipulation is done manually by the welder.

o. surface

_____ 16. In an _____ operation, a machine has been programmed to perform operations repeatedly without interaction of the operator.

p. coalescence

_____ 17. Welder assemblers, or _____, position all the parts in their proper places and make them ready for the tack welders.

q. consumable

_____ 18. Welding inspectors are often required to hold a special certification such as the one supervised by the American Welding Society known as _____.

r. nonconsumable

_____ 19. To become a skilled welder, both _____ and on-the-job experience are required.

s. plasma arc cutting (PAC)

_____ 20. In addition to welding skills, an entry-level welder must possess workplace skills such as teamwork, leadership, _____, honesty, organizational skills, time management, understand the importance of workplace diversity, and the Equal Employment Opportunity law.

t. welding school

CHAPTER 1: QUIZ 2

Name _____   Date _____

Class _____   Instructor _____   Grade _____

INSTRUCTIONS

Carefully read Chapter 1 in the textbook and answer each question.

MATCHING

In the space provided to the left of Column A, write the letter from Column B that best answers or completes the statement in Column A.

Column A

Column B

_____ 1. Beginners in welding who have no training often start in manual welding production jobs that require _____ skill.

a. Scanning

_____ 2. Many more skilled welders will be needed for maintenance and _____ work in the expanding metalworking industries.

b. learn

_____ 3. Before being assigned a job where service requirements of the weld are critical, welders usually must pass a _____ given by an employer.

c. thermal cutting

_____ 4. Job-related personal skills are important because people are hired based on their welding skills, but too often they may be later fired for not having good job-related _____.

d. Welding Procedure Specification

_____ 5. Each individual working on a project has to work efficiently and effectively with the other _____.

e. challenges

_____ 6. An example of a detailed written communication is a _____.

f. welding skill test

_____ 7. Each welding project may have unique _____ that must be overcome in order to complete the task or work, requiring welders to have the ability to adapt their thinking to solve different and unique problems.

g. performance (skills)

_____ 8. _____ is a reading technique where you quickly look over printed material to gain an overview while looking for a specific piece of information.

h. standard and metric

_____ 9. It is important that you be _____ for work and arrive at least 15 or 20 minutes before work begins so that you can be ready to start work promptly.

i. personal skills

_____ 10. Failing to notify the company or accruing excessive absences can result in your _____.

j. practical knowledge

_____ 11. After you graduate or complete your welding program, you need to continue to _____.

k. SkillsUSA

_____ 12. The AWS _____ certifications have gained widespread acceptance by the industry and they allow welders to demonstrate their skills on a standard welding test.

l. repair

_____ 13. The first three modules of the Entry Level Welder SENSE program relate to _____ that is common to all areas of welding and that welders must have to succeed in the welding field.

m. settings

_____ 14. Modules 4 through 7 relate to welding _____ in each of the major welding processes.

n. testing

_____ 15. Module 8, Thermal Cutting Principles and Practices, is divided into four units, with each covering different types of _____.

o. punctual

_____ 16. Module 9 covers two main areas of inspection and _____.

p. minimum

_____ 17. Level II Advanced Welding Qualifications are divided into two sections: Knowledge Subjects, which require students to be tested, and Performance Testing, which requires students to pass a _____.

q. team members

_____ 18. Each year _____ sponsors a series of welding skill competitions for its student members.

r. certification test

_____ 19. Often as you make a weld, it will be necessary for you to make changes in your equipment _____ or your technique to ensure you are making an acceptable weld.

s. SENSE

_____ 20. Many calculators today have built-in _____ conversions; of course, it is a good idea to know how to make these conversions with and without these aids.

t. termination

CHAPTER 1: QUIZ 3

Name _____   Date _____

Class _____   Instructor _____   Grade _____

INSTRUCTIONS

Carefully read Chapter 1 in the textbook and answer each question.

IDENTIFICATION

Identify the numbered items on the drawing by writing the letter next to the identifying term in the space provided.

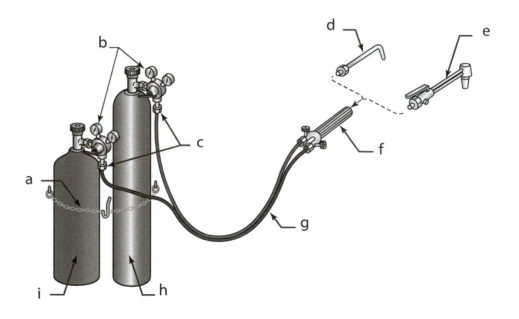

1. _____   GAS HOSES

2. _____   WELDING OR BRAZING TIP

3. _____   CUTTING HEAD

4. _____   REVERSE FLOW CHECK VALVES

5. _____   PRESSURE REGULATORS

6. _____   SAFETY CHAIN

7. _____   OXYGEN CYLINDER

8. _____   FUEL GAS CYLINDER

9. _____   TORCH BODY

CHAPTER 1: QUIZ 4

Name _____   Date _____

Class _____   Instructor _____   Grade _____

INSTRUCTIONS

Carefully read Chapter 1 in the textbook and answer each question.

IDENTIFICATION

Identify the numbered items on the drawing by writing the letter next to the identifying term in the space provided.

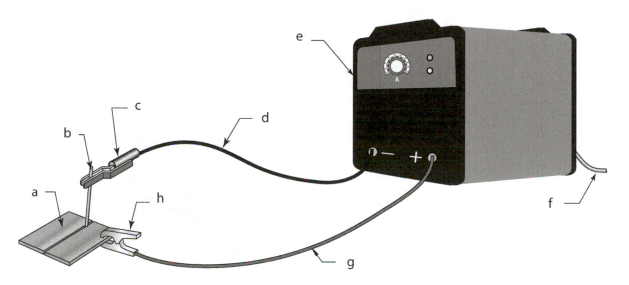

1. _____   ELECTRODE CABLE

2. _____   ELECTRODE

3. _____   WELDING MACHINE

4. _____   ELECTRODE HOLDER

5. _____   WORK

6. _____   WORK CLAMP

7. _____   MAIN POWER SUPPLY CABLE

8. _____   WORK CABLE

CHAPTER 1: QUIZ 5

Name _____   Date _____

Class _____   Instructor _____   Grade _____

INSTRUCTIONS

Carefully read Chapter 1 in the textbook and answer each question.

IDENTIFICATION

Identify the numbered items on the drawing by writing the letter next to the identifying term in the space provided.

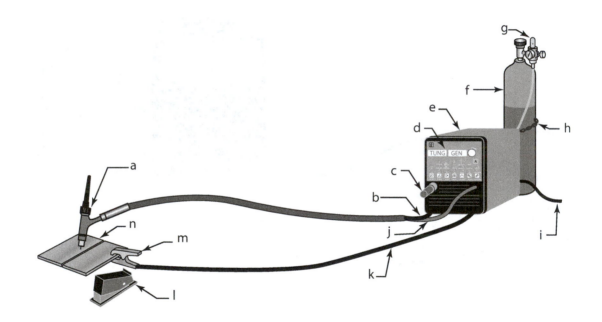

1. _____ WORK

2. _____ COMBINATION REGULATOR AND FLOWMETER

3. _____ WELDING MACHINE

4. _____ DIGITAL CONTROL PANEL AND DISPLAY

5. _____ WORK CLAMP

6. _____ MAIN POWER SUPPLY CABLE

7. _____ SAFETY CHAIN

8. _____ WIRELESS FOOT CONTROL

9. _____ WELDING POWER LEAD

10. _____ SHIELDING GAS TO TORCH

11. _____ GTA WELDING TORCH

12. _____ SHIELDING GAS CYLINDER

13. _____ WIRELESS RECEIVER

14. _____ WORK LEAD

CHAPTER 1: QUIZ 6

Name _____ Date _____

Class _____ Instructor _____ Grade _____

INSTRUCTIONS

Carefully read Chapter 1 in the textbook and answer each question.

IDENTIFICATION

Identify the numbered items on the drawing by writing the letter next to the identifying term in the space provided.

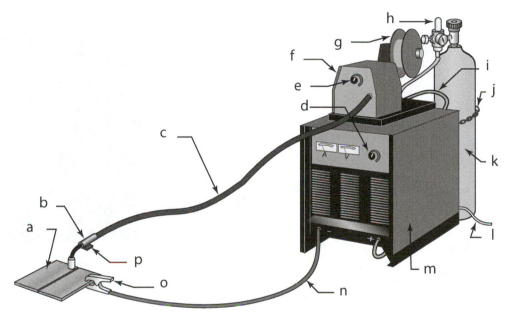

1. _____ WORK

2. _____ WIRE SPOOL

3. _____ WORK LEAD

4. _____ WELDING GUN

5. _____ GUN START/STOP TRIGGER

6. _____ MAIN POWER SUPPLY CABLE

7. _____ COMBINATION REGULATOR AND FLOWMETER

8. _____ WIRE SPEED ADJUSTMENT

9. _____ WELDING CABLE ASSEMBLY

10. _____ WELDING MACHINE CONTACTOR CONNECTION

11. _____ CYLINDER SAFETY CHAIN

12. _____ WELDING POWER LEAD

13. _____ WIRE FEED AND CONTROL UNIT

14. _____ WELDING VOLTAGE ADJUSTMENT

15. _____ SHIELDING GAS CYLINDER

16. _____ WORK CLAMP

CHAPTER 1: QUIZ 7

Name _____ Date _____

Class _____ Instructor _____ Grade _____

INSTRUCTIONS

Carefully read Chapter 1 in the textbook and answer each question.

IDENTIFICATION

Identify the numbered items on the drawing by writing the letter next to the identifying term in the space provided.

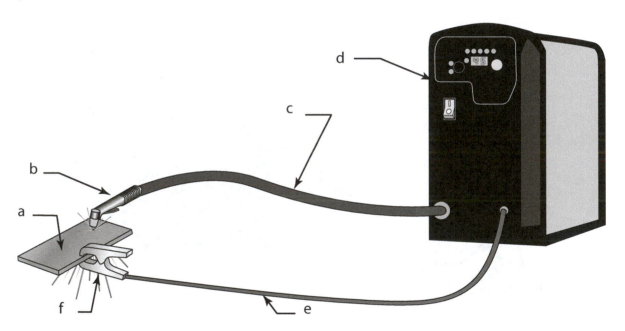

1. _____ POWER SUPPLY

2. _____ POWER & COMPRESSED AIR CABLE

3. _____ WORK CABLE

4. _____ WORK

5. _____ WORK CLAMP

6. _____ PLASMA ARC CUTTING TORCH

CHAPTER 1: QUIZ 8

Name _____ Date _____

Class _____ Instructor _____ Grade _____

INSTRUCTIONS
Carefully read Chapter 1 in the textbook and answer each question.

IDENTIFICATION
Identify the numbered items on the drawing by writing the letter next to the identifying term in the space provided.

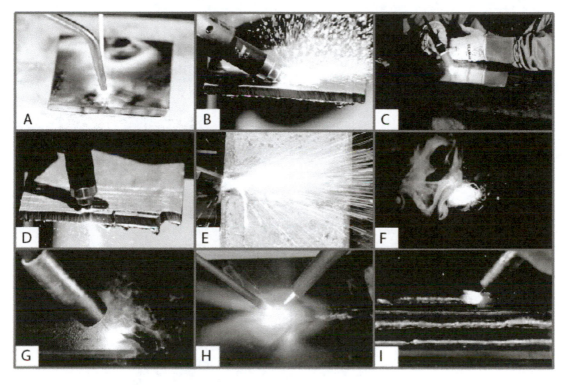

1. _____ Which image shows oxyfuel welding (OFC)?

2. _____ Which image shows plasma cutting (PAC)?

3. _____ Which image shows gas metal arc welding (GMAW)?

4. _____ Which image shows torch brazing (TB)?

5. _____ Which image shows shielded metal arc welding (SMAW)?

6. _____ Which image shows oxyfuel gouging (OFG)?

7. _____ Which image shows flux core arc welding (FCAW)?

8. _____ Which image shows gas tungsten arc welding (GTAW)?

9. _____ Which image shows plasma arc gouging (PAG)?

CHAPTER 2

Safety in Welding

1. A welder will only weld with proper eye protection: One of the primary concerns in welding is the protection of the eyes.

2. Welders will always wear protective clothing and other personal protective equipment (PPE): Protective clothing and PPE for your personal protection are required for all welding.

3. Welders may only weld with proper ventilation: Welding fumes can be hazardous to the welding operator. Fumes will vary depending on the type of welding or cutting operation along with the type of filler metal, fluxes, coating, and the base metal being welded on.

4. Welders must never weld on any type of vessel or tank that has contained any type of flammable material: Welding or cutting should not be performed on drums, barrels, tanks, or other containers until they have been cleaned thoroughly, eliminating all flammable materials and all substances (such as detergents, solvents, greases, tars, or acids), which might produce flammable, toxic, or explosive vapors when heated.

5. Welders are not to lubricate pressure regulators: Do not use oil, grease, or any type of hydrocarbon product on any torch, regulator, or cylinder fitting. Oil and grease in the presence of oxygen may burn with explosive force.

6. Welders must always use acetylene cylinders in the vertical (upright) position: Acetylene cylinders that have been lying on their side must stand upright for four hours or more before they can be used safely.

7. Welders must always use less than 15 psig acetylene working pressure: Acetylene becomes unstable at pressures higher than 30 psig and could explode. For safety, the red line on the acetylene working pressure gauge is at 15 psig, which is halfway to the point of explosion.

8. Welders will always demonstrate safe handling and storing of cylinders: Cylinders must be secured with a chain or other device so that they cannot be knocked over accidentally when in use, being moved, or stored.

9. Welders are to grind only iron, steel, or stainless steel on a grinding stone designed for ferrous material. The stone will become glazed (the surface clogs with metal) and may explode due to frictional heat buildup if nonferrous metals are ground.

10. Welders are always to keep a clean workstation with tools cleaned and properly stored: Always keep a safe clean work area and properly cleaned and stored tools when they are not in use.

11. Welders must always protect others from the welding arc's light: Welding curtains must always be used to protect other workers in an area that might be exposed to the welding light.

12. Welders must protect their hearing: Damage to your hearing caused by high sound levels may not be detected until later in life, and the resulting loss in hearing is nonrecoverable. It will not get better with time. Each time you are exposed to high levels of sound your hearing will become worse.

13. Welders must always guard against problems caused by welding on used metal: Extreme caution must be taken to avoid the fumes produced when welding is done on dirty, painted, plated, or used metal. Any chemicals that might be on the metal will become mixed with the welding fumes, and this combination can be extremely hazardous. All metal must be cleaned and ground to bare metal before welding to avoid this potential problem.

14. Welders must avoid electrical shock: Welding cables must never be spliced within 10 ft (3 m) of the electrode holder. All extension cords and equipment power cords must be in good repair. Never use equipment not properly grounded.

15. Welders must not carry flammable items in their pockets: There is no safe place to carry butane lighters and matches while welding or cutting. They may catch fire or explode if they are subjected to welding heat or sparks. Butane lighters may explode with the force of 1/4 stick of dynamite. Matches can erupt into a ball of fire. Both butane lighters and matches must always be removed from the welder's pockets and placed a safe distance away before any work is started.

16. Welders must always be well informed about safe and proper equipment usage and maintenance: Before operating any power equipment for the first time, you must read the manufacturer's safety and operating instructions and should be assisted by your welding instructor or by your supervisor or someone who has experience with the equipment. After receiving a safe operating demonstration on any new equipment you will be using in the welding shop, you must demonstrate the safe assembly, use, shutdown, service, and disassembly of the equipment to your instructor or supervisor.

17. Welders must always keep their hands clear before any equipment is started. Always "Lock-Out and Tag-Out" power before working on any part of the equipment.

18. Welders must remove or secure any long hair, rings, watches, or other jewelry so that it does not pose a safety hazard to them in the welding shop.

19. Welders must never engage in unsafe activities: Acts such as horseplay or distractions such as texting create safety problems.

20. Welders must know where safety equipment is located and how it is used: In an emergency, time is important so knowing where items such as fire extinguishers, eye wash stations, first aid kits, emergency exits, safety data sheets, etc., are located will save time.

21. Welders must observe safety markings: There are areas in welding shops that have safety markings and signs such as "No Storage Within __ Feet." Restricted areas include space in front of electrical panels, emergency exits, fire extinguishers, eye wash stations, walkways, or other similarly restricted areas.

22. Welders must be cautious when welding in a confined space: Confined spaces are defined as relatively small or restricted areas such as storage bins, pits, manholes, tunnels, etc. In addition confined spaces have restricted entry or exits and may have limited ventilation. Someone on the outside of the confined space must be assigned to maintain contact with anyone working inside and be available to aid them in an emergency.

23. Welders must always be alert: You must be watchful and report potential safety hazards that might occur in the shop. The protection of welders from hazards is a major industry concern. The safety information in these rules, Chapter 2 "Safety in Welding" in Welding Principles and Applications, and all of the additional safety instructions given by your instructor, because of the diversity of the welding industry are by no means all-inclusive, on the topic of safe practices. You must keep yourself updated with the latest regulations, codes, standards, manufacturers' guidelines, and safety data sheets for current information. I have read and understand all of the information and promise to abide by all safety rules.

Name:_____ Class:_____Instructor:_____Date:_____

■ PRACTICE 2-1

Name _____ Date _____

Class _____ Instructor _____ Grade _____

OBJECTIVES: After completing this practice, you should be able to explain:

- Why welding curtains are used in a welding environment
- What can happen to your hearing, short-term and long-term, if proper hearing protection is not used
- Why caution must be taken to avoid the fumes produced when welding is done on dirty, painted, plated, or used metal
- Why professional welders NEVER carry butane or propane lighters or matches on their person when welding or cutting
- When and why it is important to read and understand the manufacturer's safety and operating instructions on any piece of equipment

EQUIPMENT AND MATERIALS NEEDED FOR THIS PRACTICE

Paper and pencil.

INSTRUCTIONS

Look carefully at Chapter 2, "Safety in Welding." Reread all the figure captions, shaded caution areas, and tables, then answer the following questions in short answer form.

1. Why are welding curtains used in a welding environment?

   _____

   _____

   _____

2. What can happen if you do not use proper hearing protection?

   _____

   _____

   _____

   _____

3. What is the one way to avoid potentially hazardous fumes when welding on metal that has been painted or has any grease, oil, or chemicals on its surface?

   _____

   _____

   _____

   _____

   _____

4. Why do professional welders never carry butane or propane lighters or matches on their person?

_____

_____

_____

_____

_____

5. When is it important to read and understand the manufacturer's safety and operating instructions regarding a piece of equipment?

_____

_____

_____

_____

_____

INSTRUCTOR'S COMMENTS _____

_____

_____

CHAPTER 2: QUIZ 1

Name _____ Date _____

Class _____ Instructor _____ Grade _____

INSTRUCTIONS

Carefully read Chapter 2 in the textbook and answer each question.

MATCHING

In the space provided to the left of Column A, write the letter from Column B that best answers or completes the statement in Column A.

Column A	Column B
_____ 1. Welding is a very large and _____ industry.	a. face
_____ 2. _____ can be caused by ultraviolet (UV) light rays as well as by contact with hot welding material.	b. Ultraviolet and infrared
_____ 3. _____ occur when the surface of the skin is reddish in color, tender, and painful and there is no involvement of any broken skin.	c. general work clothing
_____ 4. _____ occur when the surface of the skin is severely damaged.	d. First-degree burns
_____ 5. _____ occur when the surface of the skin and possibly the tissue below the skin appear white or charred.	e. Type B
_____ 6. _____ are types of light that can cause burns.	f. full face shield
_____ 7. Personal protection equipment is commonly referred to as _____.	g. Second-degree burns
_____ 8. It is important to choose _____ that will minimize the possibility of getting burned because of the high temperature and amount of hot sparks, metal, and slag produced during welding, cutting, or brazing.	h. respirators
_____ 9. _____ is often the best material to use because it is lightweight, flexible, resists burning, and is readily available.	i. safety data sheets (SDS)
_____ 10. _____ protection must be worn in the shop at all times.	j. PPE
_____ 11. For heavy grinding, chipping, or overhead work, a _____ should be worn in addition to safety glasses.	k. separately
_____ 12. The sound level in the welding environment is at times high enough to cause pain and some _____ if the welder's ears are unprotected.	l. diverse

_____ 13. When welders must work in an area where effective general controls to remove airborne welding by-products are not feasible, _____ shall be provided by their employers when this equipment is necessary to protect their health.

m. planned maintenance (PM)

_____ 14. Any system of ventilation should draw the fumes or smoke away before it rises past the level of the welder's _____.

n. hot work permit

_____ 15. All manufacturers of potentially hazardous materials must provide _____ to the users of their products.

o. Third-degree burns

_____ 16. Oxygen and fuel gas cylinders or other flammable materials must be stored _____.

p. four

_____ 17. Acetylene cylinders that have been lying on their sides must stand upright for _____ hours or more before they are used.

q. loss of hearing

_____ 18. When performing welding outside of a shop, the welder may be required to obtain a _____ from the local fire marshal.

r. burns

_____ 19. _____ extinguishers are used for combustible liquids, such as oil, gas, and paint thinner.

s. eye

_____ 20. A routine schedule for _____ of equipment will aid in detecting potential problems such as leaking coolant, loose wires, poor grounds, frayed insulation, or split hoses.

t. Leather

CHAPTER 2: QUIZ 2

Name _____   Date _____

Class _____   Instructor _____   Grade _____

INSTRUCTIONS

Carefully read Chapter 2 in the textbook and answer each question.

MATCHING

In the space provided to the left of Column A, write the letter from Column B that best answers or completes the statement in Column A.

Column A	Column B
_____ 1. Hand tools are used by the welder to do necessary assembly and disassembly of parts for welding as well as to perform routine _____.	a. outlet
_____ 2. A hammer blow should always be struck _____, with the hammer face parallel to the surface being struck.	b. double insulation
_____ 3. Electrical resistance is lowered in the presence of _____, so welders must take special precautions when working under damp or wet conditions, including perspiration.	c. spark
_____ 4. The workpiece being welded and the frame or chassis of all electrically powered machines must be connected to a good _____.	d. cracks
_____ 5. For protection from electrical shock, the standard portable power tool is built with either of two equally safe systems: external grounding or _____.	e. legs
_____ 6. Before connecting a tool to a power supply, be sure the _____ supplied is the same as that specified on the nameplate of the tool.	f. electrically non-conductive
_____ 7. Always connect the cord of a portable electric power tool into the extension cord before the extension cord is connected to the _____.	g. equipment maintenance
_____ 8. Extension cords should be checked frequently while in use to detect unusual _____.	h. each time
_____ 9. Because electric tools _____, portable electric tools should never be started or operated in the presence of propane, natural gas, gasoline, paint thinner, acetylene, or other flammable vapors that could cause a fire or explosion.	i. heating
_____ 10. Welders use a wide variety of _____ to help them produce welded products.	j. voltage
_____ 11. Before a grinding stone is put on the machine, it should be tested for _____.	k. band saws

_____ 12. Before starting to drill, secure the workpiece as necessary and fasten it in a _____.

_____ 13. Many types of mechanical metal cutting machines are used in the welding shop—for example, shears, punches, cut-off machines, and _____.

_____ 14. When you are lifting a heavy object, the weight of the object should be distributed evenly between both hands, and you should use your _____, not your back, to lift.

_____ 15. Keep any load as close to the ground as possible while a _____ is moving it.

_____ 16. Improper _____ of ladders is often a factor in falls.

_____ 17. An advantage of a fiberglass ladder is that it is _____.

_____ 18. Over time, ladders can become worn or damaged and should be inspected _____ they are used.

_____ 19. Stepladders must be _____ in the full opened position with the spreaders.

_____ 20. Straight or extension ladders must be used at the proper _____.

l. use

m. locked

n. hoist or crane

o. squarely

p. angle

q. water or moisture

r. vise or clamp

s. electrical ground

t. power tools

CHAPTER 2: QUIZ 3

Name _____ Date _____

Class _____ Instructor _____ Grade _____

INSTRUCTIONS

Carefully read Chapter 2 in the textbook and answer each question.

IDENTIFICATION

Identify the numbered items on the drawing by writing the letter next to the identifying term in the space provided.

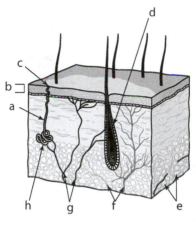
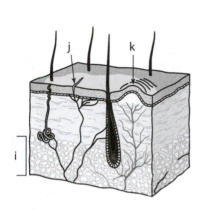
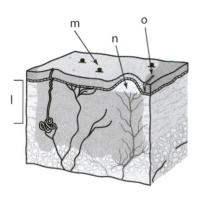

1. _____ BLISTER

2. _____ NERVES

3. _____ DUCT

4. _____ HAIR BURNED OFF

5. _____ BREAK

6. _____ SERUM

7. _____ SWEAT GLAND

8. _____ FAT CELLS

9. _____ EPIDERMIS

10. _____ SWEAT PORE

11. _____ DERMIS

12. _____ HAIR FOLLICLE

13. _____ SUBCUTANEOUS LAYER

14. _____ CRACK IN THE SKIN

15. _____ BLOOD VESSELS

CHAPTER 2: QUIZ 4

Name _____   Date _____

Class _____   Instructor _____   Grade _____

INSTRUCTIONS

Carefully read Chapter 2 in the textbook and answer each question.

IDENTIFICATION

Identify the numbered items on the drawing by writing the letter next to the identifying term in the space provided.

1. _____   SAFETY GLASSES WITH SIDE SHIELDS

2. _____   WELDER CAP

3. _____   100% COTTON SHIRT

4. _____   LEATHER OR FLAME-RETARDANT COTTON WELDING JACKET

5. _____   EARPLUGS

6. _____   WELDING HELMET

7. _____   100% COTTON PANTS

8. _____   LEATHER STEEL-TOED SHOES

9. _____   GAUNTLET WELDING GLOVES

CHAPTER 2: QUIZ 5

Name _____   Date _____

Class _____   Instructor _____   Grade _____

INSTRUCTIONS

Carefully read Chapter 2 in the textbook and answer each question.

IDENTIFICATION

Identify the numbered items on the drawing by writing the letter next to the identifying term in the space provided.

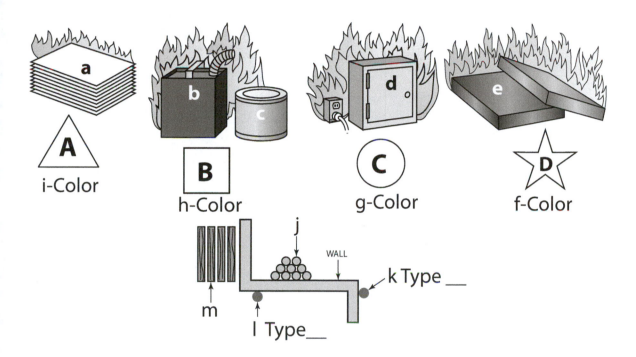

1. _____ OIL	10. _____ PAPER
2. _____ YELLOW	11. _____ GREEN
3. _____ MAGNESIUM & ZINC	12. _____ FUSE BOX
4. _____ WOOD	13. _____ A
5. _____ B	
6. _____ PAINT	
7. _____ RED	
8. _____ GAS	
9. _____ BLUE	

Shielded Metal Arc Equipment, Setup, and Operation

■ PRACTICE 3-1

Name _____ Date _____

Class _____ Instructor _____ Grade _____

OBJECTIVE: After completing this practice, you will be able to accurately calculate the amperage when the knob is at various settings.

EQUIPMENT AND MATERIALS NEEDED FOR THIS PRACTICE

1. Paper and pencil.

2. Amperage ranges as given for this practice (or from a machine in your shop).

INSTRUCTIONS

Estimate the amperage when the knob is at the 1/4, 1/2, and 3/4 settings.

INSTRUCTOR'S COMMENTS _____
_____
_____

■ PRACTICE 3-2

Name _____   Date _____

Class _____   Instructor _____   Grade _____

OBJECTIVES: After completing this practice, you will be able to accurately calculate the amperages when the knob is set at 1, 4, 7, 9, 2.3, 5.7, and 8.5.

EQUIPMENT AND MATERIALS NEEDED FOR THIS PRACTICE

1. Paper and pencil or a calculator.

2. Amperage ranges as given for this practice (or from a machine in your shop).

INSTRUCTIONS

Calculate the amperages for each of the following knob settings: 1, 4, 7, 9, 2.3, 5.7, and 8.5.

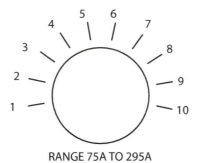

RANGE 75A TO 295A

INSTRUCTOR'S COMMENTS _____

_____

_____

■ PRACTICE 3-3

Name _____ Date _____

Class _____ Instructor _____ Grade _____

OBJECTIVE: After completing this practice, you should be able to read a duty cycle chart.

EQUIPMENT AND MATERIALS NEEDED FOR THIS PRACTICE

1. Paper and pencil.

2. The duty cycle chart shown in the textbook as Figure 3-37 on page 70.

INSTRUCTIONS

1. Determine the maximum welding amperage for Welder 1.

2. Determine the percent duty cycle at the maximum amperage for Welder 1.

3. Determine the maximum welding amperage at 100% duty cycle for Welder 1.

4. Determine the maximum welding amperage for Welder 2.

5. Determine the percent duty cycle at the maximum amperage for Welder 2.

6. Determine the maximum welding amperage at 100% duty cycle for Welder 2.

7. Determine the maximum welding amperage for Welder 3.

8. Determine the percent duty cycle at the maximum amperage for Welder 3.

9. Determine the maximum welding amperage at 100% duty cycle for Welder 3.

INSTRUCTOR'S COMMENTS _____

_____

_____

■ PRACTICE 3-4

Name _____ Date _____

Class _____ Instructor _____ Grade _____

OBJECTIVE: After completing this practice, you should be able to determine welding lead sizes.

EQUIPMENT AND MATERIALS NEEDED FOR THIS PRACTICE

1. Paper and pencil.

2. Table 3-6, shown on textbook page 71.

INSTRUCTIONS

1. Determine the minimum copper welding lead size for a 200 amp welder with 100 ft (30 m) leads.

2. Determine the minimum copper welding lead size for a 125 amp welder with 225 ft (69 m) leads.

3. Determine the maximum length aluminum welding lead that can carry 300 amps.

INSTRUCTOR'S COMMENTS _____

_____

_____

■ PRACTICE 3-5

Name _____ Date _____

Class _____ Instructor _____ Grade _____

OBJECTIVE: After completing this practice, you should be able to properly repair electrode holders.

EQUIPMENT AND MATERIALS NEEDED FOR THIS PRACTICE

1. Manufacturer's instructions for your type of electrode holder.

2. Replacement parts for your type of electrode holder.

3. A set of hand tools.

INSTRUCTIONS

Before starting any work, make sure that the power to the welder is off and locked off or the welding lead has been removed from the machine.

1. Remove the electrode holder from the welding cable.

2. Remove the jaw insulating covers.

3. Replace the jaw insulating covers with new ones.

4. Reconnect the electrode holder to the welding cable.

5. Reconnect the welding cable to the welder (if it is not already connected) and turn on the welding power.

6. Make a weld to ensure that the repair was made correctly.

INSTRUCTOR'S COMMENTS _____

_____

_____

CHAPTER 3: QUIZ 1

Name _____   Date _____

Class _____   Instructor _____   Grade _____

INSTRUCTIONS

Carefully read Chapter 3 in the textbook and answer each question.

MATCHING

In the space provided to the left of Column A, write the letter from Column B that best answers or completes the statement in Column A.

Column A	Column B
_____ 1. Shielded metal arc welding is abbreviated as _____.	a. electric current
_____ 2. SMAW can be used to weld and _____ almost any metal or alloy.	b. Amperage
_____ 3. Metal as thin as _____, approximately 1/16 in. (2 mm) thick can be SMA welded.	c. heat
_____ 4. The source of heat for arc welding is an _____.	d. CV
_____ 5. _____ is the measurement of electrical pressure.	e. Closed circuit
_____ 6. _____ is the measurement of the total number of electrons flowing.	f. Wattage
_____ 7. _____ is a measurement of the amount of electrical energy or power in the arc.	g. 16 gauge
_____ 8. The term _____ refers to the degree or level of thermal energy in a material.	h. Arc blow
_____ 9. The term _____ refers to the quantity of thermal energy in a material.	i. positive
_____ 10. An arc's temperature is dependent on the voltage, arc length, and the _____.	j. welding arc
_____ 11. The higher the amperage setting, the higher the heat produced by the _____.	k. decreases
_____ 12. In DCEN, the electrode is _____.	l. negative
_____ 13. In DCEP, the electrode is _____.	m. Open circuit
_____ 14. The rapid reversal of AC causes the welding heat to be _____ distributed on both the work and the electrode.	n. Voltage

——— 15. On _____ type welding machines, the arc voltage remains   o. hardface
constant at the selected setting even if the arc length and
amperage increase or decrease.

——— 16. On _____ type welding machines, the total welding current   p. CC
(watts) remains the same.

——— 17. The shielded metal arc welding machine's voltage output   q. SMAW
_____ as current increases.

——— 18. _____ voltage is the voltage at the electrode before striking   r. atmosphere
an arc.

——— 19. _____ voltage is the voltage at the arc during welding.   s. evenly

——— 20. _____ makes the arc drift like a string would drift in the   t. temperature
wind.

CHAPTER 3: QUIZ 2

Name _____   Date _____

Class _____   Instructor _____   Grade _____

INSTRUCTIONS

Carefully read Chapter 3 in the textbook and answer each question.

MATCHING

In the space provided to the left of Column A, write the letter from Column B that best answers or completes the statement in Column A.

Column A	Column B
_____ 1. Because transformer-type welding machines are quieter, are more energy efficient, require less maintenance, and are less expensive, they are now the _____.	a. alternating current
_____ 2. Transformers use the _____ supplied to the welding shop at a high voltage to produce the low voltage welding power.	b. movable coil
_____ 3. What part of a transformer is connected to the welding lead and work leads? _____	c. tap-type
_____ 4. A _____ transformer takes a high-voltage, low-amperage current and changes it into a low-voltage, high-amperage current.	d. to save fuel
_____ 5. A welding transformer machine that has multiple-coils is also called a _____ machine.	e. tied
_____ 6. What type of welding machine is adjusted by turning a handwheel to move the internal parts closer together or farther apart? _____	f. DC
_____ 7. The use of _____ in the inverter-type welders allows them to produce any desired type of welding power.	g. maintenance
_____ 8. An alternator-type welding machine produces _____ only.	h. AC
_____ 9. A generator-type welding machine produces _____.	i. secondary coil
_____ 10. Why do some engine-driven generators reduce their speed to an idle when welding stops? _____	j. sparks
_____ 11. Portable engine-driven welders require more _____ than do the other types of welding machines.	k. splice
_____ 12. A _____ allows current to flow in one direction only.	l. loose
_____ 13. The _____ is the percentage of time a welding machine can be used continuously.	m. industry standards
_____ 14. Most welding cables are made from stranded _____ wire.	n. touched

—— 15. A _____ in a cable should not be within 10 ft (3 m) of the electrode.

o. copper

—— 16. Higher amperages than the electrode holder's rating will cause the holder to _____.

p. rectifier

—— 17. A properly sized electrode holder can overheat if the jaws are _____.

q. overheat

—— 18. The work clamp should be carefully _____ occasionally to find out if it is getting hot.

r. step-down

—— 19. Arc welding machines should be located near the welding site, but far enough away so that they are not covered with _____.

s. duty cycle

—— 20. The cable should never be _____ to scaffolding or ladders.

t. computer boards

CHAPTER 3: QUIZ 3

Name _____   Date _____

Class _____   Instructor _____   Grade _____

INSTRUCTIONS

Carefully read Chapter 3 in the textbook and answer each question.

IDENTIFICATION

Identify the numbered items on the drawing by writing the letter next to the identifying term in the space provided.

1. _____ WORK CABLE

2. _____ ELECTRODE

3. _____ WORK

4. _____ WORK CLAMP

5. _____ ELECTRODE HOLDER

6. _____ FLOW OF ELECTRONS

7. _____ WELDING MACHINE

8. _____ IRON CORE

9. _____ AMPERAGE SCALE

10. _____ HIGH CURRENT POSITION

11. _____ PRIMARY COIL

12. _____ LOW CURRENT POSITION

13. _____ AC POWER INPUT SIDE

14. _____ WELDING POWER OUTPUT SIDE

15. _____ ADJUSTING CRANK

16. _____ MOVABLE COIL

17. _____ MOVABLE CORE

18. _____ AMPERAGE POINTER

19. _____ TAP TYPE

20. _____ SECONDARY COIL

Shielded Metal Arc Welding of Plate

Operation	Electrode Size – inch (mm)	Arc Current (Amperes)	OSHA Minimum Protective Shade Number	ANSI & AWS Shade Number Recommendations*
Shielded Metal Arc Welding (SMAW)	Less than 3/32 (2.4)	Fewer than 60	7	-
	3/32-5/32 (2.4-4.0)	60-160	8	10
	More than 5/32-1/4 (4.0-6.4)	More than 160-250	10	12
	More than 1/4 (6.4)	More than 250-550	11	14

Filter Lenses for Protection during Shielded Metal Arc Welding

Welding Current Adjustment

The selection of the proper welding current depends on the electrode size, plate thickness, welding position, and the welder's skill. Electrodes of the same size can be used with a higher current in the flat position than they can in the vertical or overhead position. Since several factors affect the current requirements, data provided by welding equipment and electrode manufacturers should be used. For initial settings, use the table below.

Electrode Size	Classification					
	E6010	E6011	E6012	E6013	E7016	E7018
3/32 in.	40–80	50–70	40–90	40–85	75–105	70–110
1/8 in.	70–130	85–125	75–130	70–120	100–150	90–165
5/32 in.	100–165	130–160	120–200	130–160	140–190	125–220

WORKSTATION CHECKLIST

OBJECTIVE: You will demonstrate your knowledge of inspecting an SMA welding station.

In the space provided to the left, place an (S) for Satisfactory if the item is in good working condition or an (R) for Repair if the site needs repair.

_____   1.   Check to see that the power to the welding machine has been turned off (if not, turn it off before starting).

_____   2.   Inspect your welding helmet for holes, cracks, and proper headband adjustment.

_____   3.   Inspect your filter lens for cracks and for the correct shade, and make sure the lens cover is clean and that there are not any light leaks around the lens.

_____   4.   Make sure you are wearing proper clothing for welding: long-sleeved shirt, leather gloves, safety glasses, leather shoes or boots, and so forth. Refer to Chapter 2 in the text if you have any questions about welding apparel.

5.   Inspect the electrode lead and work lead for:

_____   a.   proper connections between the electrode holder and cable

_____   b.   proper connections between the work lead and the ground clamp

_____   c.   broken insulation, exposed wire over the full length of the electrode and work lead

_____   d.   tight connections at both the electrode and work lugs on the welding machine

_____   6.   Proper ventilation. Refer to Chapter 2 in the text if you have questions about proper ventilation.

_____   7.   Inspect the welding area for the safety of others who may be working around you. Check for holes in curtains, cracks in walls, or even doors that are to be closed while you are welding. Always be concerned with the safety of others.

_____   8.   Inspect for flammable materials in the work area such as paper, rags, wood, oil, and other flammable liquids.

_____   9.   Make sure you have pliers, a wire brush, and chipping hammer that are in good working condition.

OPERATING INSTRUCTIONS FOR SMAW EQUIPMENT

1. Check to see that your work area is safe for welding.

2. Select the proper electrode for the work to be done.

3. Set the welding machine on the correct current and polarity for the electrode you have chosen. Some welding machines will have a switch for the polarity, while others only have welding cables that must sometimes be disconnected from the positive and negative poles of the machine and then reconnected again. Remember that on reverse polarity the electrode holder is connected to the positive lug of the welding machine and the work (ground) clamp is connected to the negative lug of the welding machine. When using straight polarity, the cables are the opposite, the electrode holder is connected to the negative lug of the welding machine and work (ground) clamp is connected to the positive lug of the welding machine. If you are using an AC welding machine, there is no straight or reverse polarity, therefore, the AC welding machine does not have a positive or negative lug. See Chapter 3 of the text for AC welding current information.

4. Connect the work clamp (ground) to the part that is to be welded. In some instances, you may wish to connect the clamp to a metal table on which the part to be welded can make direct contact.

5. Place the range selector switch in the desired position. Check the electrode amperage settings for the desired settings.

6. Place the amperage adjustment control to the desired setting.

7. Turn on the welding machine's power switch.

8. Place the electrode in the electrode holder.

9. Start welding.

10. Readjust the amperage adjustment control if necessary.

SMAW TROUBLESHOOTING	
Welding Problem	**How to Correct**
Arc Blow	1. adjust electrode angle 2. move ground clamp 3. use AC current instead of DC 4. inspect the part to see if it has become magnetized and if so, demagnetize (explain how to do this.)
Brittle Welds	1. use proper pre and post-heat on the metal to be welded 2. use low-hydrogen electrodes 3. make sure parts being welded are not cooled too quickly, as by dipping in water
Cracks in Weld	1. reduce welding speed, use an electrode that produces a more convex bead 2. use low-hydrogen electrodes 3. use pre and post-heat on the weld
Distortion	1. reduce current 2. use chill plates 3. increase welding speed 4. clamp or fix parts being welded 5. weld thick sections first
Incomplete Penetration	1. increase amperage 2. use a larger root opening 3. decrease the electrode diameter 4. reduce the welding speed
Inferior Appearance	1. use correct electrode 2. use electrodes that have not been wet 3. use the correct polarity 4. increase or decrease current 5. manipulate the electrode differently
Porosity	1. use dry electrodes 2. do not weld on wet metal 3. clean paint, grease, oil, etc., from the metal being welded 4. shorten arc length 5. use low-hydrogen electrodes
Slag Inclusions	1. increase current 2. decrease welding speed 3. do not allow the welding pool to get ahead of the arc 4. change polarities 5. if welding multipass welds, always chip and wire brush between passes
Spatter	1. decrease current 2. shorten arc length 3. weld on dry metal with dry electrodes
Undercutting	1. decrease current 2. reduce welding speed 3. shorten arc length 4. change electrode angle

■ PRACTICE 4-1

Name _____   Date _____   Electrode Used _____

Class _____   Instructor _____   Grade _____

OBJECTIVE: After completing this practice, you should be able to both demonstrate and explain the safe way to set-up a welding station and how to use all of the required PPE.

EQUIPMENT AND MATERIALS NEEDED FOR THIS PRACTICE

1. A welding workstation, welding machine, welding electrodes, welding helmet, eye and ear protection, welding gloves, proper work clothing, and any special protective clothing that may be required.

INSTRUCTIONS

1. Demonstrate to your instructor and other students the safe way to prepare yourself and the welding workstation for welding.

2. Include in your demonstration appropriate references to burn protection, eye and ear protection, material specification data sheets, ventilation, electrical safety, general work clothing, special protective clothing, and area cleanup.

INSTRUCTOR'S COMMENTS _____

_____

_____

■ PRACTICE 4-2

Name _____ Date _____ Electrode Used _____

Class _____ Instructor _____ Grade _____

OBJECTIVE: After completing this practice, you should be able to an arc and make a short weld bead using an E6011 electrode.

EQUIPMENT AND MATERIALS NEEDED FOR THIS PRACTICE

1. Using a properly set-up and adjusted arc welding machine, the proper safety protection, as demonstrated in Practice 4-1, E6011 welding electrodes with 1/8-in. (3 mm) diameter, and one piece of mild steel plate 1/4 in. (6 mm) thick, you will practice striking an arc, Figure 4-1.

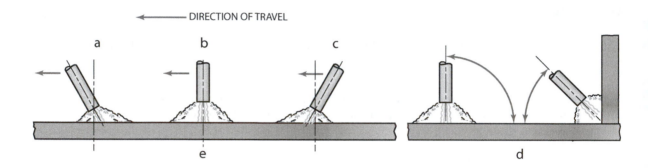

INSTRUCTIONS

1. With the electrode held over the plate, lower your helmet. Scratch the electrode across the plate (like striking a large match), Figure 4-2.

2. As the arc is established, slightly raise the electrode to the desired arc length.

3. Hold the arc in one place until the molten weld pool builds to the desired size.

4. Slowly lower the electrode as it burns off and moves it forward to start the bead.

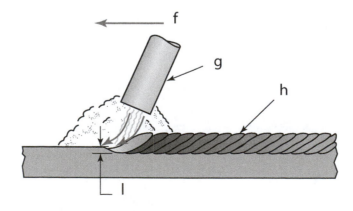

NOTE: If the electrode sticks to the plate, quickly squeeze the electrode holder lever to release the electrode. Allow the electrode to briefly cool, and then break it free by bending it back and forth a few times. Do not touch the electrode without gloves because it will still be hot. Sometimes the flux will break away from the end of the electrode as the electrode is broken free from the plate. Restarting a partially used electrode is more difficult than restarting a new electrode, and restarting one with the flux broken off can be very difficult. While you are first beginning to make your practice welds, only use new electrodes or used electrodes that have the flux all the way to the electrode's end. Save the damaged electrodes so you can practice restarting them once you have developed your welding skills. With practice you should be able to strike these damaged electrodes as easily as you strike a new electrode, Figure 4-3.

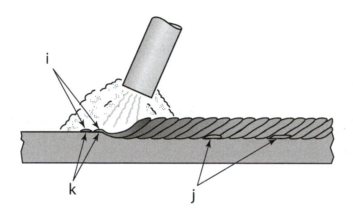

5. Once you are able to easily strike an arc and make a weld, try to strike the arc where it will be remelted by the weld you are making. Arc strikes on the metal's surface that are not covered up by the weld are considered to be weld defects by most codes.

6. Break the arc by rapidly raising the electrode back over the weld after completing a 1-in. (25-mm)-long weld bead.

7. Breaking the arc back over the weld prevents arc marks in front of the weld and results in the arc ending back away from the leading edge of the weld bead. Both are good practices to start learning early in your welding career.

8. Restart the arc as you did before and make another short weld. Repeat this process until you can easily start the arc each time.

9. Turn off the welding machine and clean your work area when you are finished welding.

INSTRUCTOR'S COMMENTS _____

_____

_____

■ PRACTICE 4-3

Name _____ Date _____ Electrode Used _____

Class _____ Instructor _____ Grade _____

OBJECTIVE: After completing this practice, you should be able to safely set-up a shielded metal arc welding workstation.

EQUIPMENT AND MATERIALS NEEDED FOR THIS PRACTICE

1. Properly set-up and adjusted arc welding machine.

2. Proper safety protection (welding hood, safety glasses, wire brush, chipping hammer, leather gloves, pliers, long-sleeved shirt, long pants, and leather boots or shoes). Refer to Chapter 2 in the text for more specific safety information.

INSTRUCTIONS

1. Demonstrate the use of proper general work clothing, special protective clothing, and eye and ear protection to prevent burns and other possible injuries.

2. Explain what material safety data sheets are and what types of information they contain.

3. Describe the ventilation provided in the area.

4. Point out electrical safety provisions.

5. List proper workstation cleanup procedures.

INSTRUCTOR'S COMMENTS _____

_____

_____

■ PRACTICE 4-4

Name _____   Date _____   Electrode Used _____

Class _____   Instructor _____   Grade _____

OBJECTIVE: After completing this practice, you should be able to make a straight stringer bead in the flat position using an E6010 or E6011 series electrode, an E6012 or E6013 series electrode, and an E7016 or E7018 series electrode.

EQUIPMENT AND MATERIALS NEEDED FOR THIS PRACTICE

1. Properly set-up and adjusted arc welding machine.

2. Proper safety protection (welding hood, safety glasses, wire brush, chipping hammer, leather gloves, pliers, long-sleeved shirt, long pants, and leather boots or shoes). Refer to Chapter 2 in the text for more specific safety information.

3. Arc welding electrodes with a 1/8-in. (3-mm) diameter.

4. One piece of mild steel plate, 6 in. (152 mm) long by 1/4 in. (6 mm) thick.

INSTRUCTIONS

1. Starting at one end of the plate, make a straight weld the full length of the plate. Watch the molten weld pool, not the end of the electrode. As you become more skillful, it is easier to watch the molten weld pool.

2. Repeat the beads with all three (F) groups of electrodes until you have consistently good quality beads that are the same height and width with good tie-in of the weld bead edges to the base metal.

3. Cool, chip, and inspect the bead for defects after completing it.

4. Turn off the welding machine and clean up your work area when you are finished welding.

INSTRUCTOR'S COMMENTS _____

_____

_____

■ PRACTICE 4-5

Name _____   Date _____   Electrode Used _____

Class _____   Instructor _____   Grade _____

OBJECTIVE: After completing this practice, you should be able to make stringer beads in the vertical up position using E6010 or E6011 electrodes, E6012 or E6013 electrodes, and E7016 or E7018 electrodes.

EQUIPMENT AND MATERIALS NEEDED FOR THIS PRACTICE

1. Properly set-up and adjusted arc welding machine.

2. Proper safety protection (welding hood, safety glasses, wire brush, chipping hammer, leather gloves, pliers, long-sleeved shirt, long pants, and leather boots or shoes). Refer to Chapter 2 in the text for more specific safety information.

3. Arc welding electrodes with a 1/8-in. (3-mm) diameter.

4. One piece of mild steel plate, 6 in. (152 mm) long by 1/4 in. (6 mm) thick.

INSTRUCTIONS

1. Start with the plate at a 45° angle. This technique is the same as that used to make a vertical weld. However, a lower level of skill is required at 45°, and it is easier to develop your skill. After the welder masters the 45° angle, the angle is increased by a few degrees successively until a vertical position is reached.

2. Before the molten metal drips down the bead, the back of the molten weld pool will start to bulge out and away from the base metal. When this happens, increase the speed of travel and the weave pattern.

3. Cool, chip, and inspect each completed weld for defects.

4. Repeat the beads as necessary with all three (F) groups of electrodes until consistently good quality beads are obtained in this position.

5. Turn off the welding machine and clean up your work area when you are finished welding.

INSTRUCTOR'S COMMENTS _____

_____

_____

■ PRACTICE 4-6

Name _____ Date _____ Electrode Used _____

Class _____ Instructor _____ Grade _____

OBJECTIVE: After completing this practice, you should be able to weld horizontal stringer beads using E6010 or E6011 electrodes, E6012 or E6013 electrodes, and E7016 or E7018 electrodes.

EQUIPMENT AND MATERIALS NEEDED FOR THIS PRACTICE

1. Properly set-up and adjusted arc welding machine.

2. Proper safety protection (welding hood, safety glasses, wire brush, chipping hammer, leather gloves, pliers, long-sleeved shirt, long pants, and leather boots or shoes). Refer to Chapter 2 in the text for more specific safety information.

3. Arc welding electrodes having a 1/8-in. (3-mm) diameter.

4. One piece of mild steel plate, 6 in. (152 mm) long by 1/4 in. (6 mm) thick.

INSTRUCTIONS

1. Start practicing these welds with the plate at a slight angle.

2. Strike the arc on the plate and build the molten weld pool.

3. The "J" weave pattern is recommended in order to deposit metal on the plate so that it can support the bead.

4. As you acquire more skill, gradually increase the plate angle until it is vertical and the weld is horizontal.

5. Cool, chip, and inspect the weld for uniformity and defects.

6. Repeat the welds with all three (F) groups of electrodes until you can consistently make welds free of defects.

7. Turn off the welding machine and clean up your work area when you are finished welding.

INSTRUCTOR'S COMMENTS _____

_____

_____

■ PRACTICE 4-7

Name _____   Date _____   Electrode Used _____

Class _____   Instructor _____   Grade _____

OBJECTIVE: After completing this practice, you should be able to weld a square butt joint in the flat position using E6010 or E6011 electrodes, E6012 or E6013 electrodes, and E7016 or E7018 electrodes.

EQUIPMENT AND MATERIALS NEEDED FOR THIS PRACTICE

1. Properly set-up and adjusted arc welding machine.

2. Proper safety protection (welding hood, safety glasses, wire brush, chipping hammer, leather gloves, pliers, long-sleeved shirt, long pants, and leather boots or shoes). Refer to Chapter 2 in the text for more specific safety information.

3. Arc welding electrodes with a 1/8-in. (3-mm) diameter.

4. Two or more pieces of mild steel plate, 6 in. (152 mm) long by 1/4 in. (6 mm) thick.

INSTRUCTIONS

1. Hold the plates together tightly.

2. Tack weld the plates together.

3. Chip the tacks before you start to weld.

4. The zigzag weave pattern works well on this joint.

5. Strike the arc and establish a molten pool directly in the joint.

6. Continue the weld along the joint.

7. Cool, chip, and inspect the weld for uniformity and defects.

8. Repeat the welds with all three (F) groups of electrodes until you can consistently make welds free of defects.

9. Turn off the welding machine and clean up your work area when you are finished welding.

INSTRUCTOR'S COMMENTS _____

_____

_____

■ PRACTICE 4-8

Name _____   Date _____   Electrode Used _____

Class _____   Instructor _____   Grade _____

OBJECTIVE: After completing this practice, you should be able to weld a vertical (3G) up-welded square butt joint using E6010 or E6011 electrodes, E6012 or E6013 electrodes, and E7016 or E7018 electrodes.

EQUIPMENT AND MATERIALS NEEDED FOR THIS PRACTICE

1. Properly set-up and adjusted arc welding machine.

2. Proper safety protection (welding hood, safety glasses, wire brush, chipping hammer, leather gloves, pliers, long-sleeved shirt, long pants, and leather boots or shoes). Refer to Chapter 2 in the text for more specific safety information.

3. Arc welding electrodes with a 1/8-in. (3-mm) diameter.

4. Two or more pieces of mild steel plate, 6 in. (152 mm) long by 1/4 in. (6 mm) thick.

INSTRUCTIONS

1. Hold the plates together tightly.

2. Tack weld the plates together.

3. Start with the plates at a 45° angle, then increase the angle as skill develops.

4. The "C," "J," or square weave pattern works well for this joint.

5. After completing the weld, cool, chip, and inspect the weld for uniformity and defects.

6. Repeat the welds using all three (F) groups of electrodes until you can consistently make welds free of defects.

7. Turn off the welding machine and clean up your work area when you are finished welding.

INSTRUCTOR'S COMMENTS _____

_____

_____

■ PRACTICE 4-9

Name _____ Date _____ Electrode Used _____

Class _____ Instructor _____ Grade _____

OBJECTIVE: After completing this practice, you will be able to weld a horizontal square butt joint in the 2G position using E6010 or E6011 electrodes, E6012 or E6013 electrodes, and E7016 or E7018 electrodes.

EQUIPMENT AND MATERIALS NEEDED FOR THIS PRACTICE

1. Properly set-up and adjusted arc welding machine.

2. Proper safety protection (welding hood, safety glasses, wire brush, chipping hammer, leather gloves, pliers, long-sleeved shirt, long pants, and leather boots or shoes). Refer to Chapter 2 in the text for more specific safety information.

3. Arc welding electrodes with a 1/8-in. (3-mm) diameter.

4. Two or more pieces of mild steel plate, 6 in. (152 mm) long by 1/4 in. (6 mm) thick.

INSTRUCTIONS

1. After the plates are tack welded together, place them on the welding table so that the weld bead will be in the horizontal position.

2. Start practicing these welds at a slight angle.

3. Strike the arc on the bottom plate and build the molten weld pool until it bridges the gap.

4. The "J" weave pattern works well on this joint.

5. When the 6-in. (152-mm) long weld is completed, cool, chip, and inspect it for uniformity and soundness.

6. Repeat the welds as needed for all these groups of electrodes until you can consistently make welds free of defects.

7. Turn off the welding machine and clean up your work area when you are finished welding.

INSTRUCTOR'S COMMENTS _____

_____

_____

■ PRACTICE 4-10

Name _____   Date _____   Electrode Used _____

Class _____   Instructor _____   Grade _____

OBJECTIVE: After completing this practice, you should be able to weld an edge joint in the flat position using E6010 or E6011 electrodes, E6012 or E6013 electrodes, and E7016 or E7018 electrodes.

EQUIPMENT AND MATERIALS NEEDED FOR THIS PRACTICE

1. Properly set-up and adjusted arc welding machine.

2. Proper safety protection (welding hood, safety glasses, wire brush, chipping hammer, leather gloves, pliers, long-sleeved shirt, long pants, and leather boots or shoes). Refer to Chapter 2 in the text for more specific safety information.

3. Arc welding electrodes with a 1/8-in. (3-mm) diameter.

4. Two or more pieces of mild steel plate, 6 in. (152 mm) long by 1/4 in. (6 mm) thick.

INSTRUCTIONS

1. Clamp the plates flat together and tack on each end.

2. Start the arc and weld the full length of the plate.

3. Make the weld bead as wide as the width of the edge joint.

4. Cool, chip, and inspect the weld for uniformity and defects.

5. Repeat the welds as necessary with all three (F) groups of electrodes until you can consistently make welds free of defects.

6. Turn off the welding machine and clean up your work area when you are finished welding.

INSTRUCTOR'S COMMENTS _____

_____

_____

■ PRACTICE 4-11

Name _____   Date _____   Electrode Used _____

Class _____   Instructor _____   Grade _____

OBJECTIVE: After completing this practice, you should be able to weld an edge joint in the vertical down position using E6010 or E6011 electrodes, E6012 or E6013 electrodes, and E7016 or E7018 electrodes.

EQUIPMENT AND MATERIALS NEEDED FOR THIS PRACTICE

1. Properly set-up and adjusted arc welding machine.

2. Proper safety protection (welding hood, safety glasses, wire brush, chipping hammer, leather gloves, pliers, long-sleeved shirt, long pants, and leather boots or shoes). Refer to Chapter 2 in the text for more specific safety information.

3. Arc welding electrodes with a 1/8-inch (3-mm) diameter.

4. Two or more pieces of mild steel plate, 6 in. (152 mm) long by 1/4 in. (6 mm) thick.

INSTRUCTIONS

1. Clamp the pieces flat together and tack on each end.

2. Start with the plates at a 45° angle and progressively increase this angle as skill develops until a vertical position is reached.

3. Start at the top and weld downward. Make the weld bead as wide as the width of the edge joint.

4. Cool, chip, and inspect the weld for uniformity and defects.

5. Repeat the welds as necessary with all three (F) groups of electrodes until you can consistently make welds free of defects.

6. Turn off the welding machine and clean up your work area when you are finished welding.

INSTRUCTOR'S COMMENTS _____

_____

_____

■ PRACTICE 4-12

Name _____   Date _____   Electrode Used _____

Class _____   Instructor _____   Grade _____

OBJECTIVE: After completing this practice, you should be able to weld an edge joint in the vertical up position using E6010 or E6011 electrodes, E6012 or E6013 electrodes, and E7016 or E7018 electrodes.

EQUIPMENT AND MATERIALS NEEDED FOR THIS PRACTICE

1. Properly set-up and adjusted arc welding machine.

2. Proper safety protection (welding hood, safety glasses, wire brush, chipping hammer, leather gloves, pliers, long-sleeved shirt, long pants, and leather boots or shoes). Refer to Chapter 2 in the text for more specific safety information.

3. Arc welding electrodes with a 1/8-in. (3-mm) diameter.

4. Two or more pieces of mild steel plate, 6 in. (152 mm) long by 1/4 in. (6 mm) thick.

INSTRUCTIONS

1. Clamp the pieces flat together and tack on each end.

2. Start with the plates at a 45° angle and progressively increase this angle as skill develops until a vertical position is reached.

3. Start at the bottom and weld upward. Make the weld bead as wide as the width of the edge joint.

4. Cool, chip, and inspect the weld for uniformity and defects.

5. Repeat the welds as necessary with all three (F) groups of electrodes until you can consistently make welds free of defects.

6. Turn off the welding machine and clean up your work area when you are finished welding.

INSTRUCTOR'S COMMENTS _____

_____

_____

■ PRACTICE 4-13

Name _____   Date _____   Electrode Used _____

Class _____   Instructor _____   Grade _____

OBJECTIVE: After completing this practice, you should be able to weld an edge joint in the horizontal position using E6010 or E6011 electrodes, E6012 or E6013 electrodes, and E7016 or E7018 electrodes.

EQUIPMENT AND MATERIALS NEEDED FOR THIS PRACTICE

1. Properly set-up and adjusted arc welding machine.

2. Proper safety protection (welding hood, safety glasses, wire brush, chipping hammer, leather gloves, pliers, long-sleeved shirt, long pants, and leather boots or shoes). Refer to Chapter 2 in the text for more specific safety information.

3. Arc welding electrodes with a 1/8-in. (3-mm) diameter.

4. Two or more pieces of mild steel plate, 6 in. (152 mm) long by 1/4 in. (6 mm) thick.

INSTRUCTIONS

1. Clamp the pieces flat together and tack on each end.

2. Start with the plates reclined at a slight angle and progressively increase this angle as skill develops until the plates are vertical and the weld bead is horizontal.

3. The "J" weave or stepped pattern works well for this joint.

4. Angle the electrode up and back toward the weld. This will cause more metal to be deposited along the top edge of the bead.

5. Cool, chip, and inspect the weld for uniformity and defects.

6. Repeat the welds as necessary with all three (F) groups of electrodes until you can consistently make welds free of defects.

7. Turn off the welding machine and clean up your work area when you are finished welding.

INSTRUCTOR'S COMMENTS _____

_____

_____

■ PRACTICE 4-14

Name _____   Date _____   Electrode Used _____

Class _____   Instructor _____   Grade _____

OBJECTIVE: After completing this practice, you should be able to weld an edge joint in the overhead position using E6010 or E6011 electrodes, E6012 or E6013 electrodes, and E7016 or E7018 electrodes.

EQUIPMENT AND MATERIALS NEEDED FOR THIS PRACTICE

1. Properly set-up and adjusted arc welding machine.

2. Proper safety protection (welding hood, safety glasses, wire brush, chipping hammer, leather gloves, pliers, long-sleeved shirt, long pants, and leather boots or shoes). Refer to Chapter 2 in the text for more specific safety information.

3. Arc welding electrodes with a 1/8-in. (3-mm) diameter.

4. Two or more pieces of mild steel plate, 6 in. (152 mm) long by 1/4 in. (6 mm) thick.

INSTRUCTIONS

1. Clamp the pieces flat together, tack on each end and position the parts in the overhead position.

2. Strike the arc and keep the electrode in a slightly trailing angle. Keep a very short arc length.

3. Use the stepped pattern and move the electrode forward slightly when the molten weld pool grows to the correct size.

4. When the weld pool cools and begins to shrink, move the arc back near the center of the weld.

5. Hold the arc in this new position until the weld pool grows to the correct size.

6. Step the electrode forward again and keep repeating this pattern as the weld bead progresses along the entire length of the joint.

7. Cool, chip, and inspect the weld for uniformity and defects.

8. Repeat the welds as necessary with all three (F) groups of electrodes until you can consistently make welds free of defects.

9. Turn off the welding machine and clean up your work area when you are finished welding.

INSTRUCTOR'S COMMENTS _____

_____

_____

■ PRACTICE 4-15

Name _____   Date _____   Electrode Used _____

Class _____   Instructor _____   Grade _____

OBJECTIVE: After completing this practice, you should be able to weld an outside corner joint in the flat position using E6010 or E6011 electrodes, E6012 or E6013 electrodes, and E7016 or E7018 electrodes.

EQUIPMENT AND MATERIALS NEEDED FOR THIS PRACTICE

1. Properly set-up and adjusted arc welding machine.

2. Proper safety protection (welding hood, safety glasses, wire brush, chipping hammer, leather gloves, pliers, long-sleeved shirt, long pants, and leather boots or shoes). Refer to Chapter 2 in the text for more specific safety information.

3. Arc welding electrodes with a 1/8-in. (3-mm) diameter.

4. Two or more pieces of mild steel plate, 6 in. (152 mm) long by 1/4 in. (6 mm) thick.

INSTRUCTIONS

1. Clamp the pieces in a 90° "V" configuration and tack on each end.

2. Working in the flat position, start at one end and make a straight bead the full length of the outside corner of the plates.

3. Remember to watch the molten weld pool as it is being made and not the arc itself.

4. Cool, chip, and inspect the weld for uniformity and defects.

5. Repeat the welds as necessary with all three (F) groups of electrodes until you can consistently make welds free of defects.

6. Turn off the welding machine and clean up your work area when you are finished welding.

INSTRUCTOR'S COMMENTS _____

_____

_____

■ PRACTICE 4-16

Name _____   Date _____   Electrode Used _____

Class _____   Instructor _____   Grade _____

OBJECTIVE: After completing this practice, you should be able to weld an outside corner joint in the vertical down position using E6010 or E6011 electrodes, E6012 or E6013 electrodes, and E7016 or E7018 electrodes.

EQUIPMENT AND MATERIALS NEEDED FOR THIS PRACTICE

1. Properly set-up and adjusted arc welding machine.

2. Proper safety protection (welding hood, safety glasses, wire brush, chipping hammer, leather gloves, pliers, long-sleeved shirt, long pants, and leather boots or shoes). Refer to Chapter 2 in the text for more specific safety information.

3. Arc welding electrodes with a 1/8-in. (3-mm) diameter.

4. Two or more pieces of mild steel plate, 6 in. (152 mm) long by 1/4 in. (6 mm) thick.

INSTRUCTIONS

1. Clamp the pieces in a 90° "V" configuration and tack on each end.

2. Begin by positioning the plates at a 45° angle and progressively increase this angle as skill develops until the vertical position is reached.

3. Start at the top and make a straight bead downward along the full length of the outside corner of the plates.

4. Remember to watch the molten weld pool as it is being made and not the arc itself.

5. Cool, chip, and inspect the weld for uniformity and defects.

6. Repeat the welds as necessary with all three (F) groups of electrodes until you can consistently make welds free of defects.

7. Turn off the welding machine and clean up your work area when you are finished welding.

INSTRUCTOR'S COMMENTS _____

■ PRACTICE 4-17

Name _____   Date _____   Electrode Used _____

Class _____   Instructor _____   Grade _____

OBJECTIVE: After completing this practice, you should be able to weld an outside corner joint in the vertical up position using E6010 or E6011 electrodes, E6012 or E6013 electrodes, and E7016 or E7018 electrodes.

EQUIPMENT AND MATERIALS NEEDED FOR THIS PRACTICE

1. Properly set-up and adjusted arc welding machine.

2. Proper safety protection (welding hood, safety glasses, wire brush, chipping hammer, leather gloves, pliers, long-sleeved shirt, long pants, and leather boots or shoes). Refer to Chapter 2 in the text for more specific safety information.

3. Arc welding electrodes with a 1/8-in. (3-mm) diameter.

4. Two or more pieces of mild steel plate, 6 in. (152 mm) long by 1/4 in. (6 mm) thick.

INSTRUCTIONS

1. Clamp the pieces in a 90° "V" configuration and tack on each end.

2. Begin by positioning the plates at a 45° angle and progressively increase this angle as skill develops until the vertical position is reached.

3. Start at the bottom and make a straight bead upward along the full length of the outside corner of the plates.

4. Remember to watch the molten weld pool as it is being made and not the arc itself.

5. Cool, chip, and inspect the weld for uniformity and defects.

6. Repeat the welds as necessary with all three (F) groups of electrodes until you can consistently make welds free of defects.

7. Turn off the welding machine and clean up your work area when you are finished welding.

INSTRUCTOR'S COMMENTS _____

_____

_____

■ PRACTICE 4-18

Name _____   Date _____   Electrode Used _____

Class _____   Instructor _____   Grade _____

OBJECTIVE: After completing this practice, you should be able to weld an outside corner joint in the horizontal position using E6010 or E6011 electrodes, E6012 or E6013 electrodes, and E7016 or E7018 electrodes.

EQUIPMENT AND MATERIALS NEEDED FOR THIS PRACTICE

1. Properly set-up and adjusted arc welding machine.

2. Proper safety protection (welding hood, safety glasses, wire brush, chipping hammer, leather gloves, pliers, long-sleeved shirt, long pants, and leather boots or shoes). Refer to Chapter 2 in the text for more specific safety information.

3. Arc welding electrodes with a 1/8-in. (3-mm) diameter.

4. Two or more pieces of mild steel plate, 6 in. (152 mm) long by 1/4 in. (6 mm) thick.

INSTRUCTIONS

1. Clamp the pieces in a 90° "V" configuration and tack on each end.

2. Position the plates so that the joint is in a horizontal position.

3. Begin by reclining the plates at a slight angle and progressively increase this angle as skill develops until a full horizontal position is reached.

4. Start at one end and make a straight bead horizontally along the full length of the outside corner of the plates.

5. The "J" weave or stepped pattern work well for this position.

6. Angling the electrode up and back toward the weld will cause more metal to be deposited along the top edge of the weld.

7. Cool, chip, and inspect the weld for uniformity and defects.

8. Repeat the welds as necessary with all three (F) groups of electrodes until you can consistently make welds free of defects.

9. Turn off the welding machine and clean up your work area when you are finished welding.

INSTRUCTOR'S COMMENTS _____

_____

_____

■ PRACTICE 4-19

Name _____   Date _____   Electrode Used _____

Class _____   Instructor _____   Grade _____

OBJECTIVE: After completing this practice, you should be able to weld an outside corner joint in the overhead position using E6010 or E6011 electrodes, E6012 or E6013 electrodes, and E7016 or E7018 electrodes.

EQUIPMENT AND MATERIALS NEEDED FOR THIS PRACTICE

1. Properly set-up and adjusted arc welding machine.

2. Proper safety protection (welding hood, safety glasses, wire brush, chipping hammer, leather gloves, pliers, long-sleeved shirt, long pants, and leather boots or shoes). Refer to Chapter 2 in the text for more specific safety information.

3. Arc welding electrodes with a 1/8-in. (3-mm) diameter.

4. Two or more pieces of mild steel plate, 6 in. (152 mm) long by 1/4 in. (6 mm) thick.

INSTRUCTIONS

1. Clamp the pieces in a 90° "V" configuration and tack on each end. Position the plates so that the joint is in the overhead position.

2. With the electrode pointed slightly into the joint, strike the arc in the joint. Keep a very short arc length.

3. Use the stepped pattern and move the electrode forward slightly when the molten weld pool grows to the correct size.

4. When the molten weld pool cools and begins to shrink, move the arc back near the center of the weld and hold it in this new position until the weld pool again grows to the correct size.

5. Step the electrode forward again and keep repeating this pattern until the weld progresses along the entire length of the joint.

6. Cool, chip, and inspect the weld for uniformity and defects.

7. Repeat the welds as necessary with all three (F) groups of electrodes until you can consistently make welds free of defects.

8. Turn off the welding machine and clean up your work area when you are finished welding.

INSTRUCTOR'S COMMENTS _____

_____

_____

■ PRACTICE 4-20

Name _____ Date _____ Electrode Used _____

Class _____ Instructor _____ Grade _____

OBJECTIVE: After completing this practice, you should be able to weld a lap joint in the flat (1F) position using E6010 or E6011 electrodes, E6012 or E6013 electrodes, and E7016 or E7018 electrodes.

EQUIPMENT AND MATERIALS NEEDED FOR THIS PRACTICE

1. Properly set-up and adjusted arc welding machine.

2. Proper safety protection (welding hood, safety glasses, wire brush, chipping hammer, leather gloves, pliers, long-sleeved shirt, long pants, and leather boots or shoes). Refer to Chapter 2 in the text for more specific safety information.

3. Arc welding electrodes with a 1/8-in. (3-mm) diameter.

4. Two or more pieces of mild steel plate, 6 in. (152 mm) long by 1/4 in. (6 mm) thick.

INSTRUCTIONS

1. Clamp the pieces tightly together with no more than a 1/4-in. (6 mm) overlap and tack on each end. Position the plates so that the joint is in the flat position.

2. A small tack weld may be added at the center to prevent distortion during welding.

3. The "J," "C," or zigzag pattern work well for this joint.

4. Strike the arc at one end of the joint and establish a weld pool directly in the joint.

5. Move the electrode out on the bottom plate and then back to move the pool to the top edge of the top plate. Proceed in this manner along the entire length of the plate.

6. Follow the surface of the plates and not the trailing edge of the weld bead so that slag will not collect in the root.

7. Cool, chip, and inspect the weld for uniformity and defects.

8. Repeat the welds as necessary with all three (F) groups of electrodes until you can consistently make welds free of defects.

9. Turn off the welding machine and clean up your work area when you are finished welding.

INSTRUCTOR'S COMMENTS _____

_____

_____

■ PRACTICE 4-21

Name _____   Date _____   Electrode Used _____

Class _____   Instructor _____   Grade _____

OBJECTIVE: After completing this practice, you should be able to weld a lap joint in the horizontal (2F) position using E6010 or E6011 electrodes, E6012 or E6013 electrodes, and E7016 or E7018 electrodes.

EQUIPMENT AND MATERIALS NEEDED FOR THIS PRACTICE

1. Properly set-up and adjusted arc welding machine.

2. Proper safety protection (welding hood, safety glasses, wire brush, chipping hammer, leather gloves, pliers, long-sleeved shirt, long pants, and leather boots or shoes). Refer to Chapter 2 in the text for more specific safety information.

3. Arc welding electrodes with a 1/8-in. (3-mm) diameter.

4. Two or more pieces of mild steel plate, 6 in. (152 mm) long by 1/4 in. (6 mm) thick.

INSTRUCTIONS

1. Clamp the pieces tightly together with no more than a 1/4-in. (6 mm) overlap and tack on each end. Position the plates so that the joint is in the horizontal position.

2. A small tack weld may be added at the center to prevent distortion during welding.

3. The "J," "C," or zigzag pattern works well for this joint.

4. Strike the arc at one end of the joint and establish a weld pool directly in the joint.

5. Make sure that the fillet is placed so that it is equally divided between both plates and continue along the entire length of the plate.

6. Follow the surface of the plates and not the trailing edge of the weld bead so that slag will not collect in the root.

7. Cool, chip, and inspect the weld for uniformity and defects.

8. Repeat the welds as necessary with all three (F) groups of electrodes until you can consistently make welds free of defects.

9. Turn off the welding machine and clean up your work area when you are finished welding.

INSTRUCTOR'S COMMENTS _____

_____

_____

■ PRACTICE 4-22

Name _____ Date _____ Electrode Used _____

Class _____ Instructor _____ Grade _____

OBJECTIVE: After completing this practice, you should be able to weld a lap joint in the vertical up (3F) position using E6010 or E6011 electrodes, E6012 or E6013 electrodes, and E7016 or E7018 electrodes.

EQUIPMENT AND MATERIALS NEEDED FOR THIS PRACTICE

1. Properly set-up and adjusted arc welding machine.

2. Proper safety protection (welding hood, safety glasses, wire brush, chipping hammer, leather gloves, pliers, long-sleeved shirt, long pants, and leather boots or shoes). Refer to Chapter 2 in the text for more specific safety information.

3. Arc welding electrodes with a 1/8-in. (3-mm) diameter.

4. Two or more pieces of mild steel plate, 6 in. (152 mm) long by 1/4 in. (6 mm) thick.

INSTRUCTIONS

1. Clamp the pieces tightly together with no more than a 1/4-in. (6 mm) overlap and tack on each end.

2. Start practicing this joint with the plates inclined at a 45° angle and progressively increase this angle as skill develops until a full vertical position is reached.

3. Strike the arc at the bottom end of the plates and establish a weld pool in the root of the joint. The "J" or "T" patterns work well for this joint.

4. Use the "T" pattern to step ahead of the molten weld pool, allowing it to cool slightly. Do not deposit weld metal ahead of the molten weld pool.

5. As the weld pool begins to cool and shrink, move the electrode back down into the molten weld pool and quickly move the electrode from side to side to fill up the joint.

6. Cool, chip, and inspect the weld for uniformity and defects.

7. Repeat the welds as necessary with all three (F) groups of electrodes until you can consistently make welds free of defects.

8. Turn off the welding machine and clean up your work area when you are finished welding.

INSTRUCTOR'S COMMENTS _____

_____

_____

■ PRACTICE 4-23

Name _____ Date _____ Electrode Used _____

Class _____ Instructor _____ Grade _____

OBJECTIVE: After completing this practice, you should be able to weld a lap joint in the overhead (4F) position using E6010 or E6011 electrodes, E6012 or E6013 electrodes, and E7016 or E7018 electrodes.

EQUIPMENT AND MATERIALS NEEDED FOR THIS PRACTICE

1. Properly set-up and adjusted arc welding machine.

2. Proper safety protection (welding hood, safety glasses, wire brush, chipping hammer, leather gloves, pliers, long-sleeved shirt, long pants, and leather boots or shoes). Refer to Chapter 2 in the text for more specific safety information.

3. Arc welding electrodes with a 1/8-in. (3-mm) diameter.

4. Two or more pieces of mild steel plate, 6 in. (152 mm) long by 1/4 in. (6 mm) thick.

INSTRUCTIONS

1. Clamp the pieces tightly together with no more than a 1/4-in. (6 mm) overlap and tack on each end. Position the plates in the overhead position.

2. With the electrode pointed slightly into the joint, strike the arc in the inside corner (root) of the lap joint. Keep a very short arc length.

3. Use the stepped pattern and move the electrode forward slightly when the molten weld pool has grown to the correct size.

4. When the molten weld pool cools and begins to shrink, move the electrode back near the center of the weld and hold it in this new position until the weld pool grows again to the correct size.

5. Step the electrode forward again keep repeating this pattern until the weld progresses along the entire length of the joint.

6. Cool, chip, and inspect the weld for uniformity and defects.

7. Repeat the welds as necessary with all three (F) groups of electrodes until you can consistently make welds free of defects.

8. Turn off the welding machine and clean up your work area when you are finished welding.

INSTRUCTOR'S COMMENTS _____

_____

_____

■ PRACTICE 4-24

Name _____   Date _____   Electrode Used _____

Class _____   Instructor _____   Grade _____

OBJECTIVE: After completing this practice, you should be able to weld a tee joint in the flat (1F) position using E6010 or E6011 electrodes, E6012 or E6013 electrodes, and E7016 or E7018 electrodes.

EQUIPMENT AND MATERIALS NEEDED FOR THIS PRACTICE

1. Properly set-up and adjusted arc welding machine.

2. Proper safety protection (welding hood, safety glasses, wire brush, chipping hammer, leather gloves, pliers, long-sleeved shirt, long pants, and leather boots or shoes). Refer to Chapter 2 in the text for more specific safety information.

3. Arc welding electrodes with a 1/8-in. (3-mm) diameter.

4. Two or more pieces of mild steel plate, 6 in. (152 mm) long by 1/4 in. (6 mm) thick.

INSTRUCTIONS

1. Clamp the pieces together in the 90° "tee" configuration and tack on each end. Position the plates so that the joint is in the true flat position with each plate at 45° to the worktable.

2. Start at one end and establish a weld pool on both plates. Allow the weld pools to flow together before starting to move the bead along the plates.

3. Any of the weave patterns will work well on this joint.

4. To prevent slag inclusions, use a slightly higher than normal current setting.

5. Progress along the joint until the entire joint is welded being careful that there are no slag inclusions.

6. Cool, chip, and inspect the weld for uniformity and defects.

7. Repeat the welds as necessary with all three (F) groups of electrodes until you can consistently make welds free of defects.

8. Turn off the welding machine and clean up your work area when you are finished welding.

INSTRUCTOR'S COMMENTS _____

_____

_____

■ PRACTICE 4-25

Name _____ Date _____ Electrode Used _____

Class _____ Instructor _____ Grade _____

OBJECTIVE: After completing this practice, you should be able to weld a tee joint in the horizontal (2F) position using E6010 or E6011 electrodes, E6012 or E6013 electrodes, and E7016 or E7018 electrodes.

EQUIPMENT AND MATERIALS NEEDED FOR THIS PRACTICE

1. Properly set-up and adjusted arc welding machine.

2. Proper safety protection (welding hood, safety glasses, wire brush, chipping hammer, leather gloves, pliers, long-sleeved shirt, long pants, and leather boots or shoes). Refer to Chapter 2 in the text for more specific safety information.

3. Arc welding electrodes with a 1/8-in. (3-mm) diameter.

4. Two or more pieces of mild steel plate, 6 in. (152 mm) long by 1/4 in. (6 mm) thick.

INSTRUCTIONS

1. Clamp the pieces together in the 90° "tee" configuration and tack on each end. Position the plates so that one plate is flat on the table and the other plate is vertical. The joint will be horizontal.

2. Start at one end and establish a weld pool on the flat plate. Allow the weld pool to flow onto the vertical as well before starting to move the bead along the plates.

3. The "J" or "C" weave pattern will work well on this joint.

4. Push the arc into the root and slightly up the vertical plate and keep the root of the joint fusing as the weld progresses.

5. Beware of undercutting on the vertical plate. If this occurs, lower the current slightly and direct the arc a little more toward the lower plate.

6. Cool, chip, and inspect the weld for uniformity and defects.

7. Repeat the welds as necessary with all three (F) groups of electrodes until you can consistently make welds free of defects.

8. Turn off the welding machine and clean up your work area when you are finished welding.

INSTRUCTOR'S COMMENTS _____

_____

_____

■ PRACTICE 4-26

Name _____ Date _____ Electrode Used _____

Class _____ Instructor _____ Grade _____

OBJECTIVE: After completing this practice, you should be able to weld a tee joint in the vertical (3F) position using E6010 or E6011 electrodes, E6012 or E6013 electrodes, and E7016 or E7018 electrodes.

EQUIPMENT AND MATERIALS NEEDED FOR THIS PRACTICE

1. Properly set-up and adjusted arc welding machine.

2. Proper safety protection (welding hood, safety glasses, wire brush, chipping hammer, leather gloves, pliers, long-sleeved shirt, long pants, and leather boots or shoes). Refer to Chapter 2 in the text for more specific safety information.

3. Arc welding electrodes with a 1/8-in. (3-mm) diameter.

4. Two or more pieces of mild steel plate, 6 in. (152 mm) long by 1/4 in. (6 mm) thick.

INSTRUCTIONS

1. Clamp the pieces together in the 90° "tee" configuration and tack on each end.

2. Start practicing this joint with the plates inclined at a 45° angle and progressively increase this angle as skill develops until a full vertical position is reached.

3. The square, "J," "C," or "T" weave patterns will work well on this joint. The "T" or stepped pattern will give the best root penetration.

4. For this weld, undercutting may be a problem on both plates. Control this by holding the arc on a side only long enough for the filler metal to run down and fill it.

5. Proceed from the bottom to the top until the entire length of the joint is welded.

6. Cool, chip, and inspect the weld for uniformity and defects.

7. Repeat the welds as necessary with all three (F) groups of electrodes until you can consistently make welds free of defects.

8. Turn off the welding machine and clean up your work area when you are finished welding.

INSTRUCTOR'S COMMENTS _____

■ PRACTICE 4-27

Name _____ Date _____ Electrode Used _____

Class _____ Instructor _____ Grade _____

OBJECTIVE: After completing this practice, you should be able to weld a tee joint in the overhead (4F) position using E6010 or E6011 electrodes, E6012 or E6013 electrodes, and E7016 or E7018 electrodes.

EQUIPMENT AND MATERIALS NEEDED FOR THIS PRACTICE

1. Properly set-up and adjusted arc welding machine.

2. Proper safety protection (welding hood, safety glasses, wire brush, chipping hammer, leather gloves, pliers, long-sleeved shirt, long pants, and leather boots or shoes). Refer to Chapter 2 in the text for more specific safety information.

3. Arc welding electrodes with a 1/8-in. (3-mm) diameter.

4. Two or more pieces of mild steel plate, 6 in. (152 mm) long by 1/4 in. (6 mm) thick.

INSTRUCTIONS

1. Clamp the pieces together in the 90° "tee" configuration and tack on each end. Position the plates in the overhead position.

2. Start the arc deep in the root of the joint. Keep a very short arc length.

3. The stepped pattern will work well for this joint and will give good root penetration.

4. For this weld, undercutting may be a problem on both plates. Control this by holding the arc on a side only long enough for the filler metal to run in and fill it.

5. Proceed from one end of the plates to the other until the entire length of the joint is welded.

6. Cool, chip, and inspect the weld for uniformity and defects.

7. Repeat the welds as necessary with all three (F) groups of electrodes until you can consistently make welds free of defects.

8. Turn off the welding machine and clean up your work area when you are finished welding.

INSTRUCTOR'S COMMENTS _____

_____

_____

CHAPTER 4: QUIZ 1

Name _____ Date _____ Electrode Used _____

Class _____ Instructor _____ Grade _____

INSTRUCTIONS

Carefully read Chapter 4 in the textbook and answer each question.

MATCHING

In the space provided to the left of Column A, write the letter from Column B that best answers or completes the statement in Column A.

Column A	Column B
_____ 1. SMAW can be used to make _____ welds which can be consistently produced on almost any type of metal, any shape, and in any position.	a. less
_____ 2. When striking the arc, scratch the electrode across the plate (like striking a _____).	b. spatter
_____ 3. Restarting a partially used electrode is more difficult than a new electrode and restarting one with the _____ can be very difficult.	c. surface
_____ 4. Breaking the arc back over the weld prevents arc marks in front of the weld and results in the arc ending back away from the _____ of the weld bead.	d. match
_____ 5. Welding with the current set too low results in _____.	e. Rutile
_____ 6. Higher amperage settings can also result in an increase in the amount of _____.	f. leading edge
_____ 7. Using smaller diameter electrodes requires _____ skill than using large diameter electrodes.	g. root opening
_____ 8. Welders often use the terms heat and amperage interchangeably when they are speaking about making changes to the _____.	h. V-groove
_____ 9. The term work angle refers to the relationship between the center of the electrode and the _____ of the work.	i. perpendicular
_____ 10. A leading electrode angle _____ molten metal and slag ahead of the weld.	j. pushes
_____ 11. A _____ angle directs all of the electrodes jetting force directly into the joint.	k. stringer bead
_____ 12. A _____ electrode angle pushes the molten metal away from the leading edge of the molten weld pool toward the back where it solidifies.	l. trailing

—— 13. Both E6010 and E6011 electrodes have _____-based fluxes.

—— 14. _____-based flux electrodes have a softer arc, smoother arc, and produce less welding fumes and sparks than cellulose-based electrodes.

—— 15. The tack weld must be small, uniform, and _____ so it does not adversely affect the finished weld.

—— 16. An outside corner joint is made by placing the plates at a 90° angle to each other, with the edges forming a _____.

—— 17. The rate that the weld metal is added to the weld is its _____.

—— 18. A large piece of metal used to absorb excessive heat is called a _____.

—— 19. A straight weld bead on the surface of a plate, with little or no side-to-side electrode movement is called a _____.

—— 20. The space between two plates to be welded is called the _____.

m. flux broken off

n. chill plate

o. poor fusion

p. free of defects

q. high-quality

r. cellulose

s. deposition rate

t. welding current

CHAPTER 4: QUIZ 2

Name _____   Date _____

Class _____   Instructor _____   Grade _____

INSTRUCTIONS

Carefully read Chapter 4 in the textbook and answer each question.

IDENTIFICATION

Identify the numbered items on the drawing by writing the letter next to the identifying term in the space provided.

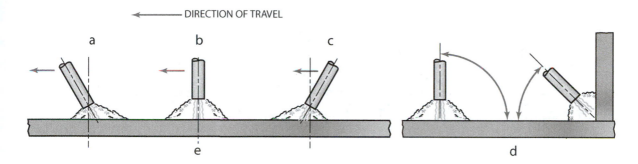

1. _____ LEADING ANGLE

2. _____ WORK ANGLE

3. _____ TRAILING ANGLE

4. _____ TRAVEL OR ELECTRODE ANGLE

5. _____ PERPENDICULAR ANGLE

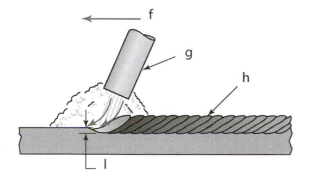

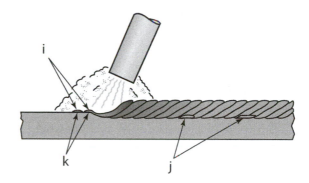

6. _____ GAS HOSES

7. _____ WELDING OR BRAZING TIP

8. _____ CUTTING HEAD

9. _____ REVERSE FLOW CHECK VALVES

10. _____ PRESSURE REGULATORS

11. _____ SAFETY CHAIN

12. _____ OXYGEN CYLINDER

CHAPTER 4: QUIZ 3

Name _____   Date _____

Class _____   Instructor _____   Grade _____

INSTRUCTIONS

Carefully read Chapter 4 in the textbook and answer each question.

IDENTIFICATION

Identify the numbered items on the drawing by writing the letter next to the identifying term in the space provided.

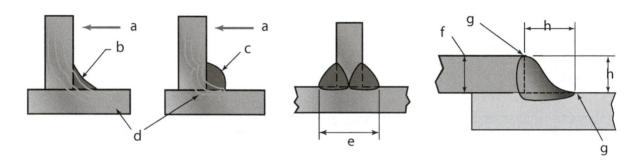

1. _____   THICKNESS

2. _____   CONVEX CONTOUR

3. _____   FORCE

4. _____   FLAT TO CONCAVE CONTOUR

5. _____   WELD SIZE

6. _____   STRESS LINES

7. _____   SMOOTH WELD TRANSITION

8. _____   LEG

Shielded Metal Arc Welding of Pipe

JOINT PREPARATION FOR PIPE WELDING

It is important that the bevel be at the correct angle, which is 37.5° for vertical up welds and 30° for vertical down welds, and that each end of the mating pipes meet squarely, Figure 5-1(A) and (B).

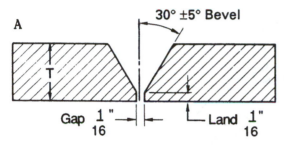

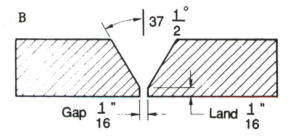

Figure 5-1A Standard joint design for vertical down welding.

Figure 5-1B Standard joint design for vertical up and horizontal welding.

Weld bead sequencing for vertical down welding is shown in Figure 5-2.

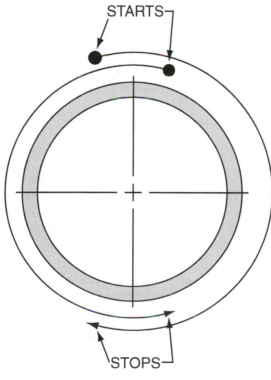

Figure 5-2

Figure 5-2 is for EXX10-type electrodes, designed for welding circumferential pipe joints in the vertical down position. The number of passes required for welding pipe will vary with the diameter of the electrode, the thickness of the pipe, the welding position, the type of current being used, and welding operators. The typical number of passes required to weld pipe using the vertical down technique is shown in Figure 5-3.

Pipe Wall Thickness	Number of Passes
1/4"	3
5/16"	4
3/8"	5
1/2"	7

Figure 5-3. Passes required for vertical down welding on pipe.

Weld bead sequencing for vertical up welding is shown in Figure 5-4.

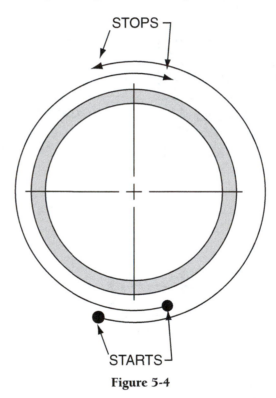

Figure 5-4

When welding carbon steel pipe, E6010 or E7010 electrodes are used for the first pass. All other passes normally are welded using an EXX18 (low-hydrogen)-type electrode. The number of passes required for welding pipe will vary with the diameter of the electrode, the thickness of the pipe, the welding position, the type of current being used, and welding operators. The typical number of passes required to weld pipe using the vertical up technique is shown in Figure 5-5.

Pipe Wall Thickness	Number of Passes
1/4"	2
5/16"	2
3/8"	3
1/2"	3
5/8"	4
3/4"	6
1"	7

Figure 5-5. Passes required for vertical down welding on pipe.

Weld bead sequencing for a horizontal butt weld on pipe is shown in Figure 5-6.

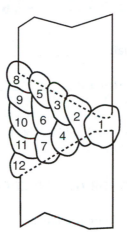

Figure 5-6

■ PRACTICE 5-1

Name _____ Date _____ Electrode Used _____

Class _____ Instructor _____ Grade _____

OBJECTIVE: After completing this practice, you should be able to weld stringer beads on carbon steel pipe in the 1G position using E6010, E6011, and E7018 electrodes.

EQUIPMENT AND MATERIALS NEEDED FOR THIS PRACTICE

1. A properly set-up and adjusted arc welding machine.

2. Proper safety protection (welding hood, safety glasses, wire brush, chipping hammer, leather gloves, long-sleeved shirt, long pants, leather boots or shoes, and a pair of pliers). Refer to Chapter 2 in the text for more specific safety information.

3. E6010 or E6011 and E7018 arc welding electrodes with a 1/8-in. (3-mm) diameter.

4. Schedule 40 mild steel pipe, 3 in. (76 mm) or larger in diameter.

5. A piece of soapstone.

6. A flexible straight edge.

INSTRUCTIONS

1. Be sure the pipe is clean and free of rust, grease, paint, or other contaminants that will affect the weld.

2. Use your flexible straightedge and soapstone and draw straight lines completely around the pipe with a spacing of 1/2 in.

3. Place the pipe horizontally on the welding table in a vee block made of angle iron. The vee block will hold the pipe steady and allow it to be moved easily between each weld bead.

4. Strike an arc on the pipe at the 11 o'clock position using an E6010 or E6011 electrode.

5. Make a stringer bead over the 12 o'clock position, stopping at the 1 o'clock position.

6. Roll the pipe until the end of the weld is at the 11 o'clock position.

7. Clean the weld crater by chipping and wire brushing.

8. Strike the arc again and establish a molten weld pool at the leading edge of the previous weld crater.

9. With the molten weld pool reestablished, move the electrode back on the weld bead just short of the last full ripple. This action will both reestablish good fusion and keep the weld bead size uniform.

10. Now that the new weld bead is tied or welded into the old weld, continue welding to the 1 o'clock position again.

11. Stop welding, roll the pipe, clean the crater, and resume welding.

12. Keep repeating this procedure until the weld is completely around the pipe.

13. Before the last weld is started, clean the beginning end of the first weld so that the end and beginning beads can be tied or welded together smoothly.

14. When you reach the beginning bead, swing your electrode around on both sides of the weld bead. The poor beginning of a weld bead is always high, narrow, and has little penetration. By swinging the weave pattern (the "C" pattern is best) on both sides of the bead, you can make the bead width uniform. The added heat will give deeper penetration at the starting point.

15. Hold the arc in the crater for a moment until the weld pool is built up.

16. Cool, chip, and inspect the bead for defects.

17. Repeat the beads as needed until they are defect free.

18. Repeat steps 1–17 using E7018 electrodes.

19. Turn off the welding machine and clean up your work area when you are finished welding.

INSTRUCTOR'S COMMENTS _____
_____
_____

■ PRACTICE 5-2

Name _____ Date _____ Electrode Used _____

Class _____ Instructor _____ Grade _____

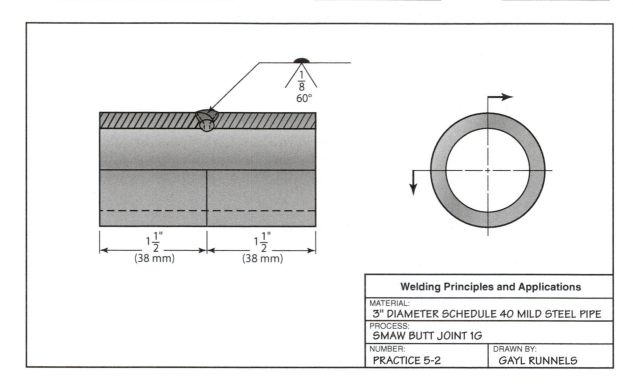

Welding Principles and Applications
MATERIAL: 3" DIAMETER SCHEDULE 40 MILD STEEL PIPE
PROCESS: SMAW BUTT JOINT 1G
NUMBER: PRACTICE 5-2   DRAWN BY: GAYL RUNNELS

OBJECTIVE: After completing this practice, you should be able to butt weld carbon steel pipe in the 1G position for low-pressure, light service applications.

EQUIPMENT AND MATERIALS NEEDED FOR THIS PRACTICE

1. A properly set-up and adjusted arc welding machine.

2. Proper safety protection (welding hood, safety glasses, wire brush, chipping hammer, leather gloves, long-sleeved shirt, long pants, leather boots or shoes, and a pair of pliers). Refer to Chapter 2 in the text for more specific safety information.

3. E6010 or E6011 arc welding electrodes with a 1/8-in. (3-mm) diameter.

4. Two or more pieces of Schedule 40 mild steel pipe, 3 in. (76 mm) or larger in diameter.

INSTRUCTIONS

1. Tack weld two pieces of pipe together.

2. Place the pipe horizontally in a vee block on the welding table.

3. Start the root pass at the 11 o'clock position.

4. Using a very short arc and high current setting, weld toward the 1 o'clock position.

5. Stop and roll the pipe, chip the slag, and repeat the weld until you have completed the root pass.

6. Clean the root pass by chipping and wire brushing. The root pass should not be ground this time.

7. Replace the pipe in the vee block on the table so that the hot pass can be done.

8. Turn up the machine amperage, enough to remelt the root weld surface, for the hot pass.

9. Use a step or whip electrode pattern, moving forward each time the molten weld pool washes out the slag, and returning each time the molten weld pool is nearly all solid.

10. Weld from the 11 o'clock position to the 1 o'clock position before stopping, rolling, and chipping the weld.

11. Repeat this procedure until the hot pass is complete. The filler pass and cover pass may be the same pass on this joint.

12. Turn down the machine amperage.

13. Use a "T," "J," "C," or zigzag pattern for the filler pass.

14. Start the weld at the 10 o'clock position and stop at the 12 o'clock position.

15. Sweep the electrode so that the molten weld pool melts out any slag trapped by the hot pass.

16. Watch the back edge of the bead to see that the molten weld pool is filling the groove completely.

17. Turn, chip, and continue the bead until the weld is complete.

18. Repeat this weld until you can consistently make welds free of defects.

19. Turn off the welding machine and clean up your work area when you are finished welding.

INSTRUCTOR'S COMMENTS _____

_____

_____

■ PRACTICE 5-3

Name _____   Date _____   Electrode Used _____

Class _____   Instructor _____   Grade _____

OBJECTIVE: After completing this practice, you should be able to butt weld carbon steel pipe in the 1G position using E6010 or E6011 electrodes for the root pass with E7018 electrodes for the filler and cover passes.

EQUIPMENT AND MATERIALS NEEDED FOR THIS PRACTICE

1. A properly set-up and adjusted arc welding machine.

2. Proper safety protection (welding hood, safety glasses, wire brush, chipping hammer, leather gloves, long-sleeved shirt, long pants, leather boots or shoes, and a pair of pliers). Refer to Chapter 2 in the text for more specific safety information.

3. E6010 or E6011 and E7018 arc welding electrodes with a 1/8-in. (3-mm) diameter.

4. Two or more pieces of Schedule 40 mild steel pipe, 3 in. (76 mm) or larger in diameter.

INSTRUCTIONS

1. Place the pipe on the arc welding table and tack weld two pieces together.

2. Strike an arc using, an E6010 or E6011 electrode and make a root weld that is as long as possible. If the root opening or gap is close and uniform, a straight, forward movement can be used. For wider gaps, a step or whip pattern must be used.

3. After completing and cleaning the root pass, make a hot pass. The hot pass need only burn out the slag to make the root pass clean. Undercut on the top of the pipe is acceptable. Use the E7018 electrode for these passes.

4. The filler and cover passes should be stringer beads or small weave beads. By keeping the molten weld pool size small, control is easier.

5. Cool, chip, and inspect the completed weld for uniformity and defects.

6. Repeat this weld until you can consistently make welds free of defects.

7. Turn off the welding machine and clean up your work area when you are finished welding.

INSTRUCTOR'S COMMENTS _____

_____

_____

■ PRACTICE 5-4

Name _____   Date _____   Electrode Used _____

Class _____   Instructor _____   Grade _____

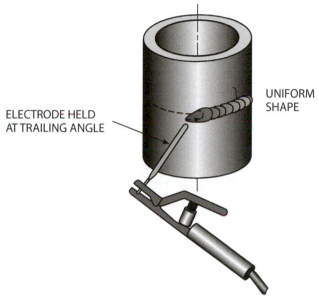

ELECTRODE HELD
AT TRAILING ANGLE

UNIFORM
SHAPE

OBJECTIVE: After completing this practice, you should be able to weld stringer beads on carbon steel pipe in the 2G position using E6010 or E6011 and E7018 electrodes.

EQUIPMENT AND MATERIALS NEEDED FOR THIS PRACTICE

1. A properly set-up and adjusted arc welding machine.

2. Proper safety protection (welding hood, safety glasses, wire brush, chipping hammer, leather gloves, long-sleeved shirt, long pants, leather boots or shoes, and a pair of pliers). Refer to Chapter 2 in the text for more specific safety information.

3. E6010 or E6011 and E7018 arc welding electrodes with a 1/8-in. (3-mm) diameter.

4. Schedule 40 mild steel pipe, 3 in. (76 mm) or larger in diameter.

INSTRUCTIONS

1. Place the pipe vertically on the welding table.

2. Hold the electrode at a 90° angle to the pipe and with a slight upward and trailing angle.

3. Use the "J" weave pattern for this practice.

4. Check the weld for uniformity and visual defects.

5. Repeat the weld until you can consistently make welds free of defects.

6. Turn off the welding machine and clean up your work area when you are finished welding.

INSTRUCTOR'S COMMENTS _____

_____

_____

■ PRACTICE 5-5

Name _____ Date _____ Electrode Used _____

Class _____ Instructor _____ Grade _____

$\frac{1}{8}$ IN. (3 MM) DI. E6010 OR E6011 ROOT PASS
E7018 FILLER AND COVER PASS

$\frac{1}{8}$
60°
G

Welding Principles and Applications

MATERIAL:	
3" DIAMETER SCHEDULE 40 MILD STEEL PIPE	
PROCESS:	
SMAW BUTT JOINT 2G	
NUMBER	DRAWN BY:
PRACTICE 5-5	LUCILLE GREENHAW

OBJECTIVE: After completing this practice, you should be able to butt weld mild steel pipe in the 2G position using E6010 or E6011 electrodes.

EQUIPMENT AND MATERIALS NEEDED FOR THIS PRACTICE

1. Properly set-up and adjusted arc welding machine.

2. Proper safety protection (welding hood, safety glasses, wire brush, chipping hammer, leather gloves, pliers, long-sleeved shirt, long pants, and leather boots or shoes). Refer to Chapter 2 in the text for more specific safety information.

3. E6010 or E6011 electrodes with a 1/8-in. (3-mm) diameter.

4. Two or more pieces of mild steel pipe, 3 in. (76 mm) or larger in diameter.

INSTRUCTIONS

1. Align the pieces of pipe and tack weld them together in at least four places.

2. Place the pipe vertically on the welding table.

3. Hold the electrode at a 90° angle to the pipe and with a slight trailing angle. The electrode should be held tightly into the joint. If a burn through occurs, quickly push the electrode back over the burn through while increasing the trailing angle. This action forces the weld metal back into the opening.

4. When the root pass is complete, chip the surface slag and then clean out the trapped slag by grinding or chipping, or use a hot pass technique on the next pass.

5. After the weld is completed, visually inspect it for 100% penetration around 80% of the root bead length.

6. Cool, chip, and inspect the weld for uniformity and visual defects on the cover pass.

7. Repeat the weld until you can consistently make welds free of defects.

8. Turn off the welding machine and clean up your work area when you are finished welding.

INSTRUCTOR'S COMMENTS _____

_____

_____

■ PRACTICE 5-6

Name _____   Date _____   Electrode Used _____

Class _____   Instructor _____   Grade _____

OBJECTIVE: After completing this practice, you should be able to make a butt weld on pipe in the 2G position.

EQUIPMENT AND MATERIALS NEEDED FOR THIS PRACTICE

Using the same setup, materials, and electrodes as listed in Practice 5-1, you will make straight stringer beads in the horizontal fixed 2G position using both groups of electrodes for the root pass and E7018 electrodes for the filler and cover passes.

INSTRUCTIONS

1. The root pass is to have 100% penetration over 80% or more of the length of the weld.

2. Place the pipe vertically on the welding table. Hold the electrode at a 90° angle to the pipe axis and with a slight trailing angle, Figure 5-42.

3. The electrode should be held tightly into the joint. If a burn through occurs, quickly push the electrode back over the burn through while increasing the trailing angle.

4. This action forces the weld metal back into the opening.

5. When the root pass is complete, chip the surface slag and then clean out the trapped slag by grinding or chipping, or use a hot pass.

6. Use E7018 electrodes for the filler and cover passes with a stringer pattern.

NOTE: The weave beads are not recommended with this electrode and position because they tend to undercut the top and overlap the bottom edge.

7. After the weld is completed, visually inspect it for 100% penetration around 80% of the root length.

8. Check the weld for uniformity and visual defects on the cover pass.

9. Repeat the weld until you can consistently make welds free of defects.

10. Turn off the welding machine and clean your work area when you are finished welding.

INSTRUCTOR'S COMMENTS _____

_____

_____

■ PRACTICE 5-7

Name _____   Date _____   Electrode Used _____

Class _____   Instructor _____   Grade _____

OBJECTIVE: After completing this practice, you should be able to make stringer welds around a pipe in the 5G position.

EQUIPMENT AND MATERIALS NEEDED FOR THIS PRACTICE

Using the same setup, materials, and electrodes as listed in Practice 5-1, you will make straight stringer beads in the horizontal fixed 5G position using both groups of electrodes.

INSTRUCTIONS

1. Clamp the pipe horizontally between waist and chest level.

2. Starting at the 11 o'clock position, make a downhill straight stringer bead through the 12 o'clock and 6 o'clock positions.

3. Stop at the 7 o'clock position, Figure 5-46. Using a new electrode, start at the 5 o'clock position and make an uphill straight stringer bead through the 6 o'clock and 12 o'clock positions.

4. Stop at the 1 o'clock position.

5. Change the electrode angle to control the molten weld pool.

6. Repeat these stringer beads as needed, with each group of electrodes, until you can consistently make welds free of defects. Turn off the welding machine and clean your work area when you are finished welding.

INSTRUCTOR'S COMMENTS _____

_____

_____

■ PRACTICE 5-8

Name _____ Date _____ Electrode Used _____

Class _____ Instructor _____ Grade _____

OBJECTIVE: After completing this practice, you should be able to make a butt weld around a pipe in the 5G position.

EQUIPMENT AND MATERIALS NEEDED FOR THIS PRACTICE

Using the same setup, materials, and procedures as listed in Practice 5-2, you will make a horizontal fixed 5G pipe weld.

INSTRUCTIONS

1. The root pass is to have 100% penetration over 80% or more of the length of the weld.

2. Mark the top of the pipe and mount it horizontally between waist and chest level.

3. Weld the root pass uphill or downhill using E6010 or E6011 electrodes.

4. Either grind the root pass or use a hot pass to clean out trapped slag.

5. Use E7018 electrodes for the filler and cover passes with stringer or weave patterns.

6. When the weld is completed, visually inspect it for 100% penetration around 80% of the root length.

7. Check the weld for uniformity and visual defects on the cover pass.

8. Repeat the weld until you can consistently make welds free of defects.

9. Turn off the welding machine and clean your work area when you are finished welding.

INSTRUCTOR'S COMMENTS _____

_____

_____

■ PRACTICE 5-9

Name _____ Date _____ Electrode Used _____

Class _____ Instructor _____ Grade _____

OBJECTIVE: After completing this practice, you should be able to make a butt weld on a pipe in the 5G position using both the uphill and downhill weld progression.

EQUIPMENT AND MATERIALS NEEDED FOR THIS PRACTICE

Using a properly set-up and adjusted arc welding machine, proper safety protection, E6010 or E6011 arc welding electrodes with a 1/8 in. (3 mm) diameter, and two or more pieces of schedule 40 mild steel pipe 3 in. (76 mm) or larger in diameter, you will make a butt joint on a horizontally fixed 5G pipe.

INSTRUCTIONS

1. Mark the top of the pipe and mount it between waist and chest level.

2. Depending on the root gap, make a root weld uphill or downhill using E6010 or E6011 electrodes.

3. Check the root penetration to determine if it is better in one area than in another area.

4. Chip and wire brush the weld and set the machine for a hot pass.

5. Start the hot pass at the bottom and weld upward on both sides.

6. The bead should be kept uniform with little buildup.

7. Using stringer or weave beads, make the filler and cover welds. If stringer beads are used, downhill welds can be made.

8. Cool, chip, and inspect the weld for uniformity and defects.

9. Repeat this weld until you can consistently make welds free of defects.

10. Turn off the welding machine and clean your work area when you are finished welding.

INSTRUCTOR'S COMMENTS _____

_____

_____

■ PRACTICE 5-10

Name _____   Date _____   Electrode Used _____

Class _____   Instructor _____   Grade _____

Welding Principles and Applications

MATERIAL:
3" DIAMETER SCHEDULE 40 MILD STEEL PIPE

PROCESS:
SMAW STRINGER BEADS 6G

NUMBER:
PRACTICE 5-10   DRAWN BY:

OBJECTIVE: After completing this practice, you should be able to weld stringer beads on mild steel pipe in the 45° fixed 6G position using E6010 or E6011 and E7018 electrodes.

EQUIPMENT AND MATERIALS NEEDED FOR THIS PRACTICE

1. Properly set-up and adjusted arc welding machine.

2. Proper safety protection (welding hood, safety glasses, wire brush, chipping hammer, leather gloves, pliers, long-sleeved shirt, long pants, and leather boots or shoes). Refer to Chapter 2 in the text for more specific safety information.

3. E6010 or E6011 and E7018 electrodes with a 1/8-in. (3-mm) diameter.

4. One or more pieces of mild steel pipe, 3 in. (76 mm) or larger in diameter.

INSTRUCTIONS

1. Clamp the pipe in a fixed 45° position and at a height that is between waist-high and chest level.

2. Begin by using the E6010 or E6011 electrodes.

3. Using the straight stepped "T" or whipping pattern, start at the bottom of the pipe and establish a molten weld pool in the overhead position. Keep the molten weld pool small and narrow for easier control.

4. Move the molten weld pool up the side of the pipe, keeping the electrode on the downhill side with a trailing or pulling angle.

5. When the weld passes the side, decrease the downhill and trailing angles. This is done so that when the weld reaches the top, the electrode is perpendicular to the top of the pipe.

6. Repeat steps 3 through 5 using the E7018 electrodes.

7. Chip and wire brush the welds and check for appearance and defects.

8. Repeat this practice as needed until you can consistently make welds free of defects.

9. Turn off the welding machine and clean up your work area when you are finished welding.

INSTRUCTOR'S COMMENTS _____

_____

_____

■ PRACTICE 5-11

Name _____ Date _____ Electrode Used _____

Class _____ Instructor _____ Grade _____

HORIZONTAL

VERTICAL

BUTT JOINT

OVERHEAD

WELDING POSITIONS CHANGE AS THE BEAD IS MADE

45° ± 5°

Welding Principles and Applications

MATERIAL:
3" DIAMETER SCHEDULE 40 MILD STEEL PIPE

PROCESS:
SMAW BUTT WELD 6G

NUMBER:
PRACTICE 5-11

DRAWN BY:

OBJECTIVE: After completing this practice, you should be able to make a butt weld joint on mild steel pipe in the 45° fixed 6G position using E6010 or E6011 electrodes.

EQUIPMENT AND MATERIALS NEEDED FOR THIS PRACTICE

1. Properly set-up and adjusted arc welding machine.

2. Proper safety protection (welding hood, safety glasses, wire brush, chipping hammer, leather gloves, pliers, long-sleeved shirt, long pants, and leather boots or shoes). Refer to Chapter 2 in the text for more specific safety information.

3. E6010 or E6011 electrodes with a 1/8-in. (3-mm) diameter.

4. Two or more pieces of mild steel pipe, 3 in. (76 mm) or larger in diameter.

INSTRUCTIONS

1. Tack weld two pieces of pipe securely together in four or more places.

2. Clamp the pipe in a fixed 45° position and at a height that is between waist-high and chest level.

3. Use the E6010 or E6011 electrodes.

4. Start at the top of the pipe, make a vertical down root pass that ends just beyond the bottom.

5. Repeat this weld on the other side.

6. Chip and wire brush the slag so that an uphill hot pass can be made.

7. Make the uphill hot pass. It must be kept small and concave so that more slag will not be trapped along the downhill side.

8. Clean the bead, turn down the amperage, and complete the joint with stringer beads.

9. Chip and wire brush the welds and check for appearance and defects.

10. Repeat this practice as needed until you can consistently make welds free of defects.

11. Turn off the welding machine and clean up your work area when you are finished welding.

INSTRUCTOR'S COMMENTS _____

_____

_____

■ PRACTICE 5-12

Name _____ Date _____ Electrode Used _____

Class _____ Instructor _____ Grade _____

OBJECTIVE: After completing this practice, you should be able to make butt weld around a pipe in the 6G position.

EQUIPMENT AND MATERIALS NEEDED FOR THIS PRACTICE

Using the same setup, materials, and procedures as listed in Practice 5-11, you will make a 45° fixed inclined pipe weld. The root pass is to have 100% penetration over 80% of the root length of the weld.

INSTRUCTIONS

1. Make the root pass as a vertical up or down weld, depending on the root opening.

2. Chip the slag and clean the weld by grinding or by using a hot pass. If a hot pass is used, chip and wire brush the joint.

3. Using the E7018 electrode, start slightly before the center on the bottom and make a small stringer bead in an upward direction.

4. Keep a trailing and a somewhat uphill electrode angle so that the weld is deposited on the bottom of the lower pipe.

5. The next pass should use a downhill electrode angle so that the bead is on the uphill pipe.

6. Alternate this process until the bead is complete.

7. Clean the weld and inspect it for 100% penetration over 80% of the root length.

8. Check for uniformity and visual defects on the cover pass. Repeat the weld until you can consistently make welds free of defects.

9. Turn off the welding machine and clean your work area when you are finished welding.

INSTRUCTOR'S COMMENTS _____

_____

_____

CHAPTER 5: QUIZ 1

Name _____ Date _____

Class _____ Instructor _____ Grade _____

INSTRUCTIONS

Carefully read Chapter 5 in the textbook and answer each question.

MATCHING

In the space provided to the left of Column A, write the letter from Column B that best answers or completes the statement in Column A.

Column A

Column B

_____ 1. Pipe is used to carry materials such as _____.

a. handrails

_____ 2. Pipe and tubing are also used for structures such as _____.

b. inside

_____ 3. Low-pressure piping systems are considered _____.

c. schedule

_____ 4. Medium-pressure piping is used for heavy-service structural items such as _____.

d. loss of life

_____ 5. High-pressure piping systems are considered critical because if they fail it could result in the possible _____.

e. penetration

_____ 6. Pipe and tubing are both available as welded (seamed) or extruded (_____).

f. less turbulence

_____ 7. Pipe sizes are given as the _____ diameter for pipe that is 12 in. (305 mm) in diameter or smaller.

g. outside

_____ 8. Tubing sizes are always given as the _____ diameter.

h. flexible

_____ 9. Pipe strength is given as a _____.

i. Machines

_____ 10. Tubing can be specified as rigid or _____.

j. petroleum

_____ 11. Rigid tubing is normally used for _____ applications.

k. root face

_____ 12. Welded piping systems, compared to pipe joined by any other method, are _____.

l. leaks

_____ 13. The thickness and strength of the pipe and fitting _____ when they are welded together.

m. overtighten

_____ 14. Over much time and use, welded pipe joints are resistant to _____.

n. noncritical

_____ 15. Small cracks between the threads on threaded pipe are likely spots for _____ to start.

o. corrosion

——— 16. The inside of a welded fitting is the same size as the pipe itself, so as material flows through the pipe, _____ is caused by unequal diameters.

p. stay the same

——— 17. Alignment of welded parts is easier because it is not necessary to _____ fittings so that they will line up.

q. stronger

——— 18. The ends of pipe must be beveled for maximum _____ and high joint strength.

r. trailer axles

——— 19. _____ designed specifically for beveling pipe can accurately cut a 37 ½° angle.

s. structural

——— 20. The _____ will help a welder control both penetration and root suck-back.

t. seamless

CHAPTER 5: QUIZ 2

Name _____   Date _____

Class _____   Instructor _____   Grade _____

INSTRUCTIONS

Carefully read Chapter 5 in the textbook and answer each question.

MATCHING

In the space provided to the left of Column A, write the letter from Column B that best answers or completes the statement in Column A.

Column A	Column B
_____ 1. _____ is caused by the surface tension of the molten metal trying to pull itself into a ball.	a. narrow grinding disk
_____ 2. One of the major problems to be overcome in pipe welding is learning how to make the _____ from one position to another.	b. At the 11 o'clock position
_____ 3. _____ outside of the weld groove are considered to be defects by much of the pipe welding industry.	c. penetration and buildup
_____ 4. A _____ is the first weld in a joint.	d. continue past
_____ 5. The _____ is the part of this weld pass that must be smooth and uniform.	e. root face
_____ 6. A grinder with a _____ can be used to repair the face of the root pass.	f. bottom
_____ 7. The hot pass is used to quickly _____ small amounts of slag trapped along the edge of the root pass.	g. cover pass or cap
_____ 8. The filler pass(es) may be a series of _____ beads.	h. Arc strikes
_____ 9. When the weld has gone completely around the pipe, it should _____ the starting point so that good fusion is ensured.	i. burn out
_____ 10. The final covering on a weld is referred to as the _____.	j. vertical
_____ 11. E7018 electrodes are highly susceptible to _____.	k. 5G
_____ 12. Where should the root pass start for the 1G position? _____.	l. 6G
_____ 13. The _____ of the weld are controlled more easily with the pipe in the 1G position.	m. Root suck-back
_____ 14. With the molten weld pool reestablished, move the electrode back on the weld bead just short of the last _____.	n. undercut

_____ 15. In the 2G vertical fixed pipe position, the pipe is _____ and the weld is horizontal.

o. transition

_____ 16. The weave beads are not recommended with E7018 electrodes and the 2G position because they tend to _____ the top edge.

p. stringer

_____ 17. The 5G _____ fixed pipe position is the most often used pipe welding position.

q. root weld

_____ 18. The _____ root pass can be performed by welding uphill or downhill.

r. moisture

_____ 19. Which welding position has the pipe mounted at a 45° angle? _____

s. horizontal

_____ 20. On a 6G weld, start the root pass at the top and make a vertical down root pass that ends just beyond the _____.

t. full ripple

CHAPTER 5: QUIZ 3

Name _____   Date _____

Class _____   Instructor _____   Grade _____

INSTRUCTIONS

Carefully read Chapter 5 in the textbook and answer each question.

IDENTIFICATION

Identify the numbered items on the drawing by writing the letter next to the identifying term in the space provided.

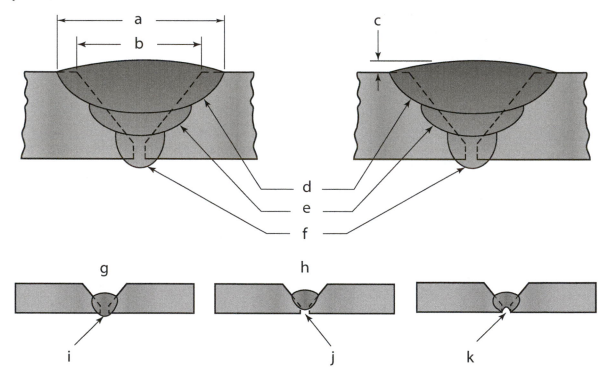

1. _____ CONCAVE ROOT FACE

2. _____ GROOVE WIDTH

3. _____ COMPLETE ROOT FACE PENETRATION

4. _____ COVER PASS

5. _____ COVER PASS MAX. WIDTH = $\frac{1}{8}''$ (3 mm) WIDER THAN THE GROOVE

6. _____ ROOT PASS

7. _____ POOR ROOT PASS

8. _____ FILLER PASS

9. _____ GOOD ROOT PASS

10. _____ MAX. COVER PASS REINFORCEMENT $\frac{1}{8}''$ (3 mm)

11. _____ INCOMPLETE ROOT FACE PENETRATION

CHAPTER 5: QUIZ 4

Name _____   Date _____

Class _____   Instructor _____   Grade _____

INSTRUCTIONS

Carefully read Chapter 5 in the textbook and answer each question.

IDENTIFICATION

Identify the numbered items on the drawing by writing the letter next to the identifying term in the space provided.

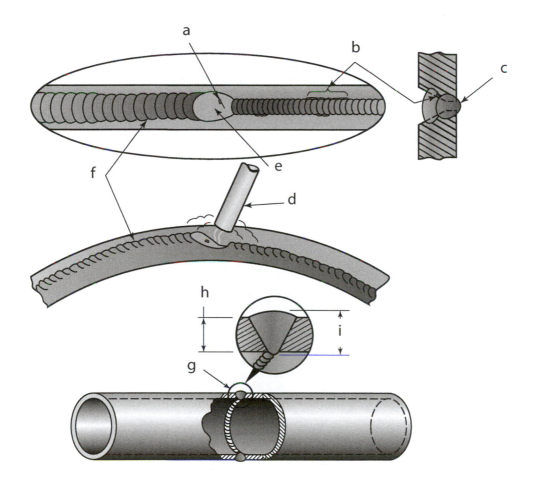

1. _____   SLAG BEING MELTED OUT

2. _____   ROOT PASS

3. _____   PIPE WALL THICKNESS

4. _____   TRAPPED SLAG

5. _____   HOT PASS

6. _____   WELDED JOINT THICKNESS

7. _____   WELDED JOINT

8. _____   ELECTRODE

9. _____   WELD CRATER

Shielded Metal Arc Welding
AWS SENSE Certification

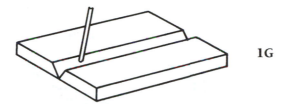

1G

1G Flat position—The welding position used to weld from the upper side of the joint; the face of the weld is approximately horizontal.

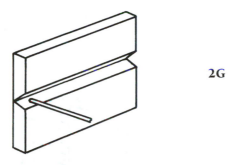

2G

2G Horizontal position—The position of welding in which the axis of the weld lies in an approximately horizontal plane and the face of the weld lies in an approximately vertical plane.

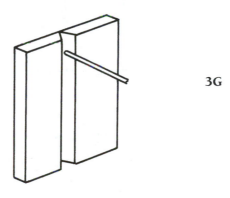

3G

3G Vertical position—The position of welding in which the axis of the weld is approximately vertical.

4G

4G Overhead position—The position in which welding is performed from the underside of the joint.

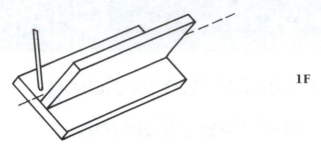

1F   Flat position—The welding position used to weld from the upper side of the joint; the face of the weld is approximately horizontal.

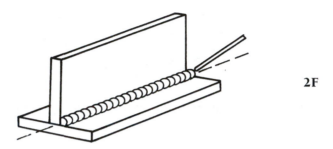

2F   Horizontal position—The position in which welding is performed on the upper side of an approximately horizontal surface and against an approximately vertical surface.

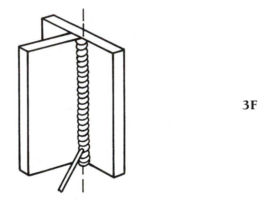

3F   Vertical position—The position of welding in which the axis of the weld is approximately vertical.

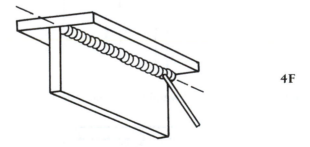

4F   Overhead position—The position in which welding is performed from the underside of the joint.

■ EXPERIMENT 6-1

Name _____   Date _____   Electrode Used _____

Class _____   Instructor _____   Grade _____

Welding Principles and Applications

MATERIAL:	
1/8" X 6" MILD STEEL PLATE	
PROCESS:	
SMAW BUTT JOINT 1G, 2G, 3G, 4G	
NUMBER:	**DRAWN BY:**
EXPERIMENT 6-1	*RONNIE AGA*

OBJECTIVE: After completing this experiment, you should be able to weld a butt joint in all positions with a backing strip using E6010 or E6011 electrodes.

EQUIPMENT AND MATERIALS NEEDED FOR THIS PRACTICE

1. A properly set-up and adjusted welding machine.

2. Proper safety protection (welding hood, safety glasses, wire brush, chipping hammer, leather gloves, pliers, long-sleeved shirt, long pants, and leather boots or shoes). Refer to Chapter 2 in the text for more specific safety information.

3. E6010 or E6011 arc welding electrodes with a 1/8-in. (3-mm) diameter.

4. Three pieces of mild steel plate: (2) 1/8 in. (3 mm) thick by 1 1/2 in. (38 mm) wide by 6 in. (152 mm) long; and one strip of mild steel, (1) 1/8 in. (3 mm) thick by 1 in. (25 mm) wide by 6 in. (152 mm) long.

INSTRUCTIONS

1. Tack weld the plates together with a 1/16-in. (2-mm) to 1/8-in. (3-mm) root opening; see the above figure. Be sure there are no gaps between the backing strip and plates when the pieces are tacked together. If there is a small gap between the backing strip and the plates, it can be removed by placing the assembled test plates on an anvil and striking the tack weld with a hammer. This will close up the gap by compressing the tack welds.

2. Chip and wire brush tack welds.

3. Adjust the welding current by welding on a scrap plate.

4. Begin with plates in the flat position.

5. Use a straight step or "T" pattern for this root weld. Push the electrode into the root opening so that there is complete fusion with the backing strip and bottom edge of the plates. Failure to push the penetration deep into the joint will result in a cold lap at the root.

6. Watch the molten weld pool and keep its size, height, and width as uniform as possible. As the molten weld pool increases in size, move the electrode out of the weld pool. When the weld pool begins to cool, bring the electrode back into the molten weld pool. Use these weld pool indications to determine how far to move the electrode and when to return to the molten weld pool. After completing the weld, cut the plate and inspect the cross section of the weld for complete fusion at the edges.

7. Chip and wire brush the weld and examine to see if you have complete fusion. Repeat the welds as necessary until you can consistently make welds free of defects.

8. Repeat above steps in the 2G, 3G, and 4G positions.

9. Turn off the welding machine and clean up your work area when you are finished welding.

INSTRUCTOR'S COMMENTS _____

_____

_____

■ EXPERIMENT 6-2

Name _____   Date _____   Electrode Used _____

Class _____   Instructor _____   Grade _____

OBJECTIVE: After completing this experiment, you should be able to weld a root pass with an open root using E6010 or E6011 electrodes in the 1G, 2G, 3G, and 4G positions.

EQUIPMENT AND MATERIALS NEEDED FOR THIS PRACTICE

1. A properly set-up and adjusted welding machine.

2. Proper safety protection (welding hood, safety glasses, wire brush, chipping hammer, leather gloves, pliers, long-sleeved shirt, long pants, and leather boots or shoes). Refer to Chapter 2 in the text for more specific safety information.

3. E6010 or E6011 arc welding electrodes with a 1/8-in. (3-mm) diameter.

4. Two pieces of mild steel plate, 1 1/2 in. (38 mm) wide by 6 in. (152 mm) long by 1/8 in. (3 mm) thick.

INSTRUCTIONS

1. Tack weld the plates together with a root opening of 0 in. (0 mm) to 1/16 in. (2 mm).

2. Chip and wire brush all tack welds.

3. Use a short arc length and high amperage to weld the joint.

4. Control penetration by the electrode angle used.

5. If burn through occurs, move back and lower the electrode angle.

6. After the weld is completed, it can be visually inspected for uniformity and complete penetration. Repeat this weld until you can consistently make it defect free.

7. Turn off the welding machine and clean up your work area when you are finished welding.

INSTRUCTOR'S COMMENTS _____

_____

_____

■ EXPERIMENT 6-3

Name _____ Date _____

Class _____ Instructor _____ Grade _____

OBJECTIVE: After completing this experiment, you should be able to weld an open root joint using the step technique with E6010 or E6011 electrodes in the 1G, 2G, 3G, and 4G positions.

EQUIPMENT AND MATERIALS NEEDED FOR THIS PRACTICE

1. A properly set-up and adjusted welding machine.

2. Proper safety protection (welding hood, safety glasses, wire brush, chipping hammer, leather gloves, pliers, long-sleeved shirt, long pants, and leather boots or shoes). Refer to Chapter 2 in the text for more specific safety information.

3. E6011 or E6011 arc welding electrodes with a 1/8-in. (3-mm) diameter.

4. Two pieces of mild steel plate 1 1/2 in. (38 mm) wide by 6 in. (152 mm) long by 1/8 in. (3 mm) thick.

INSTRUCTIONS

1. Tack weld the plates together with a root opening of 0 in. (0 mm) to 1/16 in. (2 mm).

2. Chip and wire brush all tack welds.

3. Begin with the plate in the flat position.

4. Starting at one end, make a stringer bead along the entire length using the straight step or "T"-weave pattern. Use a medium amperage setting.

5. As the bead is being made, push the electrode deeply into the root to establish a keyhole.

6. Once the keyhole is established, the electrode is moved out and back in the molten weld pool at a steady, rhythmic rate.

7. After the weld is completed, visually inspect it for uniformity and penetration. Repeat this weld until you can consistently make it defect free.

8. Repeat the above steps in the 2G, 3G, and 4G positions.

9. Turn off the welding machine and clean up your work area when you are finished welding.

INSTRUCTOR'S COMMENTS _____

_____

_____

■ PRACTICE 6-1

AWS SENSE LEVEL I

Name _____   Date _____

Class _____   Instructor _____   Grade _____

WELDING PROCEDURE SPECIFICATION (WPS)

Welding Procedure Specification No: Practice 6-1

TITLE:

Welding SMAW of plate to plate.

SCOPE:

This procedure is applicable for V-groove plate with a backing strip within the range of 1/8 in. (3 mm) through 1-1/2 in. (38 mm).

Welding may be performed in the following positions: 3G.

BASE METAL:

The base metal shall conform to Carbon Steel M-1 or P-1, Group 1 or 2.

Backing Material Specification Carbon Steel M-1 or P-1, Group 1, 2, or 3.

FILLER METAL:

The filler metal shall conform to AWS specification no. E7018 from AWS specification A5.1. This filler metal falls into F-number F-4 and A-number A-1.

SHIELDING GAS:

The shielding gas, or gases, shall conform to the following compositions and purity:

N/A.

JOINT DESIGN AND TOLERANCES:

Joint Details:

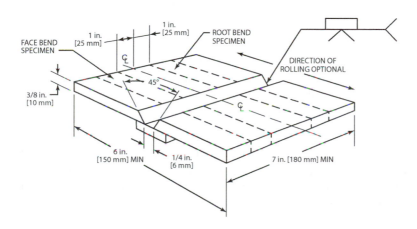

PREPARATION OF BASE METAL:

The bevel is to be flame or plasma cut on the edge of the plate before the parts are assembled. The beveled surface must be smooth and free of notches. Any roughness or notches that are deeper than 1/64 in. (0.4 mm) must be ground smooth.

All hydrocarbons and other contaminants, such as cutting fluids, grease, oil, and primers, must be cleaned off of all parts and filler metals before welding. This cleaning can be done with any suitable solvents or detergents. The backing strip, groove face, and inside and outside plate surface within 1 in. (25 mm) of the joint must be mechanically cleaned of slag, rust, and mill scale. Cleaning must be done with a wire brush or grinder down to bright metal.

ELECTRICAL CHARACTERISTICS:

The current shall be <u>direct-current electrode positive (DCEP)</u>. The base metal shall be on the <u>negative</u> side of the line.

WELDS	FILLER METAL DIA.	CURRENT	AMPERAGE RANGE
Tack	3/32 in. (2.4 mm)	DCEP	70 to 115
Root	1/8 in. (3.2 mm)	DCEP	115 to 165
Filler	5/32 in. (4 mm)	DCEP	150 to 220

PREHEAT:

The parts must be heated to a temperature higher than 50°F (10°C) before any welding is started.

BACKING GAS:

N/A.

SAFETY:

Proper protective clothing and equipment must be used. The area must be free of all hazards that may affect the welder or others in the area. The welding machine, welding leads, work clamp, electrode holder, and other equipment must be in safe working order.

WELDING TECHNIQUE:

Tack weld the plates together with the backing strip. There should be about a 1/4-in. (6-mm) root gap between the plates. Use the E7018 arc welding electrodes to make a root pass to fuse the plates and backing strip together. Clean the slag from the root pass, being sure to remove any trapped slag along the sides of the weld. Using the E7018 arc welding electrodes, make a series of stringer or weave filler welds, no thicker than 1/4 in. (6.4 mm), in the groove until the joint is filled.

INTERPASS TEMPERATURE:

The plate should not be heated to a temperature higher than 500°F (260°C) during the welding process. After each weld pass is completed, allow it to cool but never to a temperature below 50°F (10°C). The weldment must not be quenched in water.

CLEANING:

The slag must cleaned off between passes. The weld beads may be cleaned by a hand wire brush, a chipping hammer, a punch and hammer, or a needle scaler. All weld cleaning must be performed with the test plate in the welding position. A grinder may not be used to remove weld control problems such as undercut, overlap, trapped slag, and so forth.

INSPECTION:

Visual Inspection Criteria for Entry Welders: There shall be no cracks, no incomplete fusion. There shall be no incomplete joint penetration in groove welds except as permitted for partial joint penetration welds.

The Test Supervisor shall examine the weld for acceptable appearance and shall be satisfied that the welder is skilled in using the process and procedure specified for the test.

Undercut shall not exceed the lesser of 10% of the base metal thickness or 1/32 in. (0.8 mm).

Where visual examination is the only criterion for acceptance, all weld passes are subject to visual examination at the discretion of the Test Supervisor.

The frequency of porosity shall not exceed one in each 4 in. (100 mm) of weld length, and the maximum diameter shall not exceed 3/32 in. (2.4 mm).

Welds shall be free from overlap.

REPAIR:

No repairs of defects are allowed.

BEND TEST:

Transverse face bend. The weld is perpendicular to the longitudinal axis of the specimen and is bent so that the weld face becomes the tension surface of the specimen.

Transverse root bend. The weld is perpendicular to the longitudinal axis of the specimen and is bent so that the weld root becomes the tension surface of the specimen.

ACCEPTANCE CRITERIA FOR BEND TEST:

For acceptance, the convex surface of the face- and root-bend specimens shall meet both of the following requirements:

1. No single indication shall exceed 1/8 in. (3.2 mm) measured in any direction on the surface.

2. The sum of the greatest dimensions of all indications on the surface that exceed 1/32 in. (0.8 mm) but are less than or equal to 1/8 in. (3.2 mm) shall not exceed 3/8 in. (9.5 mm).

Cracks occurring at the corner of the specimens shall not be considered unless there is definite evidence that they result from slag or inclusions or other internal discontinuities.

INSTRUCTOR'S COMMENTS _____

_____

_____

■ PRACTICE 6-2

Name _____   Date _____

Class _____   Instructor _____   Grade _____

WELDING PROCEDURE SPECIFICATION (WPS)

Welding Procedure Specification No: <u>Practice 6-2</u>

TITLE:

Welding <u>SMAW</u> of <u>plate</u> to <u>plate</u>.

SCOPE:

This procedure is applicable for bevel and V-groove plates with a backing strip within the range of 3/8 in. (9.5 mm) through 3/4 in. (19 mm).

Welding may be performed in the following positions: <u>1G, 2G, 3G, and 4G</u>.

BASE METAL:

The base metal shall conform to <u>Carbon Steel M-1 or P-1, Group 1 or 2</u>.

Backing Material Specification <u>Carbon Steel M-1 or P-1, Group 1, 2, or 3</u>.

FILLER METAL:

The filler metal shall conform to AWS specification no. <u>E7018</u> from AWS specification <u>A5.1</u>. This filler metal falls into F-number <u>F-4</u> and A-number <u>A-1</u>.

SHIELDING GAS:

The shielding gas, or gases, shall conform to the following compositions and purity: <u>N/A</u>.

JOINT DESIGN AND TOLERANCES:

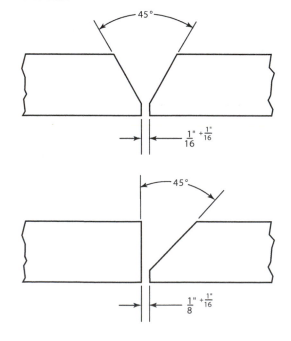

PREPARATION OF BASE METAL:

The bevel is to be flame or plasma cut on the edge of the plate before the parts are assembled. The beveled surface must be smooth and free of notches. Any roughness or notches that are deeper than 1/64 in. (0.4 mm) must be ground smooth.

All hydrocarbons and other contaminants, such as cutting fluids, grease, oil, and primers, must be cleaned off of all parts and filler metals before welding. This cleaning can be done with any suitable solvents or detergents. The backing strip, groove face, and inside and outside plate surface within 1 in. (25 mm) of the joint must be mechanically cleaned of slag, rust, and mill scale. Cleaning must be done with a wire brush or grinder down to bright metal.

ELECTRICAL CHARACTERISTICS:

The current shall be <u>direct-current electrode positive (DCEP)</u>. The base metal shall be on the <u>negative</u> side of the line.

AWS CLASSIFICATION AND POLARITY	ELECTRODE DIAMETER AND AMPERAGE RANGE		
	$\frac{3"}{32}$	$\frac{1"}{8}$	$\frac{5"}{32}$
E6010 - DCEP	40-80	70-130	110-165
E6011 - DCEP	50-70	85-125	130-160
E7018 - DCEP	70-110	90-165	125-220

PREHEAT:

The parts must be heated to a temperature higher than 50°F (10°C) before any welding is started.

BACKING GAS:

N/A.

SAFETY:

Proper protective clothing and equipment must be used. The area must be free of all hazards that may affect the welder or others in the area. The welding machine, welding leads, work clamp, electrode holder, and other equipment must be in safe working order.

WELDING TECHNIQUE:

Tack weld the plates together with the backing strip. There should be about a 1/8-in. (3.2-mm) root gap between the plates. Use the E7018 arc welding electrodes to make a root pass to fuse the plates and backing strip together. Clean the slag from the root pass, being sure to remove any trapped slag along the sides of the weld.

Using the E7018 arc welding electrodes, make a series of filler welds in the groove until the joint is filled.

INTERPASS TEMPERATURE:

The plate, outside of the heat-affected zone, should not be heated to a temperature higher than 500°F (260°C) during the welding process. After each weld pass is completed, allow it to cool; the weldment must not be quenched in water.

CLEANING:

The slag must cleaned off between passes. The weld beads may be cleaned by a hand wire brush, a chipping hammer, a punch and hammer, or a needle scaler. All weld cleaning must be performed with the test plate in the welding position. A grinder may not be used to remove weld control problems such as undercut, overlap, trapped slag, and so forth.

INSPECTION:

Visual Inspection Criteria for Entry Welders: There shall be no cracks, no incomplete fusion. There shall be no incomplete joint penetration in groove welds except as permitted for partial joint penetration welds.

The Test Supervisor shall examine the weld for acceptable appearance and shall be satisfied that the welder is skilled in using the process and procedure specified for the test.

Undercut shall not exceed the lesser of 10% of the base metal thickness or 1/32 in. (0.8 mm).

Where visual examination is the only criterion for acceptance, all weld passes are subject to visual examination at the discretion of the Test Supervisor.

The frequency of porosity shall not exceed one in each 4 in. (100 mm) of weld length, and the maximum diameter shall not exceed 3/32 in. (2.4 mm).

Welds shall be free from overlap.

REPAIR:

No repairs of defects are allowed.

SKETCHES:

1G, 2G, 3G, and 4G test plate drawing.

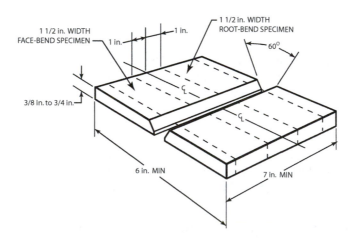

BEND TEST:

Transverse face bend. The weld is perpendicular to the longitudinal axis of the specimen and is bent so that the weld face becomes the tension surface of the specimen.

Transverse root bend. The weld is perpendicular to the longitudinal axis of the specimen and is bent so that the weld root becomes the tension surface of the specimen.

ACCEPTANCE CRITERIA FOR BEND TEST:

For acceptance, the convex surface of the face- and root-bend specimens shall meet both of the following requirements:

1. No single indication shall exceed 1/8 in. (3.2 mm) measured in any direction on the surface.

2. The sum of the greatest dimensions of all indications on the surface that exceed 1/32 in. (0.8 mm) but are less than or equal to 1/8 in. (3.2 mm) shall not exceed 3/8 in. (9.5 mm).

Cracks occurring at the corner of the specimens shall not be considered unless there is definite evidence that they result from slag or inclusions or other internal discontinuities.

INSTRUCTOR'S COMMENTS _____

_____

_____

■ PRACTICE 6-3

Name _____ Date _____

Class _____ Instructor _____ Grade _____

WELDING PROCEDURE SPECIFICATION (WPS)

Welding Procedure Specification No.: <u>Practice 6-3</u>

TITLE:

Welding SMAW of pipe to pipe.

SCOPE:

This procedure is applicable for V-groove pipe within the range of 3 in. (76 mm) schedule 40 through 8 in. (203 mm) schedule 40.

Welding may be performed in the following positions: 1G, 2G, 5G, and 6G.

BASE METAL:

The base metal shall conform to Carbon Steel M-1/P/1S-1, Group 1 or 2. Backing material specification: N/A.

FILLER METAL:

The filler metal shall conform to AWS specification No. E6010 or E6011 root pass and E7018 for the cover pass from AWS specification A5.1. This filler metal falls into F-numbers F3 and F4 and A-number A-1.

SHIELDING GAS:

The shielding gas, or gases, shall conform to the following compositions and purity: N/A.

JOINT DESIGN AND TOLERANCES:

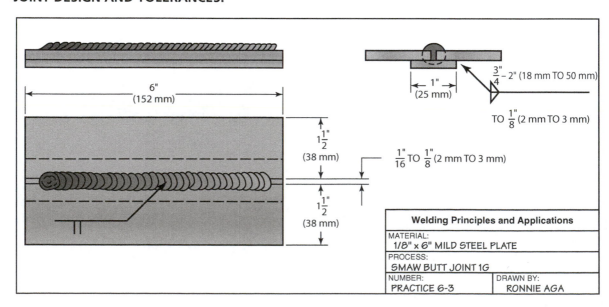

PREPARATION OF BASE METAL:

All workmanship part must prepared according to the AWS Workmanship Standard for Preparation of Base Metal.

ELECTRICAL CHARACTERISTICS:

The current shall be DCEP. The base metal shall be on the work lead or negative side of the line.

PREHEAT:

The parts must be heated to a temperature higher than 70°F (21°C) before any welding is started.

BACKING GAS:

N/A

WELDING TECHNIQUE:

Tack weld the pipes together; there should be about a 1/8-in. (3-mm) root gap between the pipe ends. Use the E6010 or E6011 arc welding electrodes to make a root pass to fuse the pipe ends together. Clean the slag from the root pass and use either a hot pass or grinder to remove any trapped slag.

Using the E7018 arc welding electrodes, make a series of filler welds in the groove until the joint is filled. Figure 6-41 shows the recommended location sequence for the weld beads for the 1G and 5G positions, Figure 6-42 for the 2G position, and Figure 6-43 for the 6G position.

INTERPASS TEMPERATURE:

The plate should not be heated to a temperature higher than 350°F (175°C) during the welding process. After each weld pass is completed, allow it to cool; the weldment must not be quenched in water.

CLEANING:

The slag can be chipped and/or ground off between passes but can only be chipped off of the cover pass.

VISUAL INSPECTION CRITERIA FOR ENTRY-LEVEL WELDERS:

The weld must pass a visual inspection by the instructor or test supervisor based on the AWS Visual Inspection Criteria.

BEND TEST:

The weld is to be mechanically tested only after it has passed the visual inspection. Be sure that the test specimens are properly marked to identify the welder, the position, and the process.

SPECIMEN PREPARATION:

The weld must prepared for bend testing in accordance to the AWS Specimen Preparation Criteria.

ACCEPTANCE CRITERIA FOR BEND TEST:

The bent specimen must not exceed any of the acceptable limits of discontinuities as listed in the AWS Acceptance Criteria for Bend Test.

REPAIR:

No repairs of defects are allowed.

SKETCHES:

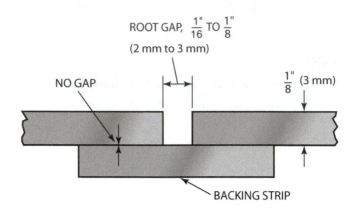

ROOT GAP, $\frac{1"}{16}$ TO $\frac{1"}{8}$
(2 mm to 3 mm)

NO GAP

$\frac{1"}{8}$ (3 mm)

BACKING STRIP

INSTRUCTOR'S COMMENTS _____
_____
_____

■ PRACTICE 6-4

AWS SENSE

Name _____ Date _____

Class _____ Instructor _____ Grade _____

WELDING PROCEDURE SPECIFICATION (WPS)

Welding Procedure Specification No.: Practice 6-4

TITLE:

Welding SMAW of pipe to pipe.

SCOPE:

This procedure is applicable for V-groove pipe with or without a back ring within the range of 6 in. (150 mm) schedule 40.

Welding may be performed in the following positions: 6G (AWS SENSE Level II).

BASE METAL:

The base metal shall conform to Carbon Steel M-1/P/1S-1, Group 1 or 2.

BACKING RING:

Backing ring to suit diameter and nominal wall thickness of pipe.

FILLER METAL:

The filler metal shall conform to AWS specification, 1/8 in. diameter, No. E7018 for root, fill, and cover passes from AWS specification A5.1. This filler metal falls into F-numbers F4 and A-number A-1.

SHIELDING GAS:

The shielding gas, or gases, shall conform to the following compositions and purity: N/A.

JOINT DESIGN AND TOLERANCES:

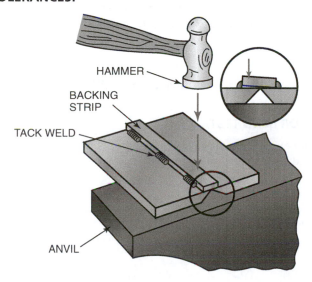

PREPARATION OF BASE METAL:

All workmanship part must prepared according to the AWS Workmanship Standard for Preparation of Base Metal.

ELECTRICAL CHARACTERISTICS:

The current shall be DCEP. The base metal shall be on the work lead or negative side of the line.

PREHEAT:

The parts must be heated to a temperature higher than 70°F (21°C) before any welding is started.

BACKING GAS:

N/A

WELDING TECHNIQUE:

Tack weld the pipes together; there should be about a 1/8-in. (3-mm) root gap between the pipe ends. Use E7018 arc welding electrodes to make a root pass to fuse the pipe ends together. Clean the slag from the root pass and use either a hot pass or grinder to remove any trapped slag.

Using the E7018 arc welding electrodes, make a series of filler welds in the groove until the joint is filled. Figure 6-40 shows the recommended location sequence for the weld beads for the 6G position.

INTERPASS TEMPERATURE:

The plate should not be heated to a temperature higher than 350°F (175°C) during the welding process. After each weld pass is completed, allow it to cool; the weldment must not be quenched in water.

CLEANING:

The slag can be chipped and/or ground off between passes but can only be chipped off of the cover pass.

VISUAL INSPECTION CRITERIA FOR ENTRY-LEVEL WELDERS:

The weld must pass a visual inspection by the instructor or test supervisor based on the AWS Visual Inspection Criteria.

BEND TEST:

The weld is to be mechanically tested only after it has passed the visual inspection. Be sure that the test specimens are properly marked to identify the welder, the position, and the process.

SPECIMEN PREPARATION:

The weld must prepared for bend testing in accordance to the AWS Specimen Preparation Criteria.

ACCEPTANCE CRITERIA FOR BEND TEST:

The bent specimen must not exceed any of the acceptable limits of discontinuities as listed in the AWS Acceptance Criteria for Bend Test.

REPAIR:

No repairs of defects are allowed.

SKETCHES:

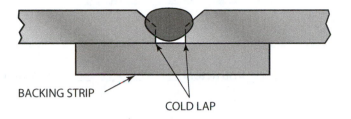

BACKING STRIP

COLD LAP

INSTRUCTOR'S COMMENTS _____
_____
_____

CHAPTER 6: QUIZ 1

Name _____ Date _____

Class _____ Instructor _____ Grade _____

INSTRUCTIONS

Carefully read Chapter 6 in the textbook and answer each question.

MATCHING

In the space provided to the left of Column A, write the letter from Column B that best answers or completes the statement in Column A.

	Column A	Column B
_____	1. The _____ pass is the first weld bead of a multiple pass weld.	a. angle
_____	2. The root pass fuses the two parts together and establishes the _____.	b. adjustment
_____	3. The backing strip used in a _____ root may remain as part of the weld, or it may be removed.	c. burn through
_____	4. You can change the electrode _____ to control penetration and burn through.	d. hot pass
_____	5. When a burn through occurs, rapidly move the electrode back to a point on the weld surface just _____ the burn through.	e. shielding gas
_____	6. If the weld metal is not fusing to one side, the slag will be _____ in color on one side.	f. ground
_____	7. Extremely large molten weld pool sizes can cause a large hole to be formed or cause _____.	g. before
_____	8. Raising the end of the electrode too high or moving it too far ahead of the molten weld pool can cause all of the _____ to be blown away from the molten weld pool.	h. root
_____	9. Changing from one welding position to another requires an _____ in timing, amperage, and electrode angle.	i. concave
_____	10. The surface of a _____ may be irregular and/or have undercut, overlap, slag inclusions, or other defects.	j. stringer beads
_____	11. The surface of a root pass can be cleaned by grinding or by using a _____.	k. filler passes
_____	12. On critical, high-strength code welds, it is usually required that the root pass as well as each filler pass be _____.	l. cover pass
_____	13. The ideal way to apply a hot pass is to _____ a large surface area so that the trapped slag can float to the surface.	m. closed

_____ 14. A very small amount of metal should be deposited during the hot pass so that the resulting weld is _____.

_____ 15. After the root pass is completed and it has been cleaned, the groove is filled with weld metal by making _____.

_____ 16. Filler passes are made with stringer beads or _____.

_____ 17. Chipping, wire brushing, and _____ are the best ways to remove slag between filler weld passes.

_____ 18. The last weld bead on a multipass weld is known as the _____.

_____ 19. A cover pass should not be more than _____ wider than the groove opening.

_____ 20. The cover pass may be made as one wide weld or as a series of two or more _____.

n. 1/8 in. (3 mm)

o. weave beads

p. brighter

q. grinding

r. root pass

s. rapidly melt

t. depth of weld metal penetration

CHAPTER 6: QUIZ 2

Name _____ Date _____

Class _____ Instructor _____ Grade _____

INSTRUCTIONS

Carefully read Chapter 6 in the textbook and answer each question.

MATCHING

In the space provided to the left of Column A, write the letter from Column B that best answers or completes the statement in Column A.

	Column A	Column B
_____	1. When welding on thick plate, it is impossible or impractical for the welder to try to get 100% _____ without preparing the plate for welding.	a. restarted
_____	2. The preparation of a thick plate is usually in the form of a _____.	b. tapered
_____	3. _____ is a process of cutting a groove in the back side of a joint that has been welded.	c. corners
_____	4. On all but short welds, the welding bead will need to be _____ after a welder stops to change electrodes.	d. visual inspection
_____	5. When a weld bead is nearing completion, it should be _____ so that when it is restarted the buildup will be more uniform.	e. cooldown
_____	6. To taper a weld bead, the travel rate should be _____ just before welding stops.	f. 1 in. (25 mm)
_____	7. Starting and stopping weld beads in _____ should be avoided.	g. 1/32 in. (0.8 mm).
_____	8. _____ is the application of heat to the metal before it is welded.	h. 4 in. (100 mm)
_____	9. Preheating helps to reduce cracking, hardness, distortion, and stresses by reducing the thermal shock from the weld and slowing the _____ rate.	i. weld groove
_____	10. _____ is the application of heat to the metal after welding and is used to slow the cooling rate and reduce hardening.	j. weld root
_____	11. _____ is the temperature of the metal during welding.	k. minimum and maximum
_____	12. The interpass temperature is given as a _____.	l. fitted
_____	13. The backing strip, groove face, and inside and outside plate surfaces within _____ of the joint must be mechanically cleaned of slag, rust, and mill scale.	m. penetration

_____ 14. All welds must pass a _____ before they are qualified to be bend tested.

n. molten weld pool

_____ 15. Visual inspection criteria require that undercut shall not exceed the lesser of 10% of the base metal thickness or _____.

o. Postheating

_____ 16. Visual inspection criteria require that the frequency of porosity shall not exceed one in each _____ of weld length, and the maximum diameter shall not exceed 3/32 in. (2.4 mm).

p. Back gouging

_____ 17. Transverse face bend. The weld is perpendicular to the longitudinal axis of the specimen and is bent so that the _____ becomes the tension surface of the specimen.

q. Interpass temperature

_____ 18. Transverse root bend. The weld is perpendicular to the longitudinal axis of the specimen and is bent so that the _____ becomes the tension surface of the specimen.

r. increased

_____ 19. Ideally, all welding will be performed on joints that are properly _____.

s. weld face

_____ 20. To make a good weld on a poorly fitted joint requires that the welder be able to watch the _____ and know how to make needed changes in amperage, current, electrode movement, electrode angle, and timing.

t. Preheating

CHAPTER 6: QUIZ 3

Name _____ Date _____

Class _____ Instructor _____ Grade _____

INSTRUCTIONS
Carefully read Chapter 6 in the textbook and answer each question.

IDENTIFICATION
Identify the numbered items on the drawing by writing the letter next to the identifying term in the space provided.

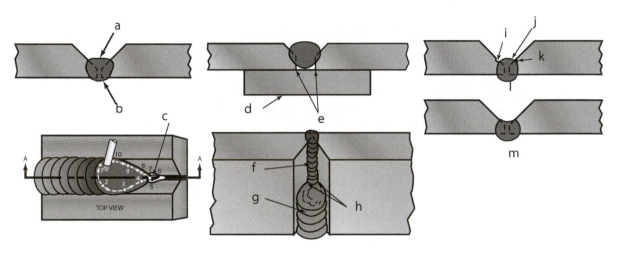

1. _____ COLD LAP

2. _____ UNDERCUT

3. _____ ROOT PASS CLEANUP; READY FOR NEXT WELD PASS

4. _____ ROOT FACE

5. _____ HOT PASS

6. _____ EXCESSIVE BUILDUP

7. _____ AS WELDED

8. _____ ROOT SURFACE

9. _____ ROOT PASS

10. _____ KEY HOLE

11. _____ WATCH THIS AREA FOR SLAG REMOVAL

12. _____ BACKING STRIP

CHAPTER 6: QUIZ 4

Name _____ Date _____

Class _____ Instructor _____ Grade _____

INSTRUCTIONS

Carefully read Chapter 6 in the textbook and answer each question.

IDENTIFICATION

Identify the numbered items on the drawing by writing the letter next to the identifying term in the space provided.

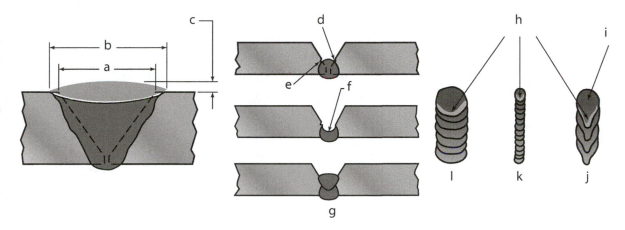

1. _____ TOO HOT

2. _____ GROOVE WIDTH

3. _____ TRAPPED SLAG

4. _____ THIS SHAPE INDICATES THAT WELD POOL IS COOLING TOO SLOWLY.

5. _____ UNDERCUT

6. _____ ROOT PASS REMAINING AFTER BACK GRINDING

7. _____ CORRECT WELD HEAT

8. _____ NEARLY 100% OF THE WELD DEPOSIT IS NOW THE HIGHER STRENGTH FILLER.

9. _____ WATCH THIS AREA

10. _____ WELD REINFORCEMENT MAXIMUM $\frac{1}{8}"$ (3 mm)

11. _____ TOO COOL

12. _____ BEAD WIDTH MAXIMUM GROOVE WIDTH + $\frac{1}{8}"$ (3 mm)

Flame Cutting

EYE PROTECTION FOR FLAME CUTTING

A general guide for the selection of eye and face protection equipment is listed below.

TYPE OF CUTTING OPERATION	HAZARD	SUGGESTED SHADE NUMBER
Light cutting, up to 1 in.	Sparks, harmful rays, molten metal, flying particles	3 or 4
Medium cutting, 1-6 in.		4 or 5
Heavy cutting, over 6 in.		5 or 6

Metal Thickness in. (mm)	Center Orifice Size No. Drill Size	Tip Cleaner No.*	Oxygen Pressure lb/in? (kPa)	Acetylene lb/in? (kPa)
1/8 (3)	60	7	10 (70)	3 (20)
1/4 (6)	60	7	15 (100)	3 (20)
3/8 (10)	55	11	20 (140)	3 (20)
1/2 (13)	55	11	25 (170)	4 (30)
3/4 (19)	55	11	30 (200)	4 (30)
1 (25)	53	12	35 (240)	4 (30)
2 (51)	49	13	45 (310)	5 (35)
3 (76)	49	13	50 (340)	5 (35)
4 (102)	49	13	55 (380)	5 (35)
5 (127)	45	**	60 (410)	5 (35)

*The tip cleaner number when counted from the small end toward the large end in a standard tip cleaner set.

**Larger than normally included in a standard tip cleaner set.

■ PRACTICE 7-1

Name _____   Date _____

Class _____   Instructor _____   Grade _____

OBJECTIVE: After completing this practice, you will be able to safely set up an oxyfuel torch set.

EQUIPMENT AND MATERIALS NEEDED

Disassembled oxyfuel torch set, two regulators, two reverse flow valves, one set of hoses, a torch body, a welding tip, two cylinders, a portable cart or supporting wall with safety chain, and a wrench.

INSTRUCTIONS

1. Safety chain the cylinders in the cart or to a wall. Then remove the valve protection caps.

2. Crack the cylinder valve on each cylinder for a second to blow away dirt that may be in the valve.

   CAUTION Always stand to one side. Point the valve away from anyone in the area and be sure there are no sources of ignition when cracking the valve.

3. Attach the regulators to the cylinder valves. The regulator nuts should be started by hand and then tightened with a wrench.

4. Attach a reverse flow valve or flashback arrestor, if the torch does not have them built in, to the hose connection on the regulator or to the hose connection on the torch body, depending on the type of reverse flow valve in the set. Occasionally, test each reverse flow valve by blowing through it to make sure it works properly.

5. Connect the hoses. The red hose has a left-hand nut and attaches to the fuel gas regulator. The green hose has a right-hand nut and attaches to the oxygen regulator.

6. Attach the torch to the hoses. Connect both hose nuts finger tight before using a wrench to tighten either one.

7. Check the tip seals for nicks or O rings, if used, for damage. Check the owner's manual, or a supplier, to determine if the torch tip should be tightened by hand only or should be tightened with a wrench.

   CAUTION Tightening a tip the incorrect way may be dangerous and might damage the equipment. Check all connections to be sure they are tight. The oxyfuel equipment is now assembled and ready for use.

   CAUTION Connections should not be overtightened. If they do not seal properly, repair or replace them.

CAUTION Leaking cylinder valve stems should not be repaired. Turn off the valve, disconnect the cylinder, mark the cylinder, and notify the supplier to come and pick up the bad cylinder.

The assembled oxyfuel welding equipment is now tested and ready to be ignited and adjusted.

INSTRUCTOR'S COMMENTS _____

_____

_____

■ PRACTICE 7-2

Name _____ Date _____

Class _____ Instructor _____ Grade _____

OBJECTIVE: After completing this practice, you should be able to safely turn on and test oxyfuel equipment for gas leaks.

EQUIPMENT AND MATERIALS NEEDED FOR THIS PRACTICE

1. Oxyfuel equipment that is properly assembled.

2. A nonadjustable tank wrench.

3. A leak-detecting solution.

INSTRUCTIONS

1. Back out the regulator pressure adjusting screws until they are loose.

2. Standing to one side of the regulator, open the cylinder valve SLOWLY so that the pressure rises on the gauge slowly.

    CAUTION **If the valve is opened quickly, the regulator or gauge may be damaged, or the gauge may explode.**

3. Open the oxygen valve all the way until it stops turning.

4. Open the acetylene or other fuel gas valve 1/4 turn or just enough to get gas pressure. If the cylinder valve does not have a handwheel, use a nonadjustable wrench and leave it in place on the valve stem while the gas is on.

    CAUTION **The acetylene valve should never be opened more than 1 1/2 turns, so that in an emergency it can be turned off quickly.**

5. Open one torch valve and point the tip away from any source of ignition. Slowly turn in the pressure adjusting screw until gas can be heard escaping from the torch. The gas should flow long enough to allow the hose to be completely purged (emptied) of air and replaced by the gas before the torch valve is closed. Repeat this process with the other gas.

6. After purging is completed, and with both torch valves off, adjust both regulators to read 5 psig (35 kPag).

7. Spray a leak-detecting solution on each hose and regulator connection and on each valve stem on the torch and cylinders. Watch for bubbles which indicate a leak. Turn off the cylinder valve before tightening any leaking connections.

CAUTION Connections should not be overtightened. If they do not seal properly, repair or
replace them.

CAUTION Leaking cylinder valve stems should not be repaired. Turn off the valve, disconnect
the cylinder, mark the cylinder, and notify the supplier to come and pick up the bad
cylinder.

The assembled oxyfuel welding equipment is now tested and ready to be ignited and adjusted.

INSTRUCTOR'S COMMENTS _____

_____

_____

■ PRACTICE 7-3

Name _____ Date _____

Class _____ Instructor _____ Grade _____

OBJECTIVE: After completing this practice, you should have the ability and knowledge to clean a cutting tip properly.

EQUIPMENT AND MATERIALS NEEDED FOR THIS PRACTICE

1. Oxygen and fuel gas cylinders, security chains or straps, a place or object to secure cylinders, fuel and oxygen regulators, reverse flow valves or flashback arrestors, torch and tip, proper fitting wrench, leak-detecting solution, and tip cleaner.

2. Proper safety protection (proper shade gas welding/cutting face shield or goggles, safety glasses, wire brush, chipping hammer, leather gloves, pliers, long-sleeved shirt, long pants, and leather boots or shoes). Refer to Chapter 2 in the text for more specific safety information.

INSTRUCTIONS

1. Turn on a small amount of oxygen. This procedure is done to blow out any dirt loosened during the cleaning.

2. The end of the tip is first filed flat, using the file provided in the tip cleaning set.

3. Try several sizes of tip cleaners in a preheat hole until the correct size cleaner is determined. It should easily go all the way into the tip.

4. Push the cleaner in and out of each preheat hole several times. Tip cleaners are small, round files. Excessive use of them will greatly increase the orifice (hole) size. Avoid this.

5. Next, depress the cutting lever and, by trial and error, select the correct size tip cleaner for the center cutting orifice.

   NOTE: A tip cleaner should never be forced. If the tip needs additional care, refer back to the section on tip care in Chapter 4.

INSTRUCTOR'S COMMENTS _____

_____

_____

■ PRACTICE 7-4

Name _____ Date _____

Class _____ Instructor _____ Grade _____

OBJECTIVE: After completing this practice, you should be able to light and extinguish a torch safely.

EQUIPMENT AND MATERIALS NEEDED FOR THIS PRACTICE

1. Oxygen and fuel gas cylinders, security chains or straps, a place or object to secure cylinders, fuel and oxygen regulators, reverse flow valves or flashback arrestors, torch and tip, proper fitting wrench, leak-detecting solution, and tip cleaner.

2. Proper safety protection (proper shade gas welding/cutting face shield or goggles, safety glasses, wire brush, chipping hammer, leather gloves, pliers, long-sleeved shirt, long pants, and leather boots or shoes). Refer to Chapter 2 in the text for more specific safety information.

INSTRUCTIONS

1. Set the regulator working pressure for the tip size. If you do not know the correct pressure setting for the tip, start with the fuel set at 5 psig (35 kPag) and the oxygen set at 25 psig (170 kPag).

2. Point the torch tip upward and away from any equipment or other students.

3. Turn on just the acetylene valve and use only a spark lighter to ignite the acetylene. The torch may not stay lit. If this happens, close the valve slightly and try to relight the torch.

4. If the flame is small, it will produce heavy black soot and smoke. In this case, turn the flame up to stop the soot and smoke. The welder need not be concerned if the flame jumps slightly away from the torch tip.

5. With the acetylene flame burning almost smoke-free, slowly open the oxygen valve, and by using only the oxygen valve, adjust the flame to a neutral setting.

6. When the cutting oxygen lever is depressed, the flame may become slightly carbonizing. This may occur because of a drop in line pressure due to the high flow of oxygen through the cutting orifice.

7. With the cutting lever depressed, readjust the preheat flame to a neutral setting. The flame will become slightly oxidizing when the cutting lever is released. Since an oxidizing flame is hotter than a neutral flame, the metal being cut will be preheated faster. When the cut is started by depressing the lever, the flame automatically returns to the neutral setting and does not oxidize the top of the plate.

8. Extinguish the flame by turning off the oxygen and then the acetylene.

    **CAUTION Turning off the acetylene first will often cause a loud POP. This can often cause
    ▰▰▰ soot to clog the tip and torch.**

INSTRUCTOR'S COMMENTS _____

_____

_____

■ PRACTICE 7-5

Name _____   Date _____

Class _____   Instructor _____   Grade _____

OBJECTIVE: After completing this practice, you should be able to turn off and disassemble oxyfuel equipment.

EQUIPMENT AND MATERIALS NEEDED FOR THIS PRACTICE

1. Oxyfuel welding equipment that is properly assembled and tested.

2. A spark lighter.

3. Gas welding goggles.

4. Gloves and proper protective clothing.

INSTRUCTIONS

1. First, quickly turn off the torch fuel-gas valve. This action blows the flame out and away from the tip, ensuring that the fire is out. In addition, it prevents the flame from burning back inside the torch. On large tips or hot tips, turning the fuel off first may cause the tip to pop. The pop is caused by a lean fuel mixture in the tip.

   CAUTION If you find that the tip pops each time you turn the fuel off first, turn the oxygen off first to prevent the pop. Be sure that the flame is out before putting the torch down.

2. After the flame is out, turn off the oxygen valve.

3. Turn off the cylinder valves.

4. Open one torch valve at a time to bleed off the pressure.

5. When all of the pressure is released from the system, back both regulator adjusting screws out until they are loose.

6. Loosen both ends of both hoses and unscrew them.

7. Loosen both regulators and unscrew them from the cylinder valves.

8. Replace the valve protection caps.

INSTRUCTOR'S COMMENTS _____

_____

_____

■ PRACTICE 7-6

Name _____ Date _____

Class _____ Instructor _____ Grade _____

OBJECTIVE: After completing this practice, you should be able to safely set the working pressure on a regulator using oxygen and acetylene.

EQUIPMENT AND MATERIALS NEEDED FOR THIS PRACTICE

1. Oxygen and fuel gas cylinders, security chains or straps, a place or object to secure cylinders, fuel and oxygen regulators, reverse flow valves or flashback arrestors, torch and tip, proper fitting wrench, leak-detecting solution, and tip cleaner.

2. Proper safety protection (oxyfuel goggles or tinted face shield, safety glasses, wire brush, chipping hammer, leather gloves, pliers, long-sleeved shirt, long pants, and leather boots or shoes). Refer to Chapter 2 in the text for more specific safety information.

INSTRUCTIONS

Setting the working pressure of the regulators can be done by following a table, or it can be set by watching the flame.

1. To set the regulator by watching the flame, first set the acetylene pressure at 2 psig to 4 psig (14 kPag to 30 kPag) and then light the acetylene flame.

2. Open the acetylene torch valve one to two turns and reduce the regulator pressure by backing out the setscrew until the flame starts to smoke.

3. Increase the pressure until the smoke stops and then increase it just a little more. This is the maximum fuel gas pressure the tip needs. With a larger tip and a longer hose, the pressure must be set higher. This is the best setting, and it is the safest one to use. With this lowest possible setting, there is less chance of a leak. If the hoses are damaged, the resulting fire will be much smaller than a fire burning from a hose with a higher pressure. There is also less chance of a leak with the lower pressure.

4. With the acetylene adjusted so that the flame almost stops smoking, slowly open the torch oxygen valve.

5. Adjust the torch to a neutral flame. When the cutting lever is depressed, the flame will become carbonizing, not having enough oxygen pressure.

6. While holding the cutting lever down, increase the oxygen regulator pressure slightly. Readjust the flame, as needed, to a neutral setting by using the oxygen valve on the torch.

7. Increase the pressure slowly and readjust the flame as you watch the length of the clear cutting stream in the center of the flame. The center stream will stay fairly long until a pressure is reached that causes turbulence disrupting the cutting stream. This turbulence will cause the flame to shorten in length considerably.

8. With the cutting lever still depressed, reduce the oxygen pressure until the flame lengthens once again. This is the maximum oxygen pressure that this tip can use without causing turbulence in the cutting stream. This turbulence will cause a very inferior cut. The lower pressure also will keep the sparks from being blown a longer distance from the work.

INSTRUCTOR'S COMMENTS _____

_____

_____

■ PRACTICE 7-7

Name _____   Date _____

Class _____   Instructor _____   Grade _____

OBJECTIVE: After completing this practice, you should be able to safely set-up and operate a track burner.

EQUIPMENT AND MATERIALS NEEDED FOR THIS PRACTICE

1. The manufacturer's instruction and safety manual.

2. Tools as listed in the manufacturer's instructions.

3. **All required PPE.**

INSTRUCTIONS

1. Make a safety check of the equipment to see that everything is in good working order with no damaged hoses, regulators, torch tip, track, electrical supply, or track burner.

2. Once the equipment has been safety checked, you can set it up according to the manufacturer's specifications.

3. Make sure the track is clean, level, parallel to the metal to be cut, and is not sitting on any spatter or debris.

4. Move the track burner along the track to see that it will move freely once it is turned on.

5. Adjust the gas pressures to match the required setting for the material thickness to be cut.

6. Move the tip off of the plate to be cut or raise it enough to make it easy to light.

7. Turn on a little acetylene, and be ready to quickly turn on a little oxygen once the torch is lit.

8. Use a flint or spark lighter to light the acetylene.

   NOTE: Because the torch is pointed downward, the flame will rise around the tip and torch until a little oxygen is turned on.

9. Adjust the flame to be neutral.

10. Turn on the cutting stream to see if it is clear and straight.

11. Move the torch up to the edge of the plate and allow the preheat flames to bring the edge up to a bright red color before turning on the cutting oxygen.

12. Start the track torch moving, and adjust the speed so that there is a smooth steady stream of sparks streaming from the underside of the cut.

13. Repeat cuts until you can consistently make clean, square cuts using the track burner.

14. Turn off the oxyfuel system and track burner.

15. Clean up your work area when finished.

INSTRUCTOR'S COMMENTS _____

_____

_____

■ PRACTICE 7-8

Name _____ Date _____

Class _____ Instructor _____ Grade _____

OBJECTIVE: After completing this practice, you should be able to make a quality straight cut on a 1/4-in. (6-mm) steel plate.

EQUIPMENT AND MATERIALS NEEDED FOR THIS PRACTICE

1. Properly set-up and adjusted oxyfuel cutting equipment.

2. Proper safety protection (oxyfuel goggles or tinted face shield, safety glasses, wire brush, chipping hammer, leather gloves, pliers, long-sleeved shirt, long pants, and leather boots or shoes). Refer to Chapter 2 in the text for more specific safety information.

3. A straightedge.

4. A piece of soapstone.

5. A piece of mild steel plate, 6 in. (152 mm) square by 1/4 in. (6 mm) thick.

INSTRUCTIONS

1. Using a straightedge and soapstone, make several straight lines 1/2 in. (13 mm) apart. Starting at one end, make a cut along the entire length of plate.

2. The cut strip must fall free, be slag-free with vertical drag lines, and be within ±3/32 in. (2 mm) of a straight line and ±5° of being square.

3. Repeat this procedure until the cut is consistently produced.

4. Turn off the cylinder valves, bleed off the trapped pressure in the hoses, back off the regulator adjusting screws, and close the torch valves.

5. Coil up the hoses to prevent them from becoming damaged.

6. Clean up your work area when finished with this practice.

INSTRUCTOR'S COMMENTS _____

_____

_____

■ PRACTICE 7-9

Name _____   Date _____

Class _____   Instructor _____   Grade _____

OBJECTIVE: After completing this practice, you should be able to make a quality straight cut on a 1/2-in. (13-mm) steel plate.

EQUIPMENT AND MATERIALS NEEDED FOR THIS PRACTICE

1. Properly set-up and adjusted oxyfuel cutting equipment.

2. Proper safety protection (oxyfuel goggles or tinted face shield, safety glasses, wire brush, chipping hammer, leather gloves, pliers, long-sleeved shirt, long pants, and leather boots or shoes). Refer to Chapter 2 in the text for more specific safety information.

3. A straightedge.

4. A piece of soapstone.

5. A piece of mild steel plate, 6 in. (152 mm) square by 1/2 in. (13 mm) thick.

INSTRUCTIONS

1. Using a straightedge and soapstone, make several straight lines 1/2 in. (13 mm) apart. Starting at one end, make a cut along the entire length of plate.

2. The strip must fall free, be slag-free with vertical drag lines, and be within ±3/32 in. (2 mm) of a straight line and ±5° of being square.

3. Repeat this procedure until the cut is consistently produced.

4. Turn off the cylinder valves, bleed off the trapped pressure in the hoses, back off the regulator adjusting screws, and close the torch valves.

5. Coil up the hoses to prevent them from becoming damaged.

6. Clean up your work area when finished with this practice.

   NOTE: Remember that starting a cut in thick plate will take longer and the cutting speed will be slower.

INSTRUCTOR'S COMMENTS _____

_____

_____

■ PRACTICE 7-10

Name _____ Date _____

Class _____ Instructor _____ Grade _____

OBJECTIVE: After completing this practice you should be able to make a flat, straight cut in sheet metal.

EQUIPMENT AND MATERIALS NEEDED FOR THIS PRACTICE

1. Properly set-up and adjusted oxyfuel cutting equipment.

2. Proper safety protection (oxyfuel goggles or tinted face shield, safety glasses, wire brush, chipping hammer, leather gloves, pliers, long-sleeved shirt, long pants, and leather boots or shoes). Refer to Chapter 2 in the text for more specific safety information.

3. A piece of soapstone.

4. A straightedge.

5. A piece of mild steel sheet 10 in. (254 mm) square by 18 gauge to 11 gauge thick.

INSTRUCTIONS

1. Using a straightedge and soapstone, mark the plate in strips 1/2 in. (13 mm) wide.

2. Hold the torch at a very sharp leading angle and cut along the line.

3. The cut must be smooth and straight with as little slag as possible.

4. Repeat this procedure until the cut can be made smooth, straight, and slag-free.

5. Turn off the cylinder valves, bleed off the trapped pressure in the hoses, back off the regulator adjusting screws, and close the torch valves.

6. Coil up the hoses to prevent them from becoming damaged.

7. Clean up your work area when finished with this practice.

INSTRUCTOR'S COMMENTS _____
_____
_____

■ PRACTICE 7-11

Name _____   Date _____

Class _____   Instructor _____   Grade _____

OBJECTIVE: After completing this practice, you should be able to cut holes in flat plate.

EQUIPMENT AND MATERIALS NEEDED FOR THIS PRACTICE

1. Properly set-up and adjusted oxyfuel cutting equipment.

2. Proper safety protection (oxyfuel goggles or tinted face shield, safety glasses, wire brush, chipping hammer, leather gloves, pliers, long-sleeved shirt, long pants, and leather boots or shoes). Refer to Chapter 2 in the text for more specific safety information.

3. A piece of soapstone.

4. A circle template or circular pattern.

5. A piece of mild steel plate 6 in. (152 mm) square by 1/4 in. (6 mm) thick.

INSTRUCTIONS

1. Using a circle template or circular pattern and soapstone, mark several 1/2-in. (13-mm) diameter and 1-in. (25-mm) diameter circles.

2. Start in the center of the circle and make an outward spiraling cut until the hole is the desired size.

3. The hole must be within ±3/32 in. (2 mm) of being round and ±5° of being square to metal.

4. The hole may have slag on the bottom.

5. Repeat this procedure until both the small and large holes can be made with consistent accuracy.

6. Turn off the cylinder valves, bleed off the trapped pressure in the hoses, back off the regulator adjusting screws, and close the torch valves.

7. Coil up the hoses to prevent them from becoming damaged.

8. Clean up your work area when finished with this practice.

INSTRUCTOR'S COMMENTS _____

_____

_____

■ PRACTICE 7-12

Name _____   Date _____

Class _____   Instructor _____   Grade _____

OBJECTIVE: After completing this practice, you should be able to bevel 3/8-inch plate.

EQUIPMENT AND MATERIALS NEEDED FOR THIS PRACTICE

1. Properly set-up and adjusted oxyfuel cutting equipment.

2. Proper safety protection (oxyfuel goggles or tinted face shield, safety glasses, wire brush, chipping hammer, leather gloves, pliers, long-sleeved shirt, long pants, and leather boots or shoes). Refer to Chapter 2 in the text for more specific safety information.

3. A piece of soapstone.

4. A straightedge.

5. A piece of mild steel plate, 6 in. (152 mm) square by 3/8 in. (10 mm) thick.

INSTRUCTIONS

1. Using a straightedge and soapstone, mark the plate in strips 1/2 in. (13 mm) wide.

2. Set the tip for beveling and cut a bevel. The bevel should be within ±3/32 in. (2 mm) of a straight line and ±5° of a 45° angle. There may be some soft slag, but no hard slag, on the beveled plate.

3. Repeat this practice until the beveled angle cut is consistently produced.

4. Turn off the cylinder valves, bleed off the trapped pressure in the hoses, back off the regulator adjusting screws, and close the torch valves.

5. Coil up the hoses to prevent them from becoming damaged.

6. Clean up your work area when finished with this practice.

INSTRUCTOR'S COMMENTS _____

_____

_____

■ PRACTICE 7-13

Name _____ Date _____

Class _____ Instructor _____ Grade _____

OBJECTIVE: After completing this practice, you should be able to cut 1/4-in. to 3/8-in. (6-mm to 10-mm) carbon steel in both the vertical up and down positions.

EQUIPMENT AND MATERIALS NEEDED FOR THIS PRACTICE

1. Properly set-up and adjusted oxyfuel cutting equipment.

2. Proper safety protection (oxyfuel goggles or tinted face shield, safety glasses, wire brush, chipping hammer, leather gloves, pliers, long-sleeved shirt, long pants, and leather boots or shoes). Refer to Chapter 2 in the text for more specific safety information.

3. A piece of soapstone.

4. A straightedge.

5. A piece of mild steel plate, 6 in. (152 mm) square by 1/4 in. (6 mm) to 3/8 in. (10 mm) thick, marked in strips 1/2 in. (13 mm) wide all along the 1/4-in. and 3/8-in. (6-mm and 10-mm) thick carbon steel plate.

INSTRUCTIONS

1. Using a straightedge and soapstone, make several straight lines 1/2 in. (13 mm) apart on the steel plate.

2. Place the plate in a vertical position.

3. Start your first cut at the top of the plate and travel downward. Then, starting at the bottom, make the next cut upward. Make sure that the sparks do not cause a safety hazard and that the metal being cut off will not fall on any person or object.

4. The cut must be free of hard to remove dross and within ±3/32 in. (2 mm) of a straight line and ±5° of being square.

5. Repeat these cuts until they are consistently produced.

6. Turn off the cylinder valves, bleed off the trapped pressure in the hoses, back off the regulator adjusting screws, and close the torch valves.

7. Coil up the hoses to prevent them from becoming damaged.

8. Clean up your work area when finished with this practice.

INSTRUCTOR'S COMMENTS _____

_____

_____

■ PRACTICE 7-14

Name _____ Date _____

Class _____ Instructor _____ Grade _____

OBJECTIVE: After completing this practice, you should be able to make an overhead straight cut.

EQUIPMENT AND MATERIALS NEEDED FOR THIS PRACTICE

1. Properly set-up and adjusted oxyfuel cutting equipment.

2. Proper safety protection (oxyfuel goggles or tinted face shield, safety glasses, wire brush, chipping hammer, leather gloves, pliers, long-sleeved shirt, long pants, and leather boots or shoes). Refer to Chapter 2 in the text for more specific safety information.

3. A piece of soapstone.

4. A straightedge.

5. A piece of mild steel sheet 10 in. (254 mm) square by 1/4 in. (6 mm) to 3/8 in. (10 mm) thick.

INSTRUCTIONS

1. Using a straightedge and soapstone, mark the plate in strips 1/2 in. (13 mm) wide.

2. Be sure that you are completely protected from the hot sparks which will fall from the cut.

3. Angle the torch so that most of the sparks and slag will be blown away from you.

4. Starting at one edge of the plate, make a cut in the overhead position. The plate should fall free when the cut is completed.

5. The cut must be smooth and straight with as little slag as possible and within 1/8 in. (3 mm) of being a straight line and within ±5° of being square to metal.

6. Repeat this procedure until the cut can be made smooth, straight, and slag-free.

7. Turn off the cylinder valves, bleed off the trapped pressure in the hoses, back off the regulator adjusting screws, and close the torch valves.

8. Coil up the hoses to prevent them from becoming damaged.

9. Clean up your work area when finished with this practice.

INSTRUCTOR'S COMMENTS _____

_____

_____

■ PRACTICE 7-15

Name _____ Date _____

Class _____ Instructor _____ Grade _____

OBJECTIVE: After completing this practice you should be able to cut out internal and external shapes.

EQUIPMENT AND MATERIALS NEEDED FOR THIS PRACTICE

1. Properly set-up and adjusted oxyfuel cutting equipment.

2. Proper safety protection (oxyfuel goggles or tinted face shield, safety glasses, wire brush, chipping hammer, leather gloves, pliers, long-sleeved shirt, long pants, and leather boots or shoes). Refer to Chapter 2 in the text for more specific safety information.

3. A piece of soapstone and a center punch.

4. A pattern of some sort, such as those shown in the textbook in Figure 7-128.

5. A piece of mild steel sheet 10 in. (254 mm) square by 1/4 in. (6 mm) to 3/8 in. (10 mm) thick.

INSTRUCTIONS

1. Using the pattern and soapstone, trace the pattern onto the plate.

2. Use the center punch and make punch marks at half-inch intervals along the line.

3. Make the cut in such a way that it is no more than 1/8 in. (3 mm) away from the punched line.

4. The cut must be smooth and straight with as little slag as possible.

5. Repeat this procedure until the cut can consistently be made correctly.

6. Turn off the cylinder valves, bleed off the trapped pressure in the hoses, back off the regulator adjusting screws, and close the torch valves.

7. Coil up the hoses to prevent them from becoming damaged.

8. Clean up your work area when finished with this practice.

INSTRUCTOR'S COMMENTS _____

_____

_____

■ PRACTICE 7-16

Name _____   Date _____

Class _____   Instructor _____   Grade _____

OBJECTIVE: After completing this practice, you should be able to make a square cut on pipe in the 1G (horizontal rolled) position while keeping the tip pointed downward.

EQUIPMENT AND MATERIALS NEEDED FOR THIS PRACTICE

1. Properly set-up and adjusted oxyfuel cutting equipment.

2. Proper safety protection (oxyfuel goggles or tinted face shield, safety glasses, wire brush, chipping hammer, leather gloves, pliers, long-sleeved shirt, long pants, and leather boots or shoes). Refer to Chapter 2 in the text for more specific safety information.

3. A piece of soapstone and a flexible straightedge which can be wrapped around the pipe.

4. A piece of schedule 40 mild steel pipe 3 in. (76 mm) in diameter by 6 in. (152 mm) long.

INSTRUCTIONS

1. Using the straightedge and soapstone, wrap the straightedge around the pipe and mark rings every 1/2 in. (13 mm) along the pipe.

2. Place the pipe horizontally on the cutting table and start the cut at the top.

3. Move the torch backward along the line and then forward to keep the slag out of the cut.

4. Keeping the tip pointed downward, make the cut as far as possible in one direction then cut in the other direction. Roll the pipe 180°, start at the top and repeat. The ring should fall free.

5. When the pipe is placed upright on a flat surface, it must stand within 5° of vertical and have no gaps greater than 1/8 in. (3 mm) between the cut edge and the flat surface.

6. Repeat this procedure until the cut can consistently be made correctly.

7. Turn off the cylinder valves, bleed off the trapped pressure in the hoses, back off the regulator adjusting screws, and close the torch valves.

8. Coil up the hoses to prevent them from becoming damaged.

9. Clean up your work area when finished with this practice.

INSTRUCTOR'S COMMENTS _____

_____

_____

■ PRACTICE 7-17

Name _____   Date _____

Class _____   Instructor _____   Grade _____

OBJECTIVE: After completing this practice, you should be able to make a square cut on pipe in the 1G (horizontal rolled) position while keeping the tip pointed toward the center of the pipe.

EQUIPMENT AND MATERIALS NEEDED FOR THIS PRACTICE

1. Properly set-up and adjusted oxyfuel cutting equipment.

2. Proper safety protection (oxyfuel goggles or tinted face shield, safety glasses, wire brush, chipping hammer, leather gloves, pliers, long-sleeved shirt, long pants, and leather boots or shoes). Refer to Chapter 2 in the text for more specific safety information.

3. A piece of soapstone and a flexible straightedge which can be wrapped around the pipe.

4. A piece of schedule 40 mild steel pipe 3 in. (76 mm) in diameter by 6 in. (152 mm) long.

INSTRUCTIONS

1. Using the straightedge and soapstone, wrap the straightedge around the pipe and mark rings every 1/2 in. (13 mm) along the pipe.

2. Place the pipe horizontally on the cutting table and start the cut at the top.

3. Move the torch backward along the line and then forward to keep the slag out of the cut.

4. Keeping the tip pointed toward the center of the pipe, make the cut as far as possible then roll the pipe and continue the cut. The ring should fall free at the end of the cut.

5. When the pipe is placed upright on a flat surface, it must stand within 5° of vertical and have no gaps greater than 1/8 in. (3 mm) between the cut edge and the flat surface.

6. Repeat this procedure until the cut can consistently be made correctly.

7. Turn off the cylinder valves, bleed off the trapped pressure in the hoses, back off the regulator adjusting screws, and close the torch valves.

8. Coil up the hoses to prevent them from becoming damaged.

9. Clean up your work area when finished with this practice.

INSTRUCTOR'S COMMENTS _____
_____
_____

■ PRACTICE 7-18

Name _____ Date _____

Class _____ Instructor _____ Grade _____

OBJECTIVE: After completing this practice, you should be able to make a square cut on pipe in the 5G (horizontal fixed) position while keeping the tip pointed toward the center of the pipe.

EQUIPMENT AND MATERIALS NEEDED FOR THIS PRACTICE

1. Properly set-up and adjusted oxyfuel cutting equipment.

2. Proper safety protection (oxyfuel goggles or tinted face shield, safety glasses, wire brush, chipping hammer, leather gloves, pliers, long-sleeved shirt, long pants, and leather boots or shoes). Refer to Chapter 2 in the text for more specific safety information.

3. A piece of soapstone and a flexible straightedge which can be wrapped around the pipe.

4. A piece of schedule 40 mild steel pipe 3 in. (76 mm) in diameter by 6 in. (152 mm) long.

INSTRUCTIONS

1. Using the straightedge and soapstone, wrap the straightedge around the pipe and mark rings every 1/2 in. (13 mm) along the pipe.

2. Place the pipe horizontally on the cutting table and start the cut at the top.

3. Move the torch backward along the line and then forward to keep the slag out of the cut.

4. Keeping the tip pointed toward the center of the pipe, make the cut as far as possible then reposition yourself and cut around the bottom. The ring should fall free at the end of the cut.

5. When the pipe is placed upright on a flat surface, it must stand within 5° of vertical and have no gaps greater than 1/8 in. (3 mm) between the cut edge and the flat surface.

6. Repeat this procedure until the cut can consistently be made correctly.

7. Turn off the cylinder valves, bleed off the trapped pressure in the hoses, back off the regulator adjusting screws, and close the torch valves.

8. Coil up the hoses to prevent them from becoming damaged.

9. Clean up your work area when finished with this practice.

INSTRUCTOR'S COMMENTS _____

_____

_____

■ PRACTICE 7-19

Name _____ Date _____

Class _____ Instructor _____ Grade _____

OBJECTIVE: After completing this practice, you should be able to make a square cut on pipe in the 2G (vertical) position.

EQUIPMENT AND MATERIALS NEEDED FOR THIS PRACTICE

1. Properly set-up and adjusted oxyfuel cutting equipment.

2. Proper safety protection (oxyfuel goggles or tinted face shield, safety glasses, wire brush, chipping hammer, leather gloves, pliers, long-sleeved shirt, long pants, and leather boots or shoes). Refer to Chapter 2 in the text for more specific safety information.

3. A piece of soapstone and a flexible straightedge which can be wrapped around the pipe.

4. A piece of schedule 40 mild steel pipe 3 in. (76 mm) in diameter by 6 in. (152 mm) long.

INSTRUCTIONS

1. Using the straightedge and soapstone, wrap the straightedge around the pipe and mark rings every 1/2 in. (13 mm) along the pipe.

2. Place the pipe vertically on the cutting table and place a plate over the open end of the pipe to contain the sparks.

3. Start at one side and make the cut horizontally around the pipe.

4. When the cut is completed the ring may have to be tapped free.

5. When the pipe is placed upright on a flat surface, it must stand within 5° of vertical and have no gaps greater than 1/8 in. (3 mm) between the cut edge and the flat surface.

6. Repeat this procedure until the cut can consistently be made correctly.

7. Turn off the cylinder valves, bleed off the trapped pressure in the hoses, back off the regulator adjusting screws, and close the torch valves.

8. Coil up the hoses to prevent them from becoming damaged.

9. Clean up your work area when finished with this practice.

INSTRUCTOR'S COMMENTS _____

_____

_____

■ PRACTICE 7-20

Name _____ Date _____

Class _____ Instructor _____ Grade _____

OBJECTIVE: After completing this practice, you should be able to make U and J Grooves in mild steel in the 1G and 2G positions.

EQUIPMENT AND MATERIALS NEEDED FOR THIS PRACTICE

1. A properly lit and adjusted cutting torch with a gouging tip.

2. All required PPE.

3. One piece of mild steel plate 6 in. (152 mm) long and 3/8 in. (10 mm) thick.

INSTRUCTIONS

1. To make a U-groove down the length of the plate and a J-groove along the side of the plate.

2. Hold the tip at a 30° angle at the edge of the plate.

3. Once the plate edge is glowing bright red, slowly depress the torch cutting lever.

4. When the groove starts, move the torch in a straight line and travel at a consistent speed.

5. The groove should be within ±3/32 in. (2 mm) of a straight line and at a consistent width and depth.

6. There may be some soft slag, but no hard slag, on the plate.

7. Repeat this practice until a groove can be made within tolerance.

8. Turn off the cylinder valves, bleed the hoses, back out the pressure regulators, and clean your work area when you are finished cutting.

INSTRUCTOR'S COMMENTS _____

_____

_____

■ PRACTICE 7-21

Name _____   Date _____

Class _____   Instructor _____   Grade _____

OBJECTIVE: After completing this practice, you should be able to remove a weld using an oxyfuel gouging torch.

EQUIPMENT AND MATERIALS NEEDED FOR THIS PRACTICE

1. A properly lit and adjusted cutting torch with a gouging tip.

2. All required PPE.

3. One or more piece of scrap welded mild steel plate.

INSTRUCTIONS

1. Hold the tip at a 30° angle at the edge of the plate.

2. Once the plate edge is glowing bright red, slowly depress the torch cutting lever.

3. When the groove forms, watch the bottom of the groove to see that the entire depth of the weld metal is being removed.

4. Repeat this practice until the welds can be completely removed and the plates separated.

5. Turn off the cylinder valves, bleed the hoses, back out the pressure regulators, and clean your work area when you are finished cutting.

INSTRUCTOR'S COMMENTS _____

_____

_____

CHAPTER 7: QUIZ 1

Name _____ Date _____

Class _____ Instructor _____ Grade _____

INSTRUCTIONS
Carefully read Chapter 7 in the textbook and answer each question.

MATCHING
In the space provided to the left of Column A, write the letter from Column B that best answers or completes the statement in Column A.

Column A

_____ 1. An oxyfuel gas flame is used to raise the temperature of the metal to its _____ before a high-pressure stream of oxygen is directed onto the metal, causing it to be cut.

_____ 2. A good oxyfuel cut not only should be straight and square but also should require little or no _____.

_____ 3. The choice of which gas to use must be based on _____.

_____ 4. The flame condition, such as neutral, oxidizing, or carbonizing, will affect the _____ of the flame.

_____ 5. Most fuel gases used for welding are _____.

_____ 6. Combustion of acetylene is divided into two separate chemical reactions; the first reaction is referred to as _____ combustion.

_____ 7. The _____ or rate of propagation of a flame is the rate or speed at which the flame burns.

_____ 8. _____ is the most frequently used fuel gas.

_____ 9. _____, which is a liquid solvent, is used inside acetylene cylinders to absorb the gas and make it more stable.

_____ 10. The maximum temperature of an acetylene flame is when it is burning as a strongly _____ flame.

_____ 11. The _____ of a liquid increases as the temperature increases, and it decreases as the temperature decreases.

_____ 12. Many different _____ gases are in use today as fuel gases for oxyfuel cutting.

_____ 13. A safety feature of MPS gases is _____.

_____ 14. Propane and natural gas are both obtained from the _____ industry.

Column B

a. postcut cleanup

b. primary

c. Acetone

d. shock stability

e. temperature

f. Goggles

g. oxygen regulators

h. rapidly oxidize

i. length

j. kindling temperature

k. kindling

l. MPS

m. colorless

n. combustion rate

_____ 15. The hydrogen flame is almost _____.

o. petroleum

_____ 16. The oxyfuel process will work easily on any reactive metal that will _____ at an elevated temperature.

p. gas pressure

_____ 17. The oxyfuel gas cutting torch works by first preheating the metal to its _____ temperature.

q. performance

_____ 18. _____ or other suitable eye protection must be used for flame cutting.

r. oxidizing

_____ 19. The added _____ of the dedicated cutting torch helps keep the operator farther away from the heat and sparks and allows thicker material to be cut.

s. Acetylene

_____ 20. The machine cutting torch may require two _____.

t. hydrocarbons

CHAPTER 7: QUIZ 2

Name _____ Date _____

Class _____ Instructor _____ Grade _____

INSTRUCTIONS

Carefully read Chapter 7 in the textbook and answer each question.

MATCHING

In the space provided to the left of Column A, write the letter from Column B that best answers or completes the statement in Column A.

Column A	Column B
_____ 1. The _____ is the most common type of oxyfuel gas torch used in industry.	a. flexibility
_____ 2. The combination welding and cutting torch offers more _____.	b. acetylene
_____ 3. Small torches can be used for _____ work.	c. constant
_____ 4. Most cutting tips are made of _____ alloy.	d. cylinder sizes
_____ 5. The high-speed cutting tip is designed to allow a _____ cutting oxygen pressure.	e. LPG
_____ 6. The amount of _____ required to make a perfect cut is determined by the type of fuel gas used and by the material thickness, shape, and surface condition.	f. oxyacetylene hand torch
_____ 7. Acetylene must be used in tips that are designed to be used with _____.	g. gas or pressure
_____ 8. _____ are used in tips with eight preheat holes or in a two-piece tip that is not recessed.	h. MPS gases
_____ 9. Although the various types of pressure regulators all work in the same way, they are not _____.	i. heat treatment
_____ 10. Two-stage regulators are able to keep the pressure _____ at very low or high flow rates.	j. working pressure
_____ 11. The _____ gauge shows the pressure at the regulator and not at the torch.	k. copper
_____ 12. Oxygen and acetylene gases are available in a variety of _____.	l. dirt and oil
_____ 13. The cylinders go through a _____ process to strengthen them so they can withstand the designed working pressure.	m. Department of Transportation
_____ 14. Most acetylene cylinders are welded, and all of them are filled with a _____ material.	n. higher

_____ 15. Excessively high withdrawal rates may cause the acetylene to boil out of the _____.

o. acetone

_____ 16. The term _____ includes propane, butane, bottled gas, and a number of other mixtures of flammable gasses.

p. jewelry

_____ 17. _____ liquids have a boiling point below –238°F (–150°C).

q. Cryogenic

_____ 18. The _____ requires all cylinders to be labeled with basic information including name of the product, hazards, precautions, etc.

r. porous

_____ 19. A variety of inlet or cylinder fittings are available to ensure that the regulator cannot be connected to the wrong _____.

s. interchangeable

_____ 20. The connections to the cylinder and to the hose must be kept free of _____.

t. preheat flame

CHAPTER 7: QUIZ 3

Name _____ Date _____

Class _____ Instructor _____ Grade _____

INSTRUCTIONS

Carefully read Chapter 7 in the textbook and answer each question.

MATCHING

In the space provided to the left of Column A, write the letter from Column B that best answers or completes the statement in Column A.

Column A

_____ 1. High-pressure valve seats on a regulator that leak result in a _____ on the working pressure gauge.

_____ 2. The outlet connection on a regulator is a _____ for oxygen.

_____ 3. How can a regulator adjusting screw be cleaned if it becomes tight and difficult to turn? _____.

_____ 4. If a torch hose fitting leaks, then the seat should be _____.

_____ 5. A _____ can cause a torch tip to backfire.

_____ 6. To stop a _____, close the oxygen valve on the torch quickly.

_____ 7. The _____ is a spring-loaded check valve that closes when gas tries to flow backward through the torch valves.

_____ 8. What do you need to follow to test a flashback arrestor? _____

_____ 9. Fuel gas hoses must be _____ and have left-hand threaded fittings.

_____ 10. When hoses are not in use, the gas must be turned off and the pressure _____.

_____ 11. A leak-detecting solution must produce a good quantity of bubbles without _____.

_____ 12. Dirty tips can be cleaned using a set of _____.

_____ 13. Manifolds must be located 20 ft (6 m) or more from the _____.

_____ 14. Set the manifold line pressure _____, still ensuring that the type of work being done at the workstations can be performed satisfactorily.

_____ 15. To make a hand cut, the torch may be moved by _____ it toward you over your supporting hand.

Column B

a. use a dry, oil-free rag

b. loose tip

c. reverse flow valve

d. leaving a film

e. as low as possible

f. oxides

g. manufacturer's procedure

h. numbering system

i. right-hand fitting

j. angle

k. repaired or replaced

l. actual work

m. chalk line

n. preheat holes

o. creep

_____ 16. A slight forward torch _____ helps the flame preheat the metal.

p. sliding

_____ 17. The proper alignment of the _____ will speed up and improve the cut.

q. tip cleaners

_____ 18. A _____ will make a long, straight line on metal and is best used on large jobs.

r. bled off

_____ 19. Each welding equipment manufacturer uses its own _____ to designate the tip size.

s. red

_____ 20. Some metals release harmful _____ when they are cut.

t. flashback

CHAPTER 7: QUIZ 4

Name _____   Date _____

Class _____   Instructor _____   Grade _____

INSTRUCTIONS
Carefully read Chapter 7 in the textbook and answer each question.

MATCHING

In the space provided to the left of Column A, write the letter from Column B that best answers or completes the statement in Column A.

Column A	Column B
_____ 1. As a cut progresses along a plate, a _____ of what happened during the cut is preserved along both sides of the kerf.	a. Slag
_____ 2. If the flame is too small, the travel speed must be _____.	b. slower
_____ 3. The cutting speed should be fast enough so that the _____ have a slight slant backward.	c. hand
_____ 4. When the _____ is too low, the cut may not go completely through the metal.	d. depth of the gouge
_____ 5. _____ is found on bad cuts due to dirty tips.	e. supported
_____ 6. _____ plate can be cut quickly and accurately ranging in thickness from thin-gauge sheet metal or thick plate sections of a foot or more.	f. melt during a cut
_____ 7. Any piece being cut should be _____ so the torch flame will not cut through the piece and into the table.	g. Low-carbon steel
_____ 8. A welder normally supports the torch weight with the _____.	h. center
_____ 9. When cutting circles, a _____ attachment is used.	i. slightly recessed
_____ 10. If the _____ is not controlled, the end product might be worthless.	j. circle cutting
_____ 11. On larger pieces, you may have to _____ to complete the cut.	k. drag lines
_____ 12. Even an ideal cut can create sparks that _____ the plate surface.	l. Gouging
_____ 13. Heat becomes a problem when it causes the top edge of the plate to _____, as if the torch tip was too large.	m. pressure setting
_____ 14. The cutting tip will catch small sparks and become _____.	n. separate from the flame
_____ 15. If the supports underneath the plate are small, _____ may not cover you with sparks, plug the cutting tip, or cause a major flaw in the cut surface.	o. bounce around

_____ 16. On large diameter pipe, the torch tip is always pointed toward the _____ of the pipe.

_____ 17. _____ does not cut all the way through the base metal.

_____ 18. The center cutting orifice of a gouging tip may be _____ to allow the cutting stream to slightly spread out as it leaves the tip.

_____ 19. As the torch-to-work angle increases, the _____ will increase.

_____ 20. Most machine cutting torches have a cutting oxygen supply hose and regulator _____ cutting oxygen hose and regulator.

p. record

q. Blowback

r. dirty or clogged

s. move

t. distortion

CHAPTER 7: QUIZ 5

Name _____ Date _____

Class _____ Instructor _____ Grade _____

INSTRUCTIONS
Carefully read Chapter 1 in the textbook and answer each question.

IDENTIFICATION
Identify the numbered items on the drawing by writing the letter next to the identifying term in the space provided.

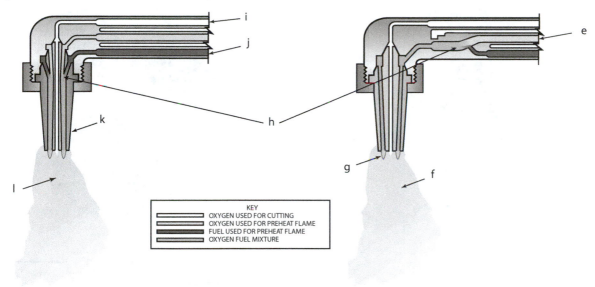

KEY
OXYGEN USED FOR CUTTING
OXYGEN USED FOR PREHEAT FLAME
FUEL USED FOR PREHEAT FLAME
OXYGEN FUEL MIXTURE

1. _____ CUTTING OXYGEN

2. _____ CUTTING TIP

3. _____ CUTTING OXYGEN VALVE LEVER

4. _____ PREHEAT FUEL

5. _____ OUTER FLAME ENVELOPE

6. _____ CUTTING OXYGEN TUBE

7. _____ MIXING CHAMBERS

8. _____ INNER FLAME CONE

9. _____ PREHEAT OXYGEN

10. _____ OXYGEN CUTTING STREAM

11. _____ HOSE CONNECTION

12. _____ NEEDLE VALVE

CHAPTER 7: QUIZ 6

Name _____ Date _____

Class _____ Instructor _____ Grade _____

INSTRUCTIONS

Carefully read Chapter 1 in the textbook and answer each question.

IDENTIFICATION

Identify the numbered items on the drawing by writing the letter next to the identifying term in the space provided.

1. _____ HOSE

2. _____ SPRING

3. _____ CYLINDER VALVE

4. _____ WORKING PRESSURE GAUGE

5. _____ THROUGH VALVE OF CYLINDER TO INPUT OF REGULATOR

6. _____ SAFETY RELEASE VALVE

7. _____ CYLINDER PRESSURE GAUGE

8. _____ HIGH-PRESSURE VALVE

9. _____ FLEXIBLE DIAPHRAGM

10. _____ ADJUSTING SPRING DESIGNED FOR PRECISE PRESET

11. _____ TORCH VALVE

12. _____ REGULATOR

13. _____ STEM-TYPE SEAT MECHANISM INLET PRESSURE AIDS IN SEALING

14. _____ PRECISION MACHINED NOZZLE

15. _____ BRONZE INLET FILTER KEEPS OUT DIRT AND FOREIGN MATERIALS

16. _____ STAINLESS STEEL DIAPHRAGM

17. _____ ADJUSTING SCREW

Plasma Arc Cutting

■ PRACTICE 8-1

Name _____ Date _____

Base Metal _____ Thickness _____

Class _____ Instructor _____ Grade _____

6"

4"

1 2"

1/8"

Welding Principles and Applications

MATERIAL: ⅛"×6" — MILD, STEEL, STAINLESS STEEL AND ALUMINUM PLATE

PROCESS: PLASMA ARC CUTTING

NUMBER: PRACTICE 8-1

DRAWN BY: BOB WISNUSKI

OBJECTIVE: After completing this practice, you should be able to safely set up and operate a PAC machine to make straight cuts on thin gauge mild steel, stainless steel, and aluminum in the flat position.

EQUIPMENT AND MATERIALS NEEDED FOR THIS PRACTICE

1. A properly set-up and adjusted PAC machine.

2. Proper safety protection (welding hood, safety glasses, leather gloves, pliers, long-sleeved shirt, long pants, and leather boots or shoes). Refer to Chapter 2 in the text for more specific safety information.

3. One or more pieces of mild steel, stainless steel, and aluminum, 6 in. (152 mm) long by 4 in. (100 mm) wide and 16 gauge and 1/8 in. (3 mm) thick.

INSTRUCTIONS

1. Starting at one end of the piece of metal that is 1/ 8 in. (3 mm) thick, hold the torch as close as possible to a 90° angle.

2. Lower your hood and establish a plasma cutting stream.

3. Move the torch in a straight line down the plate toward the other end.

4. If the width of the kerf changes, speed up or slow down the travel rate to keep the kerf the same size for the entire length of the plate.

5. Repeat the cut using both thicknesses of all three types of metals until you can make consistently smooth cuts that are within ±3/32 in. (2.3 mm) of a straight line and ±5° of being square to metal.

6. Turn off the PAC equipment and clean up your work area when you have finished cutting.

INSTRUCTOR'S COMMENTS _____

_____

_____

■ PRACTICE 8-2

Name _____   Date _____

Base Metal _____   Thickness _____

Class _____   Instructor _____   Grade _____

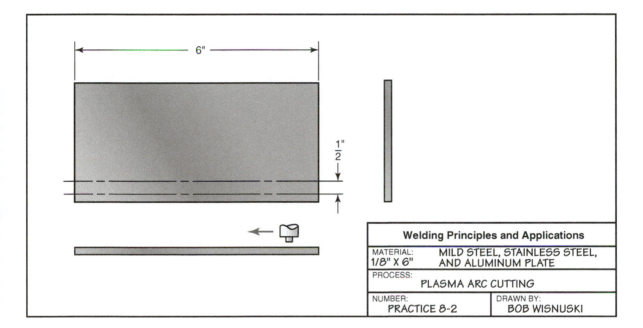

Welding Principles and Applications

MATERIAL: 1/8" X 6"	MILD STEEL, STAINLESS STEEL, AND ALUMINUM PLATE	
PROCESS: 	PLASMA ARC CUTTING	
NUMBER: PRACTICE 8-2		DRAWN BY: BOB WISNUSKI

OBJECTIVE: After completing this practice, you should be able to safely set up and operate a PAC machine to make straight cuts on thicker mild steel, stainless steel, and aluminum plates ranging in thickness from 1/4 to 1/2 in. (6 to 13 mm) in the flat position.

EQUIPMENT AND MATERIALS NEEDED FOR THIS PRACTICE

1. A properly set-up and adjusted PAC machine.

2. Proper safety protection (welding hood, safety glasses, leather gloves, pliers, long-sleeved shirt, long pants, and leather boots or shoes). Refer to Chapter 2 in the text for more specific safety information.

3. One or more pieces of mild steel, stainless steel, and aluminum, 6 in. (152 mm) long by 4 in. (100 mm) wide and 1/4 in. (6 mm) and 1/2 in. (13 mm) thick.

INSTRUCTIONS

1. Starting at one end of the piece of metal that is 1/4 in. (6 mm) thick, hold the torch as close as possible to a 90° angle.

2. Lower your hood and establish a plasma cutting stream.

3. Move the torch in a straight line down the plate toward the other end.

4. If the width of the kerf changes, speed up or slow down the travel rate to keep the kerf the same size for the entire length of the plate.

5. Repeat the cut using both thicknesses of all three types of metals until you can make consistently smooth cuts that are within ±3/32 in. (2.3 mm) of a straight line and ±5° of being square to metal.

INSTRUCTOR'S COMMENTS _____

_____

_____

■ PRACTICE 8-3

Name _____ Date _____

Base Metal _____ Thickness _____

Class _____ Instructor _____ Grade _____

OBJECTIVE: After completing this practice, you should be able to safely set up and operate a PAC machine to cut 1/2-in. (13-mm) and 1-in. (25-mm) diameter holes in mild steel, stainless steel, and aluminum plates ranging in thickness from 16 gauge sheet metal to 1/2 in. (6 to 13 mm) plate in the flat position.

EQUIPMENT AND MATERIALS NEEDED FOR THIS PRACTICE

1. A properly set-up and adjusted PAC machine.

2. Proper safety protection (welding hood, safety glasses, leather gloves, pliers, long-sleeved shirt, long pants, and leather boots or shoes). Refer to Chapter 2 in the text for more specific safety information.

3. One or more pieces of mild steel, stainless steel, and aluminum, 16 gauge, 1/8 in. (3 mm), 1/4 in. (6 mm), and 1/2 in. (13 mm) thick.

INSTRUCTIONS

1. Starting with the piece of metal that is 1/8 in. (3 mm) thick, hold the torch as close as possible to a 90° angle.

2. Lower your hood and establish a plasma cutting stream.

3. Move the torch in an outward spiral until the hole is the desired size.

4. Repeat the hole-cutting process until both sizes of holes are made using all the thicknesses of all three types of metals until you can make consistently smooth cuts that are within ±3/32 in. (2.3 mm) of being round and ±5° of being square to metal.

INSTRUCTOR'S COMMENTS _____

_____

_____

■ PRACTICE 8-4

Name _____   Date _____

Base Metal _____   Thickness _____

Class _____   Instructor _____   Grade _____

Welding Principles and Applications

MATERIAL: ¼"AND ½" MILD STEEL, STAINLESS STEEL, ALUMINUM PLATE

PROCESS: PLASMA ARC CUTTING

NUMBER: PRACTICE 8-4   DRAWN BY: LEN HEBERT

OBJECTIVE: After completing this practice, you should be able to safely set up and operate a PAC machine to bevel mild steel, stainless steel, and aluminum plates ranging from 1/4 to 1/2 in. (6 to 13 mm) thick plate.

EQUIPMENT AND MATERIALS NEEDED FOR THIS PRACTICE

1. A properly set-up and adjusted PAC machine.

2. Proper safety protection (welding hood, safety glasses, leather gloves, pliers, long-sleeved shirt, long pants, and leather boots or shoes). Refer to Chapter 2 in the text for more specific safety information.

3. One or more pieces of mild steel, stainless steel, and aluminum, 6 in. (152 mm) long by 4 in. (100 mm) wide and 1/4 in. (6 mm) and 1/2 in. (13 mm) thick. You will cut a 45° bevel down the length of the plate.

INSTRUCTIONS

1. Starting at one end of the piece of metal that is 1/4 in. (6 mm) thick, hold the torch as close as possible to a 45° angle.

2. Lower your hood and establish a plasma cutting stream.

3. Move the torch in a straight line down the plate toward the other end.

4. Repeat the cut using both thicknesses of all three types of metals until you can make consistently smooth cuts that are within ±3/32 in. (2.3 mm) of a straight line and ±5° of a 45° angle.

INSTRUCTOR'S COMMENTS _____

_____

_____

■ PRACTICE 8-5

Name _____ Date _____

Class _____ Instructor _____ Grade _____

Welding Principles and Applications

MATERIAL: 1/4" AND 1/2" MILD STEEL, STAINLESS STEEL, AND ALUMINUM PLATE

PROCESS: PLASMA ARC GOUGING

NUMBER: PRACTICE 8-5   DRAWN BY: MARY HEBERT

OBJECTIVE: After completing this practice, you should be able to safely set up and operate a PAC machine to make a U-groove in steel, stainless steel, and aluminum.

EQUIPMENT AND MATERIALS NEEDED FOR THIS PRACTICE

1. A properly set-up and adjusted PAC machine.

2. Proper safety protection (welding hood, safety glasses, leather gloves, pliers, long-sleeved shirt, long pants, and leather boots or shoes). Refer to Chapter 2 in the text for more specific safety information.

3. One or more pieces of mild steel, stainless steel, and aluminum, 6 in. (152 mm) long by 4 in. (100 mm) wide and 1/4 in. (6 mm) or 1/2 in. (13 mm) thick.

INSTRUCTIONS

1. Starting at one end of the piece of metal, hold the torch as close as possible to a 30° angle.

2. Lower your hood and establish a plasma cutting stream.

3. Move the torch in a straight line down the plate toward the other end.

4. If the width of the U-groove changes, speed up or slow down the travel rate to keep the groove the same width and depth for the entire length of the plate.

5. Repeat the gouging of the U-groove using all three types of metals until you can make consistently smooth grooves that are within ±3/32 in. (2.3 mm) of a straight line and uniform in width and depth.

INSTRUCTOR'S COMMENTS _____

_____

_____

■ PRACTICE 8-6

Name _____ Date _____

Class _____ Instructor _____ Grade _____

OBJECTIVE: After completing this practice, you should be able to safely set up and operate a PAC machine to cut a 1/2-in. (13 mm) wide cut on pipe or round stock.

EQUIPMENT AND MATERIALS NEEDED FOR THIS PRACTICE

1. A properly set-up and adjusted PAC machine.

2. Proper safety protection (welding hood, safety glasses, leather gloves, pliers, long-sleeved shirt, long pants, and leather boots or shoes). Refer to Chapter 2 in the text for more specific safety information.

3. A piece of soapstone and a flexible straightedge, which can be wrapped around the pipe or round stock.

4. One or more pieces of pipe or round stock 6 in. (152 mm) long.

INSTRUCTIONS

1. Using the straightedge and soapstone, wrap the straightedge around the pipe or round stock and mark rings every 1/2 inch (13 mm).

2. Hold the torch so it is pointed parallel to the round piece and in line with the line to be cut.

3. Lower your hood, and pull the trigger to start the pilot arc and the cutting plasma stream.

4. Use one of the piercing techniques to begin the cut.

5. If the round stock is 1/2 in. (13 mm) or smaller, keep the torch pointed in the same direction, and move the torch across the piece.

6. If the round stock is 1/2 in. (13 mm) or thicker than the PAC torch can cut through easily, then you must move the torch back and forth to make the kerf wider to allow the cut to go all the way through the round stock.

7. Repeat the cut using both thicknesses of all three types of metals until you can make consistently smooth cuts that are within ±3/32 in. (2.3 mm) of a straight line and ±5° of being square. Turn off the PAC equipment and clean up your work area when you are finished cutting.

INSTRUCTOR'S COMMENTS _____

_____

_____

■ PRACTICE 8-7

Name _____ Date _____

Class _____ Instructor _____ Grade _____

OBJECTIVE: After completing this practice, you should be able to safely set up and operate a PAC pipe beveling machine to cut a 45° angle bevel on the pipe.

EQUIPMENT AND MATERIALS NEEDED FOR THIS PRACTICE

1. A properly set-up and adjusted PAC machine.

2. Proper safety protection (welding hood, safety glasses, leather gloves, pliers, long-sleeved shirt, long pants, and leather boots or shoes). Refer to Chapter 2 in the text for more specific safety information.

3. A piece of soapstone and a flexible straightedge, which can be wrapped around the pipe or round stock.

4. One or more pieces of 6 to 8 in. (150 to 400 mm) diameter pipe.

INSTRUCTIONS

1. Attach the pipe beveling machine to the pipe.

2. Attach a machine plasma torch to the beveling machine carriage.

3. Set the torch to a 45° angle.

4. Run the carriage all the way around the pipe to check that it has freedom of movement and that the torch nozzle tip is traveling at the same height all the way around the pipe.

5. Clear any obstacles to the torch moving freely.

6. Change the beveling machine spacers if the tip height changes as the torch is moved all the way around the pipe.

7. Using the running start technique of piercing the pipe, start the carriage moving at the same time the plasma stream is started.

8. Watch the spark stream to determine that the cut is progressing at the correct speed.

9. Make any adjustments in speed to keep the spark stream flowing outward in a steady stream.

10. If the beveled piece is to be used as a welding test coupon, you should hold it with a pair of pliers so that it does not drop to the floor. The hot pipe piece can easily be bent out of round if it hits the floor.

11. Repeat the cut until you can make consistently smooth beveled cuts. Turn off the PAC equipment and clean up your work area when you are finished cutting.

INSTRUCTOR'S COMMENTS _____

_____

_____

CHAPTER 8: QUIZ 1

Name _____   Date _____

Class _____   Instructor _____   Grade _____

INSTRUCTIONS

Carefully read Chapter 8 in the textbook and answer each question.

MATCHING

In the space provided to the left of Column A, write the letter from Column B that best answers or completes the statement in Column A.

	Column A	Column B
_____	1. Plasma arc _____ is the most often used plasma process.	a. body
_____	2. Plasma cutters have the unique ability to cut metals without making them very _____.	b. head
_____	3. Plasma machines can cut _____ of metal, including aluminum, stainless steel, and cast iron.	c. plasma arc
_____	4. The term _____ is the term most often used in the welding industry when referring to the arc plasma used in welding and cutting processes.	d. heat up
_____	5. The plasma is created in both the cutting and welding torches in the same basic manner by _____ the arc into a very small gap between the electrode and nozzle.	e. consumable
_____	6. The torch _____ is a place that provides a good grip area and protects the cable and hose connections to the head.	f. any type
_____	7. The torch _____ is attached to the torch body where the cables and hoses attach to the electrode tip, nozzle tip, and nozzle.	g. cup or deflector
_____	8. The trigger is used to start and stop the power source, gas, and _____.	h. pollutants
_____	9. The _____ parts of a plasma torch include the shield, retaining cap, nozzle, electrode, and swirl ring.	i. damage
_____	10. Copper is used as the base metal for the electrode because it has low electrical resistance, so it does not _____ while carrying the high current needed for the plasma.	j. constricting
_____	11. The nozzle _____ is between the electrode tip and the nozzle tip.	k. flexibility
_____	12. The nozzle _____ has a small, cone-shaped, constricting orifice in the center.	l. shielding gas
_____	13. The shield or nozzle, sometimes called the _____, is made of ceramic or any other high-temperature–resistant substance.	m. hot

_____ 14. A water shroud nozzle tip is used to control the potential hazards of light, fumes, noise, or other _____ produced by the process.

         n. sized

_____ 15. As parts become worn the quality of the cut will decrease, and if left their complete failure can result in significant _____ to the plasma cutting torch and the metal being cut.

         o. insulator

_____ 16. A number of power and control cables and gas and cooling water hoses may be used to connect the _____ with the torch.

         p. coolant

_____ 17. Some larger high-amperage machine cutting torches may have two hoses—one hose carries the gas used to produce the plasma, and the other provides the _____ coverage.

         q. cutting

_____ 18. The power cable must have a high-voltage–rated insulation, and it is made of finely stranded copper wire to allow for maximum _____ of the torch.

         r. tip

_____ 19. The _____ system for a plasma workstation consists of a water pump, coolant reservoir, and hoses to carry the coolant to and from the torch.

         s. power supply

_____ 20. Cooling hoses must be _____ to allow the system pump to circulate the required volume of coolant to keep the torch cool.

         t. cooling water

CHAPTER 8: QUIZ 2

Name _____ Date _____

Class _____ Instructor _____ Grade _____

INSTRUCTIONS

Carefully read Chapter 8 in the textbook and answer each question.

MATCHING

In the space provided to the left of Column A, write the letter from Column B that best answers or completes the statement in Column A.

	Column A	Column B
_____	1. The _____ is a two-conductor, low-voltage, stranded copper wire that allows the welder to start and stop the plasma power and gas as needed during the cut or weld.	a. inverter-type
_____	2. Most small shop plasma arc cutting torches use _____ to form the plasma and to make the cut.	b. voltage
_____	3. Most small plasma power supplies are _____ machines capable of supplying the voltage and amperages required for plasma cutting.	c. lower
_____	4. The production of the plasma requires a _____, high-voltage, power supply.	d. conductive
_____	5. Although the _____ is higher, the current (amperage) flow is much lower than it is with most other welding processes.	e. alternating
_____	6. _____ are a measure of a plasma machine's power to do the job.	f. major
_____	7. A high travel speed with plasma cutting will result in a heat input that is much _____ than that of the oxyfuel cutting process.	g. Oxygen
_____	8. The heat of a torch cut causes metal to bend, but the _____ cut is so fast little or no bending occurs.	h. direct-current (DC)
_____	9. Any material that is electrically _____ can be cut using the PAC process.	i. C-clamps
_____	10. The torch standoff distance is the distance from the _____ to the work.	j. plasma torch
_____	11. The most common starting method uses a high-frequency _____ current carried through the conductor, the electrode, and back from the nozzle tip.	k. kerf
_____	12. The _____ is the space left in the workpiece as the metal is removed during a cut.	l. Water tables
_____	13. The type of gas or gases used will have a _____ effect on the cutting performance, although almost any gas or gas mixture can be used today for the PAC process.	m. control wire
_____	14. _____ produces the fastest and cleanest cuts on mild steels because the stream reacts with the steel to produce a very fluid fine spray of molten droplets.	n. Dross
_____	15. The type of plasma cutting gas along with the number of _____ affects the life of the stack.	o. safety

_____ 16. Stacked sheets can be held together for cutting by using standard _____. p. Watts

_____ 17. _____ is the metal compound that resolidifies and attaches itself to the bottom of a cut. q. starts

_____ 18. Almost any _____ can be attached to some type of semiautomatic or automatic device to allow it to make machine cuts. r. compressed air

_____ 19. _____ either support the metal just above the surface of the water or they submerge the metal approximately 3 in. (76 mm) below the water's surface. s. plasma

_____ 20. Low-power plasma machines are limited to low power, 200 amperes or less, primarily for _____ reasons. t. nozzle tip

CHAPTER 8: QUIZ 3

Name _____   Date _____

Class _____   Instructor _____   Grade _____

INSTRUCTIONS
Carefully read Chapter 8 in the textbook and answer each question.

IDENTIFICATION
Identify the numbered items on the drawing by writing the letter next to the identifying term in the space provided.

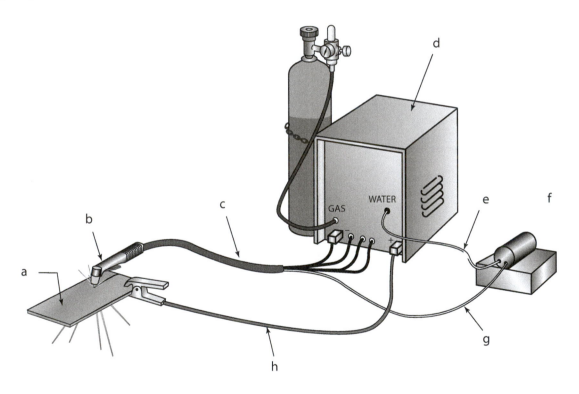

1. _____ WORK CABLE

2. _____ WATER RECIRCULATION

3. _____ WORK

4. _____ POWER SUPPLY

5. _____ WATER OUT

6. _____ PLASMA TORCH

7. _____ WATER IN

8. _____ POWER & COMPRESSED GAS OR AIR CABLE

CHAPTER 8: QUIZ 4

Name _____   Date _____

Class _____   Instructor _____   Grade _____

INSTRUCTIONS

Carefully read Chapter 8 in the textbook and answer each question.

IDENTIFICATION

Increasing or decreasing the travel speed, stando distance, torch angle, or current will increase or decrease a plasma groove's width or depth. In the space provided below put an X beside the effect these changes will have on the groove.

CHANGE	INCREASE/DECREASE	GROOVE WIDTH	GROOVE DEPTH	EFFECT
TRAVEL SPEED	INCREASE	1	2	
	DECREASE	3	4	
STANDOFF DISTANCE	INCREASE	5	6	
	DECREASE	7	8	
TORCH ANGLE	INCREASE (MORE VERTICAL)	9	10	
	DECREASE (LESS VERTICAL)	11	12	
CURRENT	INCREASE	13	14	
	DECREASE	15	16	

1. _____ Increase _____ Decrease

2. _____ Increase _____ Decrease

3. _____ Increase _____ Decrease

4. _____ Increase _____ Decrease

5. _____ Increase _____ Decrease

6. _____ Increase _____ Decrease

7. _____ Increase _____ Decrease

8. _____ Increase _____ Decrease

9. _____ Increase _____ Decrease

10. _____ Increase _____ Decrease

11. _____ Increase _____ Decrease

12. _____ Increase _____ Decrease

13. _____ Increase _____ Decrease

14. _____ Increase _____ Decrease

15. _____ Increase _____ Decrease

16. _____ Increase _____ Decrease

CHAPTER 8: QUIZ 5

Name _____   Date _____

Class _____   Instructor _____   Grade _____

INSTRUCTIONS

Carefully read Chapter 8 in the textbook and answer each question.

IDENTIFICATION

Identify the numbered items on the drawing by writing the letter next to the identifying term in the space provided.

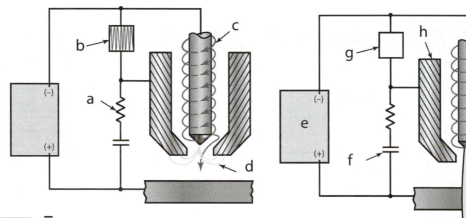

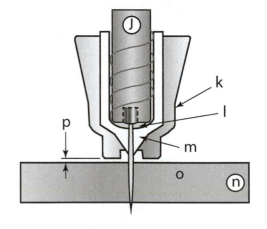

1. _____ **–**

2. _____ **+**

3. _____ RESISTOR

4. _____ NOZZLE TIP

5. _____ PILOT ARC RELAY CONTACTS

6. _____ TORCH STANDOFF

7. _____ PLASMA GAS

8. _____ HIGH FREQUENCY GENERATOR OFF

9. _____ SHIELD

10. _____ ELECTRODE

11. _____ HIGH FREQUENCY GENERATOR ON

12. _____ WORKPIECE

13. _____ PLENUM CHAMBER

14. _____ NOZZLE

15. _____ PILOT ARC

16. _____ DC POWER SUPPLY

CHAPTER 8: QUIZ 6

Name _____   Date _____

Class _____   Instructor _____   Grade _____

INSTRUCTIONS

Carefully read Chapter 8 in the textbook and answer each question.

IDENTIFICATION

Identify the numbered items on the drawing by writing the letter next to the identifying term in the space provided.

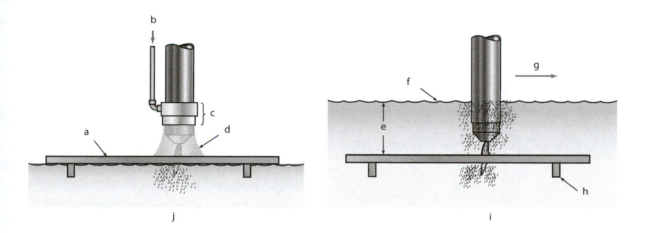

1. _____   WORKPIECE

2. _____   WATER LEVEL

3. _____   LIGHT, SOUND, AND FUME POLLUTION CONTROL

4. _____ $2\frac{1}{2}" - 3"$

5. _____   DIRECTION OF CUT

6. _____   WATER SHROUD NOZZLE

7. _____   UNDERWATER PLASMA CUTTING

8. _____   WATER SHROUD

9. _____   (15 TO 20 gpm)

10. _____   WORK SUPPORT

Related Cutting Processes

Air Carbon Arc Electrode Current Recommendations									
Electrode Diameter									
Current	5/32	3/16	1/4	5/16	3/8	1/2	5/8	3/4	FLAT
Min DC	90	200	300	350	450	800	1000	1250	300
Max DC	150	250	400	450	600	1000	1250	1600	500
Min AC	—	200	300	325	350	500	—	—	—
Max AC	—	250	400	425	450	600	—	—	—

Air Carbon Arc Cutting Electrode Recommendations			
Metal	Electrode	Current	Polarity
Steel	DC	DC	DCRP
Cast Iron:			
Gray, Ductile,	DC	DC	DCRP
Malleable	DC	DC	DCSP
Copper Alloys	DC	DC	DCSP
	DC	DC	—
Nickel Alloys	AC	AC or DC	DCSP

■ PRACTICE 9-1

Name _____ Date _____

Class _____ Instructor _____ Grade _____

OBJECTIVE: After completing this practice, you should be able make a safety inspection of a CAC-A workstation.

EQUIPMENT AND MATERIALS NEEDED FOR THIS PRACTICE

Use an air carbon arc cutting torch, welding power supply dry air supply, the manufacturer's owner's and operator's manual, safety glasses, welding helmet, welding gloves, ear protection, and any other required PPE, you will safely set up the CAC-A cutting equipment.

INSTRUCTIONS

1. Demonstrate to your instructor and fellow students how to assemble the CAC-A equipment according to the manufacturers.

2. Check the airline and fittings for any possible damage because most air hoses are not flame resistance and can be damaged by CAC-A sparks.

3. Make sure there are no flammable materials within the area where the CAC-A sparks will be scattered when making cuts.

4. Place welding curtains around the area to protect others in the area from the sparks and arc light.

INSTRUCTOR'S COMMENTS _____

_____

_____

■ PRACTICE 9-2

Name _____ Date _____

Class _____ Instructor _____ Grade _____

OBJECTIVE: After completing this practice, you should be able make minor repairs on CAC-A equipment.

EQUIPMENT AND MATERIALS NEEDED FOR THIS PRACTICE

Using the same equipment setup as in Practice 9-1 and an assortment of hand tools, CAC-A torch replacement parts make minor repairs on CAC-A equipment.

1. Use a leak detection solution to check the air fitting to the CAC-A torch to locate any possible leaks.

   - If the air fitting is leaking, remove it, clean off the thread sealant, and reapply a thread material such as Teflon® tape.
   - Reinstall the air fitting and again check it for leaks.

2. Use a small wire brush and file to remove any spatter stuck on the electrode jaws or insulators.

3. Make any other repairs as indicated by your instructor.

INSTRUCTOR'S COMMENTS _____

_____

_____

■ PRACTICE 9-3

Name _____   Date _____

Class _____   Instructor _____   Grade _____

OBJECTIVE: After completing this practice, you should be able to safely set up and operate an air carbon arc cutting torch to make a U-groove gouge on carbon steel plate.

EQUIPMENT AND MATERIALS NEEDED FOR THIS PRACTICE

1. An air carbon arc cutting torch and welding power supply that has been safely set up in accordance with the manufacturer's specific instructions in the owner's manual.

2. Proper safety protection equipment (safety glasses, welding helmet, leather gloves, pliers, long-sleeved shirt, long pants, and leather boots or shoes). Refer to Chapter 2 and Chapter 9 in the text for more specific safety information.

3. A mild steel plate, 3/8 in. (9 mm) by 4 in. (100 mm) by 6 in. (152 mm).

INSTRUCTIONS

1. Adjust the air pressure to approximately 80 psi.

2. Set the amperage within the range for the diameter electrode you are using by referring to the box that the electrodes came in.

3. Check to see that the stream of sparks will not start a fire or cause any damage to anyone or anything in the area.

4. Make sure the area is safe and turn on the welding machine.

5. Using a suitable dry leather glove to avoid electrical shock, insert the electrode in the torch jaws so that about 6 in. (152 mm) is extending outward. Be sure not to touch the electrode to any metal parts because it may short out.

6. Turn on the air at the torch head.

7. Lower your arc welding helmet.

8. Slowly bring the electrode down at about a 30° angle so it will make contact with the plate near the starting edge. Be prepared for a loud, sharp sound when the arc starts.

9. Once the arc is struck, move the electrode in a straight line down the plate toward the other end. Keep the speed and angle of the torch constant. You should be attempting to skim the surface of the plate at a depth of 1/16 in. (2 mm) to 3/32 in. (2.3 mm) with each skimming pass made with the carbon arc electrode. Keep skimming the gouged surface until you have reached the depth required.

10. When you reach the other end, lift the torch so the arc will stop.

11. Raise your helmet and turn off the air.

12. Remove the remaining electrode from the torch so it will not accidently touch anything.

13. When the metal is cool, chip or brush any slag or dross off of the plate. This material should remove easily. The groove must be within ±1/8 in. (3 mm) of being straight and within ±3/32 in. (2.4 mm) of uniformity in width and depth.

14. Repeat this cut until it can be made within these tolerances.

15. Turn off the welding machine and clean up your work area when you are finished welding.

INSTRUCTOR'S COMMENTS _____

_____

_____

■ PRACTICE 9-4

Name _____ Date _____

Class _____ Instructor _____ Grade _____

OBJECTIVE: After completing this practice, you should be able to safely set up and operate an air carbon arc cutting torch to make a J-groove along the edge of a plate.

EQUIPMENT AND MATERIALS NEEDED FOR THIS PRACTICE

1. An air carbon arc cutting torch and welding power supply that has been safely set up in accordance with the manufacturer's specific instructions in the owner's manual.

2. Proper safety protection equipment (safety glasses, welding helmet, leather gloves, pliers, long-sleeved shirt, long pants, and leather boots or shoes). Refer to Chapter 2 and Chapter 9 in the text for more specific safety information.

3. A mild steel plate, 3/8 in. (9 mm) by 4 in. (100 mm) by 6 in. (152 mm).

INSTRUCTIONS

1. Adjust the air pressure to approximately 80 psi.

2. Set the amperage within the range for the diameter electrode you are using.

3. Check to see that the stream of sparks will not start a fire or cause any damage to anyone or anything in the area.

4. Make sure the area is safe and turn on the welding machine.

5. Using a suitable dry leather glove to avoid electrical shock, insert the electrode in the torch jaws so that about 6 in. (152 mm) is extending outward. Be sure not to touch the electrode to any metal parts because it may short out.

6. Turn on the air at the torch head.

7. Lower your arc welding helmet.

8. Slowly bring the electrode down at about a 30° angle so it will make contact with the plate near the starting edge.

9. Once the arc is struck, move the electrode in a straight line down the edge of the plate toward the other end. Keep the speed and angle of the torch constant. You should be trying to just skim the edge of the plate until you have gouged a J-groove along the plate edge.

10. When you reach the other end, lift the torch so the arc will stop.

11. Raise your helmet and turn off the air.

12. Remove the remaining electrode from the torch so it will not accidently touch anything.

13. When the metal is cool, chip or brush any slag or dross off of the plate. This material should remove easily. The groove must be within ±1/8 in. (3 mm) of being straight and within ±3/32 in. (2.4 mm) of uniformity in width and depth.

14. Repeat this cut until it can be made within these tolerances.

15. Turn off the welding machine and clean up your work area when you are finished welding.

INSTRUCTOR'S COMMENTS _____

_____

_____

■ PRACTICE 9-5

Name _____   Date _____   Electrode Used _____

Class _____   Instructor _____   Grade _____

OBJECTIVE: After completing this practice, you should be able to make a U-groove along the root face of a weld joint on a plate (back side).

EQUIPMENT AND MATERIALS NEEDED FOR THIS PRACTICE

1. An air carbon arc cutting torch and welding power supply that has been safely set up in accordance with the manufacturer's specific instructions in the owner's manual.

2. Proper safety protection equipment (safety glasses, welding helmet, leather gloves, pliers, long-sleeved shirt, long pants, and leather boots or shoes). Refer to Chapter 2 and Chapter 9 in the text for more specific safety information.

3. Two plates 6 in. (152 mm) in length by 4 in. (100 mm) in width by 3/8 in. (9 mm) in thick that have been joined with a suitable partial joint penetration groove-type weld.

INSTRUCTIONS

1. Adjust the air pressure to approximately 80 metric?

2. Set the amperage within the range for the diameter electrode you are using by referring to the box that the electrodes came in.

3. Check to see that the stream of sparks will not start a fire or cause any damage to anyone or anything in the area.

4. Make sure the area is safe and turn on the welding machine.

5. Using a suitable dry leather glove to avoid electrical shock, insert the electrode in the torch jaws so that about 6 in. (152 mm) is extending outward. Be sure not to touch the electrode to any metal parts because it may short out.

6. Take the groove welded plate and place it with the root face side up. (Back side of the plate up.)

7. Turn on the air at the torch head.

8. Lower your arc welding helmet.

9. Start the arc at the joint between the two plates on the side opposite where the weld bead is.

10. Once the arc is struck, move the electrode in a straight line down the edge of the plate toward the other end. Watch the bottom of the cut to see that it is deep enough. If there is a line along the bottom of the groove, it needs to be deeper. Once the groove depth is determined, keep the speed and angle of the torch constant. Try to gouge a depth of 3/32 in. (2.3 mm) with each skim pass of the carbon arc electrode.

11. When you reach the other end, break the arc off.

12. Raise your helmet and turn off the air.

13. Remove the remaining electrode from the torch so it will not accidently touch anything.

14. When the metal is cool, chip or brush any slag or dross off of the plate. This material should remove easily. The groove must be within ±1/8 in. (3 mm) of being straight, but it may vary in depth so long as all of the unfused root of the weld has been removed.

15. Repeat this cut until it can be made within these tolerances.

16. Turn off the welding machine and clean up your work area when you are finished welding.

INSTRUCTOR'S COMMENTS _____
_____
_____

■ PRACTICE 9-6

Name _____ Date _____ Electrode Used _____

Class _____ Instructor _____ Grade _____

OBJECTIVE: After completing this practice, you should be able to make a U-groove to remove a weld from a plate.

EQUIPMENT AND MATERIALS NEEDED FOR THIS PRACTICE

1. An air carbon arc cutting torch and welding power supply that has been safely set up in accordance with the manufacturer's specific instructions in the owner's manual.

2. Proper safety protection equipment (safety glasses, welding helmet, leather gloves, pliers, long-sleeved shirt, long pants, and leather boots or shoes). Refer to Chapter 2 and Chapter 9 in the text for more specific safety information.

3. Two steel plates that have been welded together, minimum length 6 in. (152 mm).

INSTRUCTIONS

1. Adjust the air pressure to approximately 80 psi.

2. Set the amperage within the range for the diameter electrode you are using by referring to the box that the electrodes came in.

3. Check to see that the stream of sparks will not start a fire or cause any damage to anyone or anything in the area.

4. Make sure the area is safe and turn on the welding machine.

5. Using a suitable dry leather glove to avoid electrical shock, insert the electrode in the torch jaws so that about 6 in. (152 mm) is extending outward. Be sure not to touch the electrode to any metal parts because it may short out.

6. Turn on the air at the torch head.

7. Lower your arc welding helmet.

8. Start the arc on one end of the weld bead.

9. Once the arc is struck, move the electrode in a straight line down the weld toward the other end. Watch the bottom of the cut to see that it is deep enough. If there is not a line along the bottom of the groove, it needs to be deeper. Once the groove depth is determined, keep the speed and angle of the torch constant.

10. When you reach the other end, break the arc off.

11. Raise your helmet and turn off the air.

12. Remove the remaining electrode from the torch so it will not accidently touch anything.

13. When the metal is cool, chip or brush any slag or dross off of the plate. This material should remove easily. The groove must be within ±1/8 in. (3 mm) of being straight, but it may vary in depth so that all of the weld metal has been removed. You should be able to completely separate the two plates.

14. Repeat this cut until it can be made within these tolerances.

15. Turn off the welding machine and clean up your work area when you are finished welding.

INSTRUCTOR'S COMMENTS _____

_____

_____

CHAPTER 9: QUIZ 1

Name _____ Date _____

Class _____ Instructor _____ Grade _____

INSTRUCTIONS

Carefully read Chapter 9 in the textbook and answer each question.

MATCHING

In the space provided to the left of Column A, write the letter from Column B that best answers or completes the statement in Column A.

	Column A	Column B
_____	1. A laser can be used to _____ holes (LBD) through the hardest materials, such as synthetic diamonds, tungsten carbide cutting tools, quartz, glass, or ceramics.	a. efficiency
_____	2. Lasers can be used to _____ (LBC) materials that must be cut without overheating delicate parts that might be located just a few thousandths of an inch away.	b. circulating
_____	3. Most lasers used for cutting, drilling, and welding produce a laser light beam in the _____ range.	c. Solid state
_____	4. Lasers can be divided into two major types: lasers using a solid material for the laser and lasers using a _____.	d. cut
_____	5. _____ lasers are most often used with a low-power continuous or high-powered pulse.	e. 35
_____	6. Gas lasers either use a gas-charged cylinder where the gas is static or use a chamber that has a means of _____ the laser gas.	f. strength
_____	7. The ability of a material's surface to absorb or reflect the laser light affects the laser's operating _____.	g. gas
_____	8. A laser weld and laser cut both bring the material to a _____ state.	h. Gouging
_____	9. The _____ laser is the best choice for drilling operations.	i. air stream
_____	10. Most laser beam welding is performed on _____ materials.	j. Washing
_____	11. Laser equipment is _____ than most of the other welding or cutting power supplies.	k. thin
_____	12. In the air carbon arc cutting process, the _____ blows the molten metal away.	l. drill
_____	13. On an air carbon arc cutting torch, the lower electrode jaw has a series of _____.	m. molten

―――― 14. The copper coating on electrodes provides increased _____ to reduce accidental breakage.

n. postcut cleanup

―――― 15. The operating voltage required for air carbon arc cutting needs to be _____ volts or higher.

o. infrared

―――― 16. The correct air pressure supplied to the torch will result in cuts that are clean, smooth, and _____.

p. repair

―――― 17. Air carbon arc cutting is most often used for _____ work.

q. air holes

―――― 18. The removal of welds can be accomplished with such success with air carbon arc cutting that often the part needs no _____.

r. pulsed

―――― 19. _____ is the most common application of the air carbon arc cutting processes.

s. larger

―――― 20. _____ is the process sometimes used to remove large areas of metal so that hard surfacing can be applied.

t. uniform

CHAPTER 9: QUIZ 2

Name _____ Date _____

Class _____ Instructor _____ Grade _____

INSTRUCTIONS

Carefully read Chapter 9 in the textbook and answer each question.

MATCHING

In the space provided to the left of Column A, write the letter from Column B that best answers or completes the statement in Column A.

Column A

_____ 1. The quantity and volume of sparks and molten metal spatter generated during this process are a major _____ hazard.

_____ 2. A piece of _____ may be placed in line with the spark stream to stop the sparks or to reflect the sparks downward.

_____ 3. The air carbon arc cutting process produces a high level of _____.

_____ 4. Because the arc has no smoke to diffuse the light and the amperages are usually much higher, the chances of receiving arc burns are much _____.

_____ 5. Because of the intense arc light, a _____ welding filter lens for the helmet should be used.

_____ 6. The combination of the air and the metal being removed results in a high volume of _____.

_____ 7. If the used parts have paint, oils, or other contaminants that might generate hazardous fumes, then they must be removed in an acceptable manner before any _____ begins.

_____ 8. _____ is used to remove defective welds and to prepare a thick metal joint so that a full penetration weld can be made.

_____ 9. The oxygen lance cutting process uses a _____ alloy tube.

_____ 10. Fire and rescue personal can use oxygen lances to cut heavy steel beams, _____, or other material that may need to be quickly removed during a rescue.

_____ 11. It allows the quick removal of thick sections of a building without the dangerous _____ caused by most conventional methods.

_____ 12. An approved ventilation system must be provided when using oxygen lances if work is to be done in a building or any other _____ area.

Column B

a. cutting

b. U-grooving

c. robotic

d. metal

e. darker

f. erosion

g. vibration

h. higher

i. defective weld

j. sparks and slag

k. above

l. abrasive powder

———— 13. Oxygen lances produce both high levels of radiant heat and plumes of molten _____.

m. safety

———— 14. Sound is produced well _____ safety levels.

n. electrode holder

———— 15. Water jet cutting does not put any _____ into the material being cut.

o. consumable

———— 16. With water jet cutting, the cut is accomplished by the rapid _____ of the material by a high-pressure jet of water.

p. concrete

———— 17. The addition of an _____ to the stream of water can speed up the cutting, allow harder materials to be cut, and improve the surface finish of a cut.

q. enclosed

———— 18. Most water jet cutting is performed by some automated or _____ system.

r. sound

———— 19. Arc cutting electrodes do not require any special equipment other than a standard SMA welding machine, and they fit into the standard _____.

s. fumes

———— 20. Arc cutting electrodes can be used to cut metal, pierce holes, or to gouge a groove for welding or gouge out a _____.

t. heat

CHAPTER 9: QUIZ 3

Name _____ Date _____

Class _____ Instructor _____ Grade _____

INSTRUCTIONS

Carefully read Chapter 9 in the textbook and answer each question.

IDENTIFICATION

Identify the numbered items on the drawing by writing the letter next to the identifying term in the space provided.

1. _____ BODY

2. _____ TORCH

3. _____ CABLE COVER

4. _____ HEAD

5. _____ ELECTRODE LEAD DCEP OR AC

6. _____ HANDLE

7. _____ INSULATOR

8. _____ CARBON ELECTRODE

9. _____ COMPRESSED AIR

10. _____ CONDUCTOR

11. _____ FEMALE CONNECTOR

12. _____ LEVER

13. _____ CONCENTRIC CABLE

14. _____ AIR VALVE

15. _____ POWER SUPPLY

16. _____ WORKPIECE

17. _____ WORKPIECE LEAD

CHAPTER 9: QUIZ 4

Name _____ Date _____

Class _____ Instructor _____ Grade _____

INSTRUCTIONS

Carefully read Chapter 9 in the textbook and answer each question.

IDENTIFICATION

Identify the numbered items on the drawing by writing the letter next to the identifying term in the space provided.

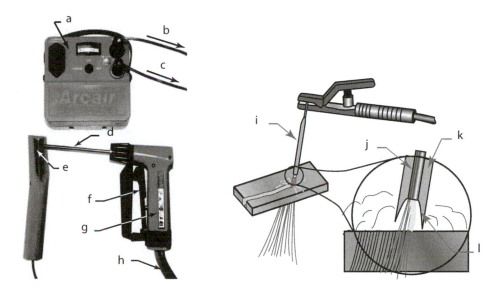

1. _____ ARC CUTTING ELECTRODE

2. _____ TO TORCH

3. _____ STRIKER

4. _____ SMALL CAVITY

5. _____ FROM OXYGEN REGULATOR

6. _____ OXYGEN LANCE ROD

7. _____ ELECTRODE CORE WIRE

8. _____ POWER SUPPLY

9. _____ TORCH

10. _____ OXYGEN LEVER

11. _____ FLUX COVERING

12. _____ TO STRIKER

Gas Metal Arc Welding
Equipment, Setup, and Operation

CHAPTER 10: QUIZ 1

Name _____   Date _____

Class _____   Instructor _____   Grade _____

INSTRUCTIONS

Carefully read Chapter 10 in the textbook and answer each question.

MATCHING

In the space provided to the left of Column A, write the letter from Column B that best answers or completes the statement in Column A.

Column A

Column B

_____ 1. Gas metal arc welding is known in the field by several other terms, such as MIG, which is short for metal inert gas welding, MAG, which is short for metal active gas welding, and _____, which describes the electrode used.

a. axial spray

_____ 2. The GMAW process may be performed as semiautomatic (SA), machine (ME), or _____ welding.

b. direct contact

_____ 3. The GMAW process is unique in that there are six modes of transferring the _____ from the wire to the weld.

c. weld metal

_____ 4. To change between short-circuiting, axial spray, globular, or buried arc transfer methods, for the most part all you have to do is make the changes in the welding machines _____ settings.

d. Modulated current

_____ 5. With the short-circuiting mode of transfer, low currents allow the liquid metal at the electrode tip to be transferred by _____ with the molten weld pool.

e. Ramp up

_____ 6. In the _____ transfer, the arc melts the end of the electrode, forming a molten ball of metal.

f. heat

_____ 7. The _____ metal transfer mode is identified by the pointing of the wire tip from which very small drops are projected (sprayed) axially across the arc gap to the molten weld pool.

g. sensor

_____   8. In the pulsed-arc metal transfer, one pulse of high current is for the axial spray transfer mode, and the other lower pulse of current should not transfer any _____.   h. wire welding

_____   9. _____ refers to how the electric current in a transformer takes a few milliseconds to build up the magnetic field to full strength once the coil is energized.   i. diameter

_____   10. _____ is the time in milliseconds from the beginning of a single pulse cycle to the beginning of the next cycle.   j. voltage and amperage

_____   11. _____ metal transfer is the newest GMA welding process; it uses an extremely modified version of GMAW-S and has a very sophisticated computer-controlled welding machine.   k. deposition rate

_____   12. In the modulated current process, a _____ is attached near the weld on the plate or pipe that provides the computer with the necessary information for it to control the metal transfer.   l. automatic (AU)

_____   13. An advantage of both modulated current and pulsed-arc metal transfer is that there is less _____ input to the base metal as a result of the periods of lower current flow.   m. crowned

_____   14. Another advantage of both modulated current and pulsed-arc metal transfer is that there is little _____ produced during welding.   n. wire melting rate

_____   15. With buried-arc transfer, the resultant welds tend to be more highly _____ than those produced with open arcs, but they are relatively free of spatter and offer a decided advantage of welding speed.   o. filler metal

_____   16. The amperage ranges for GMA welding electrodes will vary depending on the method of metal transfer, type of shielding gas, and base metal _____.   p. thickness

_____   17. The _____, measured in inches per minute (in./min) or pounds per hour (lb/hr), is the rate at which the arc consumes the wire.   q. globular

_____   18. The _____, the measure of weld metal deposited, is nearly always less than the melting rate because not all of the wire is converted to weld metal.   r. deposition efficiency

_____   19. The amount of weld metal deposited in ratio to the wire used is called the _____.   s. Pulse width

_____   20. Welders can control the deposition rate by changing the current, electrode extension, and _____ of the wire.   t. fume

CHAPTER 10: QUIZ 2

Name _____ Date _____

Class _____ Instructor _____ Grade _____

INSTRUCTIONS

Carefully read Chapter 10 in the textbook and answer each question.

MATCHING

In the space provided to the left of Column A, write the letter from Column B that best answers or completes the statement in Column A.

Column A	Column B
_____ 1. _____ is a measurement of electrical pressure in the same way that pounds per square inch is a measurement of water pressure.	a. lighter
_____ 2. Electrical potential means the same thing as voltage and is usually expressed by using the term _____.	b. Amperage, or amps (A)
_____ 3. _____ is the measurement of the total number of electrons flowing in the same way that gallons are a measurement of the amount of water flowing.	c. stop
_____ 4. Electrical current means the same thing as amperage and is usually expressed by using the term _____.	d. weld
_____ 5. GMAW _____ are the constant-voltage, constant potential (CV, CP) type of machines.	e. power supplies
_____ 6. The wire-feed speed is generally recommended by the electrode manufacturer and is selected in _____, or how fast the wire exits the contact tube.	f. Helium
_____ 7. The slope should be adjusted so that a proper _____ metal transfer occurs.	g. penetration
_____ 8. The shielding gas selected for a weld has a definite effect on the _____ produced.	h. spatter-free
_____ 9. Argon is denser than air so it effectively shields welds by pushing the _____ air away.	i. Voltage, or volts (V)
_____ 10. Adding reactive gases (oxidizing), such as oxygen or carbon dioxide, to argon tends to _____ the arc, promote favorable metal transfer, and minimize spatter.	j. wire-feed speed
_____ 11. _____ is lighter than air; thus, its flow rates must be approximately twice as high as that of argon for acceptable stiffness in the gas stream to be able to push air away from the weld.	k. increase

_____ 12. _____ is a compound made up of one carbon atom (C) and two oxygen atoms (O_2), and its molecular formula is CO_2.

_____ 13. Nitrogen is often used in blended gases to _____ the arc's heat and temperature.

_____ 14. Making an acceptable weld requires a _____ of the voltage and amperage.

_____ 15. On a GMA welding machine, the amperage at the arc is adjusted by changing the _____.

_____ 16. During the short-circuiting process, if the arc is directed to the base metal and outside the molten weld pool, then the welding process may _____.

_____ 17. Because the location of the arc inside the molten weld pool is important, the welding travel speed cannot exceed the ability of the arc to melt the _____.

_____ 18. The _____ (stickout) is the distance from the contact tube to the arc measured along the wire.

_____ 19. The *gun angle*, *work angle*, and *travel angle* are names used to refer to the relation of the gun to the _____.

_____ 20. Changes in this angle will affect the weld bead shape and _____.

l. electrode extension

m. base metal

n. current (C)

o. work surface

p. inches per minute (ipm)

q. stabilize

r. balancing

s. potential (P)

t. Carbon dioxide

CHAPTER 10: QUIZ 3

Name _____   Date _____

Class _____   Instructor _____   Grade _____

INSTRUCTIONS

Carefully read Chapter 10 in the textbook and answer each question.

MATCHING

In the space provided to the left of Column A, write the letter from Column B that best answers or completes the statement in Column A.

	Column A	Column B
_____	1. Forehand, perpendicular, and backhand are the terms most often used to describe the _____ as it relates to the work and the direction of travel.	a. gun
_____	2. The forehand technique is sometimes referred to as pushing the weld bead, and _____ may be referred to as pulling or dragging the weld bead.	b. generator
_____	3. The term _____ is used when the gun angle is at approximately 90° to the work surface.	c. smoke
_____	4. The lower percentage of nonmetallic compounds in metal cored arc welding electrodes results in less slag, less postweld cleanup, and less _____ in the welding shop.	d. rate
_____	5. Metal core electrodes can be used in _____ position and on joints that have larger or uneven root openings.	e. cooling
_____	6. There are no changes in _____ and very little change in setup to change from solid wire GMA to metal core GMA welding.	f. strength
_____	7. The basic GMAW equipment consists of the _____, electrode (wire) feed unit, electrode (wire) supply, power source, shielding gas supply with flowmeter/regulator, control circuit, and related hoses, liners, and cables.	g. backhand
_____	8. Larger, more complex systems may have water for _____, solenoids for controlling gas flow, and carriages for moving the work or the gun or both.	h. wire
_____	9. The power source may be either a transformer-rectifier or _____ type.	i. power cable
_____	10. Because of the long periods of continuous use, GMA welding machines have a 100% _____.	j. pull-type
_____	11. The purpose of the electrode feeder is to provide a steady and reliable supply of _____ to the weld.	k. gun angle

_____ 12. Slight changes in the _____ at which the wire is fed have distinct effects on the weld.

l. burns

_____ 13. In the push-type system, the electrode must have enough _____ to be pushed through the conduit without kinking.

m. feed rollers

_____ 14. In _____ systems, a smaller but higher-speed motor is located in the gun to pull the wire through the conduit.

n. compact

_____ 15. _____ feed systems use a synchronized system with feed motors located at both ends of the electrode conduit.

o. perpendicular

_____ 16. The advantage of a linear system is that the bulky system of gears is eliminated, thus _____ weight, size, and wasted power.

p. any

_____ 17. A spool gun is a _____, self-contained system consisting of a small drive system and a wire supply.

q. equipment

_____ 18. The electrode conduit or liner guides the welding wire from the _____ to the gun.

r. Push-pull

_____ 19. The welding gun attaches to the end of the _____, electrode conduit, and shielding gas hose.

s. reducing

_____ 20. The GMAW spot weld starts on one surface of one member and _____ through to the other member.

t. duty cycle

CHAPTER 10: QUIZ 4

Name _____   Date _____

Class _____   Instructor _____   Grade _____

INSTRUCTIONS

Carefully read Chapter 10 in the textbook and answer each question.

IDENTIFICATION

Identify the numbered items on the drawing by writing the letter next to the identifying term in the space provided.

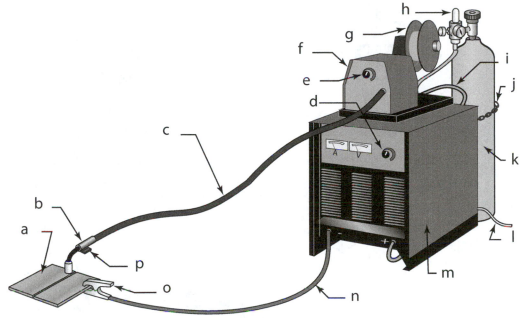

1. _____   WORK CLAMP

2. _____   WORK

3. _____   WORK LEAD

4. _____   WELDING POWER LEAD

5. _____   WELDING VOLTAGE ADJUSTMENT

6. _____   MAIN POWER SUPPLY CABLE

7. _____   WIRE SPOOL

8. _____   WIRE SPEED ADJUSTMENT

9. _____   WELDING CABLE ASSEMBLY

10. _____   WELDING MACHINE

11. _____   CYLINDER SAFETY CHAIN

12. _____   GUN START/STOP TRIGGER

13. _____   WIRE FEED AND CONTROL UNIT

14. _____   WELDING GUN

15. _____   SHIELDING GAS CYLINDER

16. _____   COMBINATION REGULATOR AND FLOWMETER

CHAPTER 10: QUIZ 5

Name _____   Date _____

Class _____   Instructor _____   Grade _____

INSTRUCTIONS
Carefully read Chapter 10 in the textbook and answer each question.

IDENTIFICATION
Identify the numbered items on the drawing by writing the letter next to the identifying term in the space provided.

1. _____ BASE METAL
2. _____ GUN TRIGGER
3. _____ GUN BODY
4. _____ CONTACT TUBE
5. _____ ELECTRODE EXTENSION
6. _____ GAS NOZZLE
7. _____ CONTACT TUBE-TO-WORK DISTANCE
8. _____ INSULATED CONDUCTOR TUBE
9. _____ WIRE GUIDE AND CONTACT TUBE

10. _____ NOZZLE-TO-WORK DISTANCE
11. _____ CONDUIT
12. _____ MOLTEN WELD POOL
13. _____ SHIELDING GAS
14. _____ GAS DIFFUSER
15. _____ ARC LENGTH
16. _____ SOLID WIRE ELECTRODE
17. _____ ARC AND METAL TRANSFER
18. _____ WELD METAL

CHAPTER 10: QUIZ 6

Name _____   Date _____

Class _____   Instructor _____   Grade _____

INSTRUCTIONS

Carefully read Chapter 10 in the textbook and answer each question.

IDENTIFICATION

Identify the numbered items on the drawing by writing the letter next to the identifying term in the space provided.

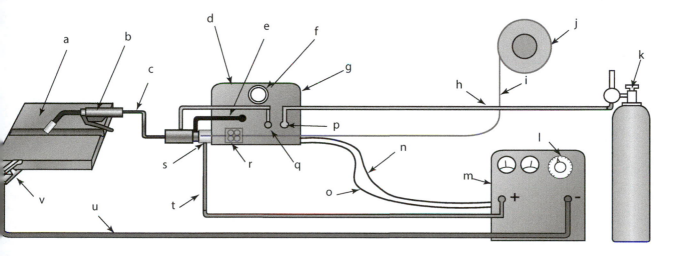

1.	_____	WIRE FEED POWER	12. _____	POWER TERMINAL
2.	_____	VOLTAGE CONTROL DIAL	13. _____	GAS OUT
3.	_____	WORK	14. _____	WIRE
4.	_____	WELDING CABLE	15. _____	WELDING MACHINE
5.	_____	WIRE FEED CONTROLLER	16. _____	WORK CABLE
6.	_____	CONTROL CABLE	17. _____	GAS HOSE
7.	_____	WELDING GUN CABLE	18. _____	WIRE REEL
8.	_____	GUN	19. _____	SHIELDING GAS SOURCE
9.	_____	GAS IN	20. _____	WIRE SPEED FEED CONTROL
10.	_____	WIRE FEED DRIVE ROLLERS	21. _____	WORK CLAMP
11.	_____	WIRE FEED UNIT	22. _____	GUN CONTROL CABLE

Gas Metal Arc Welding

Arc Welding Operation	Suggested Shade Number
Gas Metal Arc Welding Nonferrous	11
Gas Metal Arc Welding Ferrous	12

Troubleshooting for GMAW	
Problem	**Correction**
Arc Blow	1. Change gun angle. 2. Move ground clamp. 3. Use backup bars, brass or copper. 4. Demagnetize part.
Cracked Welds	1. Check filler wire compatibility with base metal. 2. Use preheat and postheat on weldment. 3. Use a convex weld bead. 4. Check design of root opening. 5. Change welding speed. 6. Change shielding gas.
Dirty Welds	1. Decrease gun angle. 2. Hold gun nozzle closer to work. 3. Increase gas flow. 4. Clean weld joint area, gas flow. 5. Check for draft that may be blowing shielding gas away. 6. Check gun nozzle for damaged or worn parts. 7. Center contact tip in gun nozzle. 8. Clean filler wire before it enters wire drive. 9. Check cables and gun for air or water leaks. 10. Keep unused filler wire in shipping containers.
Wide Weld Bead	1. Increase welding speed. 2. Reduce current. 3. Use a different welding technique. 4. Shorten arc length.
Incomplete Penetration	1. Increase current. 2. Reduce welding speed. 3. Shorten arc length. 4. Increase root opening. 5. Change gun angle.

Problem	Correction
Irregular Arc Start	1. Use wire cutters to cut off the end of the filler wire before starting new weld. 2. Check ground. 3. Check contact tip. 4. Check polarity. 5. Check for drafts. 6. Increase gas flow.
Irregular Wire-Feed Burn-back	1. Check contact tip. 2. Check wire-feed speed. 3. Increase drive roll pressure. 4. Check voltage. 5. Check polarity. 6. Check wire spool for kinks or bends. 7. Clean or replace worn conduit liner.
Welding Cables Overheating	1. Check for loose cable connections. 2. Use larger cables. 3. Use shorter cables. 4. Decrease welding time.
Porosity	1. Check for drafts. 2. Check shielding gas. 3. Increase gas flow. 4. Decrease gun angle. 5. Hold nozzle close to work. 6. Do not weld if metal is wet. 7. Clean weld joint area. 8. Center contact tip with gun nozzle. 9. Check gun nozzle for damage. 10. Check gun and cables for air or water leaks.
Spatter	1. Change gun angle. 2. Shorten arc length. 3. Decrease wire speed. 4. Check for draft.
Undercutting	1. Reduce current. 2. Change gun angle. 3. Use different welding technique. 4. Reduce welding speed. 5. Shorten arc length.
Incomplete Fusion	1. Increase current. 2. Change welding technique. 3. Shorten arc length. 4. Check joint preparation. 5. Clean weld joint area.
Unstable Arc	1. Clean weld area. 2. Check contact tip. 3. Check for loose cable connections.

SHIELDING GASES AND GAS MIXTURES USED FOR GAS METAL ARC WELDING		
Shielding Gas	**Chemical Behavior**	**Uses and Usage Notes**
1. Argon	Inert	Welding virtually all metals except steel
2. Helium	Inert	Al and Cu alloys for greater heat and to minimize porosity
3. Ar & He (20–80 to 50–50%)	Inert	Al and Cu alloys for greater heat and to minimize porosity, but with quieter, more readily controlled arc action
4. N_2	Reducing	On Cu, very powerful arc
5. Ar + 25–30% N_2	Reducing	On Cu, powerful but smoother operating, more readily controlled arc than with N_2
6. Ar + 1–2% O_2	Oxidizing	Stainless and alloy steels, also for some deoxidized copper alloys
7. Ar + 3–5% O_2	Oxidizing	Plain carbon, alloy, and stainless steels (generally requires highly deoxidized wire)
8. Ar + 3–5% O_2	Oxidizing	Various steels using deoxidized wire
9. Ar + 20–30% O_2	Oxidizing	Various steels, chiefly with short-circuiting arc
10. Ar + 5% O_2 + 15% CO_2	Oxidizing	Various steels using deoxidized wire
11. CO_2	Oxidizing	Plain-carbon and low-alloy steels, deoxidized wire essential
12. CO_2 + 3–10% O_2	Oxidizing	Various steels using deoxidized wire
13. CO_2 + 20% O_2	Oxidizing	Steels

GENERALLY RECOMMENDED FILLER METALS AND SHIELDING GASES FOR GAS METAL ARC WELDING VARIOUS BASE METALS

Base Metal Type	Shielding Gas Composition	Specific Alloy to Be Welded	Filler Metal Type Electrode
Aluminum and its alloys	Pure argon or helium-argon (75–25%)	1100 2219 3003, 3004 5050 5052 5154, 5254 5083, 5084, 5486 6061 7039	1100 or 4043 4145 or 2319 319 4043 4043 or 5554 5554 or 5154 5554 or 5154 5556 or 5356 4043 or 5556 5556, 5356, or 5183
Magnesium alloys	Pure argon	AZ31B, 61A, 81A ZE10XA ZK20XA AZ31B, 61A, 63A 80A, 81A, 91C, 92A AM80A, 100A ZE10XA XK20XA	AZ61A AZ92A
Copper	Helium-argon mixture 75–25%) pure argon on thin sections	Deoxidized copper	Deoxidized copper Silicon-0.25% Tin-0.75% Mn-0.15%
Copper-nickel alloy	Pure argon	Cu-Ni alloy: 70–30 90–10	Titanium deoxidized 70–30 Cu-Ni 70–30 or 90–10
Bronzes	Pure argon Argon + 5% O_2	Manganese bronze Aluminum bronze Nickel-aluminum bronze Tin bronze	Aluminum bronze Aluminum bronze Aluminum bronze Phosphor bronze
Nickel and nickel alloys	Helium-argon mixture (75–25) or pure argon	Nickel Nickel-copper (Monel) Nickel–chromium (Inconel)	Similar to base metal, titanium deoxidized (see supplier)
Plain low-carbon steel	CO_2; argon + 10 to 30% CO_2; or argon + 2 to 5% O_2	Hot or cold rolled sheet or plate ASTM A7, A36 A285, A373, or equivalent	Deoxidized plain carbon steel
Low-alloy carbon steel	Argon + 1–2% O_2 or argon + 10–30% CO_2	Hot or cold rolled sheet of various grades	Deoxidized low-alloy steel
Stainless steel	Argon + 1–5% O_2	302, 304, 321, 347, 309, 310 316, etc.	Electrode to match base alloy

TYPICAL SHIELDING GAS FLOW RATES* FOR GAS METAL ARC WELDING VARIOUS MATERIALS

Spray-Type Arc — 1/16 in. dia. filler wire

Materials	Argon	Helium	Argon-Helium (25–75%)	Argon-Oxygen (1–5 O_2)	CO_2
Aluminum	50	100	80	—	—
Magnesium	50	100	80	—	—
Plain C Steel	—	—	—	40	40
Low-alloy steel	—	—	—	40	40
Stainless steel	40	—	—	40	—
Nickel	50	100	80	—	—
Ni-Cu alloy	50	100	80	—	—
Ni-Cr Fe alloy	50	100	80	—	—
Copper	50	100	80	—	—
Cu-Ni alloy (70–30%)	50	100	80	—	—
Si bronze	40	80	60	—	—
Al bronze	50	100	80	—	—
Phos. bronze	40	80	80	40	—

Short-Circuiting Arc — 0.035 dia filler wire

Materials	Argon	Argon-Oxygen (1–5% O_2)	Argon-CO_2 (75–25%)	CO_2
Aluminum	35	—	—	—
Magnesium	35	—	—	—
Plain C Steel	25	25	25	35
Low-alloy steel	25	25	25	35
Stainless steel	25	25	—	35
Nickel	35	—	—	—
Ni-Cu alloy	35	—	—	—
Ni-Cr-Fe alloy	35	—	—	—
Copper	30	—	—	—
Cu-Ni alloy (70–30)	30	—	—	—
Si bronze	25	25	—	—
Al bronze	25	—	—	—
Phos. bronze	25	25	—	—

*All rates are in cubic feet per hour and are plus or minus 40%. The lower rates would be most suitable for indoor work and moderate amperage welding. The higher rates would be more suitable for high current, maximum speed, and outdoor welding.

OPERATING INSTRUCTIONS FOR GMAW

1. Set polarity.

   Making sure that the machine is turned off, connect the electrode cable to either the positive stud for reverse polarity or to the negative stud for straight polarity. Check with your instructor for the correct polarity. Some machines have a polarity switch instead of changing cables; this makes the changing of the polarity much easier.

2. Start the welder.

   Place the "power switch" to the "on" position. On some machines, you will also have to place the "power switch" on the wire feeder to the "on" position.

3. Open the shielding gas valve.

   Slowly open the flowmeter valve, simultaneously squeezing the gun trigger until the flowmeter reads approximately 15 to 20 cubic feet per hour.

4. Set voltage.

   Set the voltage by either increasing or decreasing the voltage control knob on the machine. The open circuit voltage should be approximately 2 volts higher than the desired welding voltage.

5. Set amperage (wire-speed control).

   Most amperage (wire-speed control) adjustment knobs are calibrated in inches per minute. Set the control knob to the desired wire speed. If you are not sure of what setting you need, place the adjustment knob in the 9 to 10 o'clock position and then fine-tune the amperage while making a sample weld.

6. The final setting for both voltage and amperage must be made while welding.

■ PRACTICE 11-1

Name _____   Date _____

Class _____   Instructor _____   Grade _____

OBJECTIVE: After completing this practice, you should be able to safely set up the GMAW equipment.

EQUIPMENT AND MATERIALS NEEDED FOR THIS PRACTICE

1. GMAW power source.

2. Welding gun.

3. Electrode feed unit.

4. Electrode supply.

5. Shielding gas supply.

6. Shielding gas regulator and flowmeter.

7. Electrode conduit.

8. Power and work leads.

9. Shielding gas hoses.

10. Assorted hand tools (see instructor).

INSTRUCTIONS

1. Be sure that the power to the machine is "off." If the shielding gas supply is a cylinder, it must be chained securely in place before the valve protection cap is removed. Standing to one side of the cylinder, quickly crack the valve to blow out any dirt in the valve before the flowmeter regulator is attached. Attach the correct hose from the regulator to the "gas-in" connection on the electrode feed unit or machine.

2. Install the reel of electrode (welding wire) on the holder and secure it.

3. Check the roller size to ensure that it matches the wire size.

4. The conduit liner size should be checked to be sure that it is compatible with the wire size.

5. Connect the conduit to the feed unit. The conduit or an extension should be aligned with the groove in the roller and set as close to the roller as possible without touching. Misalignment at this point can contribute to a bird's nest. Bird nesting of the electrode wire results when the feed roller pushes the wire into a tangled ball because the wire would not go through the outfeed side conduit. The welding wire appears to look like a bird's nest.

6. Be sure that the power is "off" before attaching the welding cables. The electrode and work leads should be attached to the proper terminals. The electrode lead should be attached to electrode or positive (+). If necessary, it is also attached to the power cable part of the gun lead. The work lead should be attached to work or negative (–).

7. The shielding "gas-out" side of the solenoid is then attached to the gun lead. If a separate splice is required from the gun switch circuit to the feed unit, it should be connected at this time.

8. Check to see that the welding contactor circuit is connected from the feed unit to the power source. The welding gun should be permanently attached to the main lead cable and conduit.

9. There should be a gas diffuser attached to the end of the conduit liner to ensure proper gas flow.

10. A contact tube (tip) of the correct size to match the electrode wire size being used should be installed.

11. A shielding gas nozzle is attached to complete the assembly.

12. Recheck all fittings and connections for tightness. Loose fittings can leak; loose connections can cause added resistance, reducing the welding efficiency.

13. Some manufacturers include detailed setup instructions with their equipment.

INSTRUCTOR'S COMMENTS _____

_____

_____

■ PRACTICE 11-2

Name _____ Date _____

Class _____ Instructor _____ Grade _____

OBJECTIVE: After completing this practice, you should be able to safely thread the GMAW wire into the equipment.

EQUIPMENT AND MATERIALS NEEDED FOR THIS PRACTICE

1. GMAW power source.

2. Welding gun.

3. Electrode feed unit.

4. Electrode supply.

5. Shielding gas supply.

6. Shielding gas regulator and flowmeter.

7. Electrode conduit.

8. Power and work leads.

9. Shielding gas hoses.

10. Assorted hand tools (see instructor).

INSTRUCTIONS

1. Check to see that the unit is assembled correctly according to the manufacturer's specifications.

2. Switch on the power and check the gun switch circuit by depressing the switch. The power source relays, feed relays, gas solenoid, and feed motor should all activate.

3. Cut the end of the electrode wire free. Hold it tightly so that it does not unwind. The wire has a natural curve that is known as its cast. The cast is measured by the diameter of the circle that the wire would make if it were loosely laid on a flat surface. The cast helps the wire make a good electrical contact as it passes through the contact tube. However, the cast can be a problem when threading the system.

4. To make threading easier, straighten about 12 in. (305 mm) of the end of the wire and cut any kinks off.

5. Separate the wire-feed rollers and push the wire first through the guides, then between the rollers, and finally into the conduit liner.

6. Reset the rollers so that there is a slight amount of compression on the wire.

7. Set the wire-feed speed control to a slow speed.

8. Hold the welding gun so that the electrode conduit and cable are as straight as possible.

9. Remove the nozzle and contact tip from the gun.

10. Press the gun switch. The wire should start feeding into the liner. Watch to make certain that the wire feeds smoothly and release the gun switch as soon as the end comes through the gun.

    CAUTION **If the wire stops feeding before it reaches the end of the contact tube, stop and check the system. If no obvious problem can be found, mark the wire with tape and remove it from the gun. It then can be held next to the system to determine the location of the problem.**

11. With the wire-feed running, adjust the feed roller compression so that the wire reel can be stopped easily by a slight pressure. Too light a roller pressure will cause the wire to feed erratically. Too high a pressure can turn a minor problem into a major disaster. If the wire jams at a high roller pressure, the feed rollers keep feeding the wire, causing it to bird nest and possibly short out. With a light pressure, the wire can stop, preventing bird nesting. This is very important with soft wires. The other advantage of a light pressure is that the feed will stop if something like clothing or gas hoses are caught in the reel.

12. With the feed running, adjust the spool drag so that the reel stops when the feed stops. The reel should not coast to a stop because the wire can be snagged easily. Also, when the feed restarts, a jolt occurs when the slack in the wire is taken up. This jolt can be enough to momentarily stop the wire, possibly causing a discontinuity in the weld or may even snap the wire.

13. When the test runs are completed, the wire can either be rewound or cut off. Some wire-feed units have a retract button. This allows the feed driver to reverse and retract the wire automatically. To rewind the wire on units without this retract feature, release the rollers and turn them backward by hand. If the machine will not allow the feed rollers to be released without upsetting the tension, you must cut the wire.

14. Replace the tip and nozzle on the gun.

    CAUTION **Do not discard pieces of wire on the floor. They present a hazard to safe movement around the machine. In addition, a small piece of wire can work its way into a filter screen on the welding power source. If the piece of wire shorts out inside the machine, it could become charged with high voltage, which may cause injury or death. Always wind the wire tightly into a ball or cut it into short lengths before discarding it in the proper waste container.**

15. Use wire nippers and nip the electrode wire to the correct stick out length.

INSTRUCTOR'S COMMENTS _____

_____

_____

■ PRACTICE 11-3

Name _____ Date _____

Class _____ Instructor _____ Grade _____

Welding Principles and Applications
MATERIAL: 16 GA AND 1/8" MILD STEEL SHEET 12" X 3"
PROCESS: GMAW STRINGER BEAD FLAT POSITION

NUMBER: PRACTICE 11-3	DRAWN BY: GAVIN DUBOIS

OBJECTIVE: After completing this practice, you should be able to make stringer beads on the 1/8-in. (3-mm) and 16-gauge carbon steel using the short-circuiting metal transfer method in the flat position.

EQUIPMENT AND MATERIALS NEEDED FOR THIS PRACTICE

1. A properly set-up and adjusted GMA welding machine.

2. Proper safety protection (welding hood, safety glasses, leather gloves, wire cutting pliers, long-sleeved shirt, long pants, and leather shoes or boots). Refer to Chapter 2 in the text for more specific safety information.

3. 0.035-in. and/or 0.045-in. (0.9-mm and/or 1.2-mm) diameter wire.

4. Two or more pieces of mild steel sheet, 12-in. (305-mm) long by 3-in. (76-mm) wide by 16-gauge to 1/8-in. (3-mm) thick.

INSTRUCTIONS

1. Starting at one end of the plate and using either a pushing or dragging technique, make a weld bead along the entire 12 in. (305 mm) length of the metal.

2. After the weld is complete, check its appearance. Make any needed changes to correct the weld.

3. Repeat the weld and make additional adjustments.

4. After the machine is set, start to work on improving the straightness and uniformity of the weld. Keeping the bead straight and uniform can be hard because of the limited visibility due to the small amount of light and the size of the molten weld pool. The welder's view is further restricted by the shielding gas nozzle. Even with limited visibility, it is possible to make a satisfactory weld by watching the edge of the molten weld pool, the sparks, and the weld bead produced. Watching the leading edge of the molten weld pool (forehand welding, push technique) will show you the molten weld pool fusion and width. Watching the trailing edge of the molten weld pool (backhand welding, drag technique) will show you the amount of buildup and the relative heat input. The quantity and size of sparks produced can indicate the relative location of the filler wire in the molten weld pool. The number of sparks will increase as the wire strikes the solid metal ahead of the molten weld pool. The gun itself will begin to vibrate or bump as the wire momentarily pushes against the cooler, unmelted base metal before it melts. Changes in weld width, buildup, and proper joint tracking can be seen by watching the bead as it appears from behind the shielding gas nozzle.

5. Repeat each type of bead on both plate thicknesses as needed until consistently good beads are obtained.

6. Turn off the welding machine and shielding gas and clean up your work area when you are finished welding.

INSTRUCTOR'S COMMENTS _____

_____

_____

■ PRACTICE 11-4

Name _____ Date _____

Class _____ Instructor _____ Grade _____

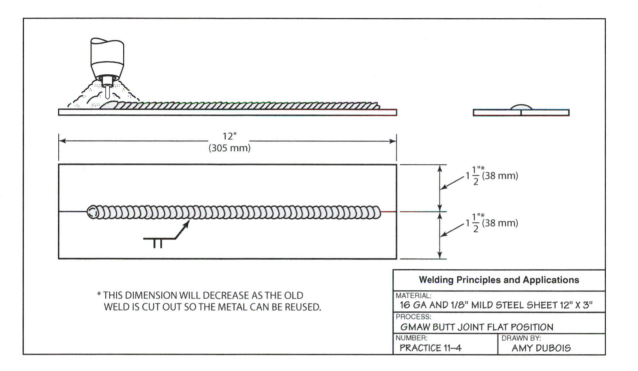

12"
(305 mm)

$1\frac{1}{2}$"* (38 mm)

$1\frac{1}{2}$"* (38 mm)

* THIS DIMENSION WILL DECREASE AS THE OLD
WELD IS CUT OUT SO THE METAL CAN BE REUSED.

Welding Principles and Applications	
MATERIAL: 16 GA AND 1/8" MILD STEEL SHEET 12" X 3"	
PROCESS: GMAW BUTT JOINT FLAT POSITION	
NUMBER: PRACTICE 11–4	DRAWN BY: AMY DUBOIS

OBJECTIVE: After completing this practice, you should be able to weld a butt, lap, and tee joint in the flat position using 1/8-in. (3-mm) and 16-gauge carbon steel.

EQUIPMENT AND MATERIALS NEEDED FOR THIS PRACTICE

1. A properly set-up and adjusted GMA welding machine.

2. Proper safety protection (welding hood, safety glasses, leather gloves, wire cutting pliers, long-sleeved shirt, long pants, and leather shoes or boots). Refer to Chapter 2 in the text for more specific safety information.

3. 0.035-in. and/or 0.045-in. (0.9-mm and/or 1.2-mm) diameter wire.

4. Two or more pieces of mild steel sheet, 12-in. (305-mm) long by 3-in. (76-mm) wide by 16-gauge to 1/8-in. (3-mm) thick.

INSTRUCTIONS

1. Tack weld the sheets together in a butt joint and place them flat on the welding table.

2. Starting at one end, run a bead along the joint. Watch the molten weld pool and bead for signs that a change in technique may be required.

3. Make any needed changes as the weld progresses. By the time the weld is complete, you should be making a nearly perfect weld.

4. Using the same technique that was established in the last weld, make another weld. This time, the entire 12 in. (305 mm) of weld should be defect free.

5. Repeat each type of joint with both thicknesses of metal as needed until consistently good quality beads are obtained.

6. Repeat the above steps for the lap joint and tee joint.

7. Turn off the welding machine and shielding gas and clean up your work area when you are finished welding.

INSTRUCTOR'S COMMENTS _____

_____

_____

■ PRACTICE 11-5

Name _____   Date _____

Class _____   Instructor _____   Grade _____

OBJECTIVE: After completing this practice, you should be able to make butt, lap, and tee joints. All with 100% penetration.

EQUIPMENT AND MATERIALS NEEDED FOR THIS PRACTICE

1. A properly set-up and adjusted GMA welding machine.

2. Proper safety protection (welding hood, safety glasses, leather gloves, wire cutting pliers, long-sleeved shirt, long pants, and leather shoes or boots). Refer to Chapter 2 in the text for more specific safety information.

3. 0.035-in. and/or 0.045-in. (0.9-mm and/or 1.2-mm) diameter wire.

4. Two or more pieces of mild steel sheet, 12-in. (305-mm) long by 3-in. (76-mm) wide by 16-gauge to 1/8-in. (3-mm) thick.

INSTRUCTIONS

1. Tack weld the sheets together in a butt joint and place them in the flat position.

2. Start at one end of the plate and run a bead along the joint.

3. Use gun angle, current, and travel speed which will produce 100% penetration.

4. Make any changes as needed as the weld progresses.

5. Repeat until consistently good welds can be made with 100% penetration on both thicknesses of metal.

6. Repeat the above steps with the lap joint and tee joint.

7. Turn off the welding machine and shielding gas and clean up your work area when you are finished welding.

INSTRUCTOR'S COMMENTS _____

_____

_____

■ PRACTICE 11-6

Name _____ Date _____

Class _____ Instructor _____ Grade _____

OBJECTIVE: After completing this practice, you should be able to make a stringer bead in the vertical up position at a 45° angle.

EQUIPMENT AND MATERIALS NEEDED FOR THIS PRACTICE

1. A properly set-up and adjusted GMA welding machine.

2. Proper safety protection (welding hood, safety glasses, leather gloves, wire cutting pliers, long-sleeved shirt, long pants, and leather shoes or boots). Refer to Chapter 2 in the text for more specific safety information.

3. 0.035-in. and/or 0.045-in. (0.9-mm and/or 1.2-mm) diameter wire.

4. One or more pieces of mild steel sheet, 12-in. (305-mm) long by 3-in. (76-mm) wide by 16-gauge to 1/8-in. (3-mm) thick.

INSTRUCTIONS

1. Brace one of the plates on the welding table at a 45° inclined angle.

2. Start at the bottom of the plate and hold the welding gun at a slight angle to the plate.

3. Brace yourself, lower your hood, and begin to weld.

4. Depending on the machine settings and type of shielding gas used, you will make a weave pattern.

5. If the molten weld pool is large and fluid (hot), use a "C" or "J" weave pattern to allow a longer time for the molten weld pool to cool.

6. Do not make the weave so long or fast that the wire is allowed to strike the metal ahead of the molten weld pool. If this happens, spatter increases and a spot or zone of incomplete fusion may occur.

7. If the molten weld pool is small and controllable, use a small "C," zigzag, or "J" weave pattern to control the width and buildup of the weld. A slower travel speed can also be used.

8. Watch for complete fusion along the leading edge of the molten weld pool.

9. A weld that is high and has little or no fusion is too "cold." Changing the welding technique will not correct this problem. The welder must stop welding and make the needed adjustments to the welding machine.

10. As the weld progresses up the plate, the back or trailing edge of the molten weld pool will cool, forming a shelf to support the molten metal. Watch the shelf to be sure that molten metal does not run over, forming a drip. When it appears that the metal may flow over the shelf, either increase the weave lengths or stop and start the current for brief moments to allow the weld to cool. Stopping for brief moments will not allow the shielding gas to be lost.

11. Continue to weld along the entire 12 in. (305 mm) length of plate.

12. Repeat this weld as needed until a straight and uniform weld bead is produced on both thicknesses of metal.

13. Turn off the welding machine and shielding gas and clean up your work area when you are finished welding.

INSTRUCTOR'S COMMENTS _____

_____

_____

■ PRACTICE 11-7

Name _____ Date _____

Class _____ Instructor _____ Grade _____

OBJECTIVE: After completing this practice, you should be able to make stringer beads on sheet steel in the vertical up position.

EQUIPMENT AND MATERIALS NEEDED FOR THIS PRACTICE

1. A properly set-up and adjusted GMA welding machine.

2. Proper safety protection (welding hood, safety glasses, wire brush, chipping hammer, leather gloves, pliers, long-sleeved shirt, long pants, and leather boots or shoes). Refer to Chapter 2 in the text for more specific safety information.

3. 0.035-in. and/or 0.045-in. (0.9-mm and/or 1.2-mm) diameter welding wire.

4. Two or more pieces of mild steel sheet, 12-in. (305-mm) long by 3-in. (76-mm) wide by 16-gauge to 1/8-in. (3-mm) thick.

INSTRUCTIONS

1. Place the sheet steel on the welding table at a 90° angle, vertical up angle.

2. Start at the bottom of the plate and hold the gun at a slight up angle to the plate.

3. Brace yourself, lower your hood, and begin to weld.

4. Depending on the machine settings and type of shielding gas, you may make a weave pattern.

5. If the weld pool is large and fluid (hot), use a "C" or "J" weave pattern to allow a longer time for the molten weld pool to cool.

6. Do not make the weave so long or so fast that the wire is allowed to strike the metal ahead of the molten weld pool. If this happens, spatter increases and a spot or zone of incomplete fusion may occur.

7. If the molten weld pool is small and controllable, use a small "C," zigzag, or "J" weave pattern to control the width and buildup of the bead. A slower travel speed can also be used.

8. Watch for complete fusion along the leading edge of the molten weld pool.

9. A weld that is too high and has little or no fusion is too "cold." Changing the welding technique will not correct the problem. The welder must stop welding and make the needed adjustments to the welding machine.

10. As the weld progresses up the plate, the back or trailing edge of the molten weld pool will cool forming a shelf to support the molten metal. Watch the shelf to be sure that molten weld metal does not run over, forming a drip. When it appears that the metal may flow over the shelf, either increase the weave lengths or stop and start the current for brief moments to allow the weld to cool. Stopping for brief moments will not allow the shielding gas to be lost.

11. Continue to weld along the entire 12 in. (305 mm) length of the plate.

12. Repeat the weld as needed on both thicknesses of plate until a straight and uniform weld bead with proper penetration can be made.

13. Turn off the welding machine and shielding gas and clean up your work area when finished with this practice.

INSTRUCTOR'S COMMENTS _____

_____

_____

■ PRACTICE 11-8

Name _____   Date _____

Class _____   Instructor _____   Grade _____

OBJECTIVE: After completing this practice, you should be able to make butt, lap, and tee joints in the vertical up position at a 45° angle.

EQUIPMENT AND MATERIALS NEEDED FOR THIS PRACTICE

1. A properly set-up and adjusted GMA welding machine.

2. Proper safety protection (welding hood, safety glasses, wire brush, chipping hammer, leather gloves, pliers, long-sleeved shirt, long pants, and leather boots or shoes). Refer to Chapter 2 in the text for more specific safety information.

3. 0.035-in. and/or 0.045-in. (0.9-mm and/or 1.2-mm) diameter welding wire.

4. Two or more pieces of mild steel sheet, 12-in. (305-mm) long by 3-in. (76-mm) wide by 16-gauge to 1/8-in. (3-mm) thick.

INSTRUCTIONS

1. Tack weld the pieces together in the butt configuration, lap configuration, and tee configuration and place them on the welding table at a 45° angle, vertical up angle.

2. Start at the bottom of the plate and hold the gun at a slight up angle to the plate.

3. Brace yourself, lower your hood, and begin to weld.

4. Depending on the machine settings and type of shielding gas, you may make a weave pattern.

5. If the weld pool is large and fluid (hot), use a "C" or "J" weave pattern to allow a longer time for the molten weld pool to cool.

6. Do not make the weave so long or so fast that the wire is allowed to strike the metal ahead of the molten weld pool. If this happens, spatter increases and a spot or zone of incomplete fusion may occur.

7. If the molten weld pool is small and controllable, use a small "C," zigzag, or "J" weave pattern to control the width and buildup of the bead. A slower travel speed can also be used.

8. Watch for complete fusion along the leading edge of the molten weld pool.

9. A weld that is too high and has little or no fusion is too "cold." Changing the welding technique will not correct the problem. The welder must stop welding and make the needed adjustments to the welding machine.

10. As the weld progresses up the plate, the back or trailing edge of the molten weld pool will cool forming a shelf to support the molten metal. Watch the shelf to be sure that molten weld metal does not run over, forming a drip. When it appears that the metal may flow over the shelf, either increase the weave lengths or stop and start the current for brief moments to allow the weld to cool. Stopping for brief moments will not allow the shielding gas to be lost.

11. Continue to weld along the entire 12 in. (305 mm) length of the plate.

12. Repeat the weld as needed on both thicknesses of plate until a straight and uniform weld bead with proper penetration can be made.

13. Turn off the welding machine and shielding gas and clean up your work area when finished with this practice.

INSTRUCTOR'S COMMENTS _____

_____

_____

■ PRACTICE 11-9

Name _____   Date _____

Class _____   Instructor _____   Grade _____

OBJECTIVE: After completing this practice, you should be able to make butt joint in the vertical up position with 100% penetration.

EQUIPMENT AND MATERIALS NEEDED FOR THIS PRACTICE

1. A properly set-up and adjusted GMA welding machine.

2. Proper safety protection (welding hood, safety glasses, wire brush, chipping hammer, leather gloves, pliers, long-sleeved shirt, long pants, and leather boots or shoes). Refer to Chapter 2 in the text for more specific safety information.

3. 0.035-in. and/or 0.045-in. (0.9-mm and/or 1.2-mm) diameter welding wire.

4. Two or more pieces of mild steel sheet, 12-in. (305-mm) long by 3-in. (76-mm) wide by 16 gauge to 1/8 in. (3-mm) thick.

INSTRUCTIONS

1. Tack weld the pieces together in the butt configuration and place it on the welding table in the vertical position.

2. Start at the bottom of the plate and hold the gun at a slight up angle to the plate.

3. Brace yourself, lower your hood, and begin to weld.

4. Depending on the machine settings and type of shielding gas, you may make a weave pattern.

5. If the weld pool is large and fluid (hot), use a "C" or "J" weave pattern to allow a longer time for the molten weld pool to cool.

6. Do not make the weave so long or so fast that the wire is allowed to strike the metal ahead of the molten weld pool. If this happens, spatter increases and a spot or zone of incomplete fusion may occur.

7. If the molten weld pool is small and controllable, use a small "C," zigzag, or "J" weave pattern to control the width and buildup of the bead. A slower travel speed can also be used.

8. Watch for complete fusion along the leading edge of the molten weld pool.

9. A weld that is too high and has little or no fusion is too "cold." Changing the welding technique will not correct the problem. The welder must stop welding and make the needed adjustments to the welding machine.

10. Repeat the weld as needed on both thicknesses of plate until a straight and uniform weld bead with 100% penetration can be made.

11. Turn off the welding machine and shielding gas and clean up your work area when you are finished welding.

INSTRUCTOR'S COMMENTS _____

_____

_____

■ PRACTICE 11-10

Name _____ Date _____

Class _____ Instructor _____ Grade _____

OBJECTIVE: After completing this practice, you should be able to make stringer beads on sheet steel in the 45° vertical down position.

EQUIPMENT AND MATERIALS NEEDED FOR THIS PRACTICE

1. A properly set-up and adjusted GMA welding machine.

2. Proper safety protection (welding hood, safety glasses, leather gloves, wire cutting pliers, long-sleeved shirt, long pants, and leather shoes or boots). Refer to Chapter 2 in the text for more specific safety information.

3. 0.035-in. and/or 0.045-in. (0.9-mm and/or 1.2-mm) diameter wire.

4. Two or more pieces of mild steel sheet, 12-in. (305-mm) long by 3-in. (76-mm) wide by 16-gauge to 1/8-in. (3-mm) thick.

INSTRUCTIONS

1. Place the sheet steel on the welding table at a 45° angle, vertical down position.

2. Start at the top of the plate and hold the welding gun at a slight dragging angle to the plate.

3. Brace yourself, lower your hood, and begin to weld.

4. Depending on the machine settings and type of shielding gas used, you may or may not make a weave pattern.

5. If the molten weld pool is large and fluid (hot), use a "C" weave pattern to allow a longer time for the molten weld pool to cool.

6. The leading edge and sides should flow into the base metal not curl over onto it. (Cold lap.)

7. Some changes in gun angle may increase penetration. Experiment with the gun angle as the weld progresses.

8. Watch for complete fusion along the leading edge and sides of the molten weld pool.

9. A weld that is high and has little or no fusion is too "cold." Changing the welding technique will not correct this problem. The welder must stop welding and make the needed adjustments to the welding machine.

10. Continue to weld along the entire 12 in. (305 mm) length of plate.

11. Repeat this weld as needed on both thicknesses of plate until a straight weld bead is produced that is uniform in height and width.

12. Turn off the welding machine and shielding gas and clean up your work area when you are finished welding.

INSTRUCTOR'S COMMENTS _____

_____

_____

■ PRACTICE 11-11

Name _____ Date _____

Class _____ Instructor _____ Grade _____

OBJECTIVE: After completing this practice, you should be able to make stringer beads on sheet steel in the vertical down position.

EQUIPMENT AND MATERIALS NEEDED FOR THIS PRACTICE

1. A properly set-up and adjusted GMA welding machine.

2. Proper safety protection (welding hood, safety glasses, leather gloves, wire cutting pliers, long-sleeved shirt, long pants, and leather shoes or boots). Refer to Chapter 2 in the text for more specific safety information.

3. 0.035-in. and/or 0.045-in. (0.9-mm and/or 1.2-mm) diameter wire.

4. Two or more pieces of mild steel sheet, 12-in. (305-mm) long by 3-in. (76-mm) wide by 16-gauge to 1/8-in. (3-mm) thick.

INSTRUCTIONS

1. Place the sheet steel on the welding table at a 90° angle, vertical down position.

2. Start at the top of the plate and hold the welding gun at a slight dragging angle to the plate.

3. Brace yourself, lower your hood, and begin to weld.

4. Depending on the machine settings and type of shielding gas used, you may or may not make a weave pattern.

5. If the molten weld pool is large and fluid (hot), use a "C" weave pattern to allow a longer time for the molten weld pool to cool.

6. Watch the sides and leading edge of the molten weld pool for proper fusion.

7. Change the gun angle as needed to get proper penetration.

8. Continue to weld along the entire 12 in. (305 mm) length of plate.

9. Repeat this weld as needed on both thicknesses of plate until a straight weld bead is produced that is uniform in height and width and is defect free.

10. Turn off the welding machine and shielding gas and clean up your work area when you are finished welding.

INSTRUCTOR'S COMMENTS _____

_____

_____

■ PRACTICE 11-12

Name _____ Date _____

Class _____ Instructor _____ Grade _____

OBJECTIVE: After completing this practice, you should be able to make butt, lap, and tee joints in the vertical down position.

EQUIPMENT AND MATERIALS NEEDED FOR THIS PRACTICE

1. A properly set-up and adjusted GMA welding machine.

2. Proper safety protection (welding hood, safety glasses, leather gloves, wire cutting pliers, long-sleeved shirt, long pants, and leather shoes or boots). Refer to Chapter 2 in the text for more specific safety information.

3. 0.035-in. and/or 0.045-in. (0.9-mm and/or 1.2-mm) diameter wire.

4. Two or more pieces of mild steel sheet, 12-in. (305-mm) long by 3-in. (76-mm) wide by 16-gauge to 1/8-in. (3-mm) thick.

INSTRUCTIONS

1. Tack weld the metal pieces of the butt, lap, and tee joints together and brace them on the welding table at a 90° angle for vertical down welding.

2. Start at the top of the plate, holding the welding gun with a slight dragging angle.

3. Brace yourself, lower your hood, and begin to weld.

4. Depending on the machine settings and type of shielding gas used, you will make a weave pattern.

5. If the molten weld pool is large and fluid (hot), use a "C" weave pattern to allow a longer time for the molten weld pool to cool.

6. Do not make the weave so long or fast that the wire is allowed to strike the metal ahead of the molten weld pool. If this happens, spatter increases and a spot or zone of incomplete fusion may occur.

7. If the molten weld pool is small and controllable, use a small "C," zigzag, or "J" weave pattern to control the width and buildup of the weld. A slower travel speed can also be used.

8. Watch for complete fusion along the leading edge and sides of the molten weld pool.

9. A weld that is high and has little or no fusion is too "cold." Changing the welding technique will not correct this problem. The welder must stop welding and make the needed adjustments to the welding machine.

10. Continue to weld along the entire 12 in. (305 mm) length of plate.

11. Repeat this weld as needed on both thicknesses of plate until a straight weld bead is produced that is uniform in height and width.

12. Turn off the welding machine and shielding gas and clean up your work area when you are finished welding.

INSTRUCTOR'S COMMENTS _____

_____

_____

■ PRACTICE 11-13

Name _____   Date _____

Class _____   Instructor _____   Grade _____

OBJECTIVE: After completing this practice, you should be able to make butt and tee joints in the vertical down position with 100% penetration.

EQUIPMENT AND MATERIALS NEEDED FOR THIS PRACTICE

1. A properly set-up and adjusted GMA welding machine.

2. Proper safety protection (welding hood, safety glasses, wire brush, chipping hammer, leather gloves, pliers, long-sleeved shirt, long pants, and leather boots or shoes). Refer to Chapter 2 in the text for more specific safety information.

3. 0.035-in. and/or 0.045-in. (0.9-mm and/or 1.2-mm) diameter welding wire.

4. Two or more pieces of mild steel sheet, 12-in. (305-mm) long by 3-in. (76-mm) wide by 16-gauge to 1/8-in. (3-mm) thick.

INSTRUCTIONS

1. Tack weld the pieces together in the butt configuration and tee configuration and place them on the welding table at a 90° vertical down position.

2. It may be necessary to adjust the root openings and machine settings to get 100% penetration.

3. Brace yourself, lower your hood and, starting at the top, begin to weld downward.

4. Depending on the machine settings and type of shielding gas, you may make a weave pattern.

5. If the weld pool is large and fluid (hot), use a "C" weave pattern to allow a longer time for the molten weld pool to cool.

6. Do not make the weave so long or so fast that the wire is allowed to strike the metal ahead of the molten weld pool. If this happens, spatter increases and a spot or zone of incomplete fusion may occur.

7. If the molten weld pool is small and controllable, use a small "C," zigzag, or "J" weave pattern to control the width and buildup of the bead. A slower travel speed can also be used.

8. Watch for complete fusion along the sides and leading edge of the molten weld pool.

9. A weld that is too high and has little or no fusion is too "cold." Changing the welding technique will not correct the problem. The welder must stop welding and make the needed adjustments to the welding machine.

10. Continue to weld along the entire 12 in. (305 mm) length of the plate.

11. Repeat the weld as needed on both thicknesses of plate until a straight and uniform weld bead with 100% penetration can be made.

12. Turn off the welding machine and shielding gas and clean up your work area when finished with this practice.

INSTRUCTOR'S COMMENTS _____

_____

_____

■ PRACTICE 11-14

Name _____   Date _____

Class _____   Instructor _____   Grade _____

OBJECTIVE: After completing this practice, you should be able to make horizontal stringer beads on a plate with a reclined angle.

EQUIPMENT AND MATERIALS NEEDED FOR THIS PRACTICE

1. A properly set-up and adjusted GMA welding machine.

2. Proper safety protection (welding hood, safety glasses, leather gloves, wire cutting pliers, long-sleeved shirt, long pants, and leather shoes or boots). Refer to Chapter 2 in the text for more specific safety information.

3. 0.035-in. and/or 0.045-in. (0.9-mm and/or 1.2-mm) diameter wire.

4. One or more pieces of mild steel sheet, 12-in. (305-mm) long by 3-in. (76-mm) wide by 16-gauge to 1/8-in. (3-mm) thick.

INSTRUCTIONS

1. Place the mild steel plate on the welding table with a 45° reclined angle.

2. Start at one end of the plate with the gun pointed in a slightly upward direction.

3. You may use a pushing or a dragging gun angle, depending on the current setting and penetration desired.

4. Undercutting along the top edge and overlap along the bottom edge are problems with both gun angles. Careful attention must be paid to the manipulation of the "weave" technique used to overcome these problems. The most successful weave patterns are the "C" and "J" patterns. The "J" pattern is the most frequently used. The "J" pattern allows weld metal to be deposited along a shelf created by the previous weave. The length of the "J" can be changed to control the weld bead size.

5. Small weld beads are easier to control than large ones.

6. The weld must be straight and uniform in size and have complete fusion.

7. Repeat these welds on both thicknesses of plate until you have established the rhythm and technique that work well for you.

8. Turn off the welding machine and shielding gas and clean up your work area when you are finished welding.

INSTRUCTOR'S COMMENTS _____

_____

_____

■ PRACTICE 11-15

Name _____   Date _____

Class _____   Instructor _____   Grade _____

OBJECTIVE: After completing this practice, you should be able to make stringer beads in the horizontal position on mild steel plate.

EQUIPMENT AND MATERIALS NEEDED FOR THIS PRACTICE

1. A properly set-up and adjusted GMA welding machine.

2. Proper safety protection (welding hood, safety glasses, leather gloves, wire cutting pliers, long-sleeved shirt, long pants, and leather shoes or boots). Refer to Chapter 2 in the text for more specific safety information.

3. 0.035-in. and/or 0.045-in. (0.9-mm and/or 1.2-mm) diameter wire.

4. One or more pieces of mild steel sheet, 12-in. (305-mm) long by 3-in. (76-mm) wide by 16-gauge to 1/8-in. (3-mm) thick.

INSTRUCTIONS

1. Place the mild steel plate in a horizontal position on the welding table.

2. Start at one end of the plate with the gun pointed in a slightly upward direction.

3. You may use a pushing or a dragging gun angle, depending on the current setting and penetration desired.

4. Undercutting along the top edge and overlap along the bottom edge are problems with both gun angles. Careful attention must be paid to the manipulation of the "weave" technique used to overcome these problems. The most successful weave patterns are the "C" and "J" patterns. The "J" pattern is the most frequently used. The "J" pattern allows weld metal to be deposited along a shelf created by the previous weave. The length of the "J" can be changed to control the weld bead size.

5. Small weld beads are easier to control than large ones.

6. The weld must be straight and uniform in size and have complete fusion.

7. Repeat these welds with both thicknesses of plate until you have established the rhythm and technique that work well for you.

8. Turn off the welding machine and shielding gas and clean up your work area when you are finished welding.

INSTRUCTOR'S COMMENTS _____

_____

_____

■ PRACTICE 11-16

Name _____   Date _____

Class _____   Instructor _____   Grade _____

OBJECTIVE: After completing this practice, you should be able to weld a butt and tee joint in the horizontal position on mild steel plate.

EQUIPMENT AND MATERIALS NEEDED FOR THIS PRACTICE

1. A properly set-up and adjusted GMA welding machine.

2. Proper safety protection (welding hood, safety glasses, leather gloves, wire cutting pliers, long-sleeved shirt, long pants, and leather shoes or boots). Refer to Chapter 2 in the text for more specific safety information.

3. 0.035-in. and/or 0.045-in. (0.9-mm and/or 1.2-mm) diameter wire.

4. Two or more pieces of mild steel sheet, 12-in. (305-mm) long by 3-in. (76-mm) wide by 16-gauge to 1/8-in. (3-mm) thick.

INSTRUCTIONS

1. Tack weld the metal pieces of the butt and tee joints together and brace them on the welding table in the horizontal position.

2. Start at one end of the plate with the gun pointed in a slightly upward direction.

3. You may use a pushing or a dragging gun angle, depending on the current setting and penetration desired.

4. Undercutting along the top edge and overlap along the bottom edge are problems with both gun angles. Careful attention must be paid to the manipulation of the "weave" technique used to overcome these problems. The most successful weave patterns are the "C" and "J" patterns. The "J" pattern is the most frequently used. The "J" pattern allows weld metal to be deposited along a shelf created by the previous weave. The length of the "J" can be changed to control the weld bead size.

5. Small weld beads are easier to control than large ones.

6. The weld must be straight and uniform in size and have complete fusion.

7. Repeat these welds on both thicknesses of metal until you have established the rhythm and technique that work well for you.

8. Turn off the welding machine and shielding gas and clean up your work area when you are finished welding.

INSTRUCTOR'S COMMENTS _____

_____

_____

■ PRACTICE 11-17

Name _____ Date _____

Class _____ Instructor _____ Grade _____

OBJECTIVE: After completing this practice, you should be able to weld a butt and tee joint in the horizontal position on mild steel plate with 100% penetration.

EQUIPMENT AND MATERIALS NEEDED FOR THIS PRACTICE

1. A properly set-up and adjusted GMA welding machine.

2. Proper safety protection (welding hood, safety glasses, leather gloves, wire cutting pliers, long-sleeved shirt, long pants, and leather shoes or boots). Refer to Chapter 2 in the text for more specific safety information.

3. 0.035-in. and/or 0.045-in. (0.9-mm and/or 1.2-mm) diameter wire.

4. Two or more pieces of mild steel sheet, 12-in. (305-mm) long by 3-in. (76-mm) wide by 16-gauge to 1/8-in. (3-mm) thick.

INSTRUCTIONS

1. Tack weld the metal pieces of the butt and tee joints together and brace them on the welding table in the horizontal position.

2. Start at one end of the plate with the gun pointed in a slightly upward direction.

3. You may use a pushing or a dragging gun angle, depending on the current setting and penetration desired.

4. Undercutting along the top edge and overlap along the bottom edge are problems with both gun angles. Careful attention must be paid to the manipulation of the "weave" technique used to overcome these problems. The most successful weave patterns are the "C" and "J" patterns. The "J" pattern is the most frequently used. The "J" pattern allows weld metal to be deposited along a shelf created by the previous weave. The length of the "J" can be changed to control the weld bead size.

5. Small weld beads are easier to control than large ones.

6. The weld must be straight and uniform in size and have 100% penetration.

7. Repeat these welds until you have established the rhythm and technique that work well for you.

8. Turn off the welding machine and shielding gas and clean up your work area when you are finished welding.

INSTRUCTOR'S COMMENTS _____

_____

_____

■ PRACTICE 11-18

Name _____ Date _____

Class _____ Instructor _____ Grade _____

OBJECTIVE: After completing this practice, you should be able to weld stringer beads in the overhead position on both 1/16-in. (2-mm) and 1/8-in. (3-mm) mild steel.

EQUIPMENT AND MATERIALS NEEDED FOR THIS PRACTICE

1. A properly set-up and adjusted GMA welding machine.

2. Proper safety protection (welding hood, safety glasses, leather gloves, wire cutting pliers, long-sleeved shirt, long pants, and leather shoes or boots). Refer to Chapter 2 in the text for more specific safety information.

3. 0.035-in. and/or 0.045-in. (0.9-mm and/or 1.2-mm) diameter wire.

4. Two or more pieces of mild steel sheet, 12-in. (305-mm) long by 3-in. (76-mm) wide by 16-gauge to 1/8-in. (3-mm) thick.

INSTRUCTIONS

1. In the overhead position, the molten weld pool should be kept small. This can be achieved by using lower current settings, traveling faster, or by pushing the molten weld pool.

2. When welding overhead, extra personal protection is required to reduce the danger of burns. Leather sleeves or leather jackets should be worn.

3. Make several short weld beads using various techniques to establish the method that is most successful and most comfortable for you. After each weld, stop and evaluate it before making a change.

4. When you have decided on the technique to be used, make a welded stringer bead that is 12 in. (305 mm) long.

5. Repeat the weld until it can be made straight, uniform in size, and free from any visual defects.

6. Turn off the welding machine and shielding gas and clean up your work area when you are finished welding.

INSTRUCTOR'S COMMENTS _____

_____

_____

■ PRACTICE 11-19

Name _____ Date _____

Class _____ Instructor _____ Grade _____

OBJECTIVE: After completing this practice, you should be able to weld a butt, lap, and tee joint in the overhead position.

EQUIPMENT AND MATERIALS NEEDED FOR THIS PRACTICE

1. A properly set-up and adjusted GMA welding machine.

2. Proper safety protection (welding hood, safety glasses, leather gloves, wire cutting pliers, long-sleeved shirt, long pants, and leather shoes or boots). Refer to Chapter 2 in the text for more specific safety information.

3. 0.035-in. and/or 0.045-in. (0.9-mm and/or 1.2-mm) diameter wire.

4. Two or more pieces of mild steel sheet, 12-in. (305-mm) long by 3-in. (76-mm) wide by 16-gauge to 1/8-in. (3-mm) thick.

INSTRUCTIONS

1. Tack weld the pieces of metal together in the butt, lap, and tee configurations and secure them in the overhead position.

2. When welding overhead, extra personal protection is required to reduce the danger of burns. Leather sleeves or leather jackets should be worn.

3. Make several short weld beads using various techniques to establish the method that is most successful and most comfortable for you. After each weld, stop and evaluate it before making a change.

4. When you have decided on the technique to be used, make butt, lap, and tee welded joints.

5. Repeat the welds on both metal thicknesses until they can be made straight, uniform in size, and free from any visual defects.

6. Turn off the welding machine and shielding gas, and clean up your work area when you are finished welding.

INSTRUCTOR'S COMMENTS _____

_____

_____

■ PRACTICE 11-20

Name _____ Date _____

Class _____ Instructor _____ Grade _____

OBJECTIVE: After completing this practice, you should be able to weld a butt and tee joint in the overhead position with 100% penetration.

EQUIPMENT AND MATERIALS NEEDED FOR THIS PRACTICE

1. A properly set-up and adjusted GMA welding machine.

2. Proper safety protection (welding hood, safety glasses, leather gloves, wire cutting pliers, long-sleeved shirt, long pants, and leather shoes or boots). Refer to Chapter 2 in the text for more specific safety information.

3. 0.035-in. and/or 0.045-in. (0.9-mm and/or 1.2-mm) diameter wire.

4. Two or more pieces of mild steel sheet, 12-in. (305-mm) long by 3-in. (76-mm) wide by 16-gauge to 1/8-in. (3-mm) thick.

INSTRUCTIONS

1. Tack weld the pieces of metal together in the butt and tee configurations and secure them in the overhead position.

2. When welding overhead, extra personal protection is required to reduce the danger of burns. Leather sleeves or leather jackets should be worn.

3. It may be necessary to adjust the root opening to allow 100% penetration.

4. Make several short weld beads using various techniques to establish the method that is most successful and most comfortable for you. After each weld, stop and evaluate it before making a change.

5. Use a dragging torch angle with a "C" or "J" weave pattern.

6. Repeat the welds on both metal thicknesses until they can be made straight, uniform in size, and free from any visual defects.

7. Turn off the welding machine and shielding gas and clean up your work area when you are finished welding.

INSTRUCTOR'S COMMENTS _____

_____

_____

■ PRACTICE 11-21

Name _____ Date _____

Class _____ Instructor _____ Grade _____

OBJECTIVE: After completing this practice, you should be able to make a stringer bead in the flat (1G) position using pulsed-arc globular metal transfer.

EQUIPMENT AND MATERIALS NEEDED FOR THIS PRACTICE

1. A properly set-up and adjusted GMA welding machine.

2. Proper safety protection (welding hood, safety glasses, wire brush, chipping hammer, leather gloves, pliers, long-sleeved shirt, long pants, and leather boots or shoes). Refer to Chapter 2 in the text for more specific safety information.

3. 0.035-in. and/or 0.045-in. (0.9-mm and/or 1.2-mm) diameter welding wire.

4. One or more pieces of mild steel plate, 12-in. (305-mm) long by 3-in. (76-mm) wide by 1/4-in. (6-mm) thick.

INSTRUCTIONS

1. Have your instructor show you how to correctly set the controls on the welding machine so that you will be welding with a pulsed-arc and globular transfer. The current and wire-feed speed as well as the frequency, amplitude, and pulse width must be correct for the metal being welded.

2. Place the metal flat on the weld table and start at one end using either a push or dragging technique to make a bead along the entire 12 in. (305 mm) of the plate.

3. Keep the arc near the center of the weld pool. If it is too far forward, the spatter will be increased and if it is too far rearward, leading edge fusion may be reduced.

4. Use a weave pattern that allows the arc to follow the leading edge of the molten weld pool.

5. Inspect the appearance of the bead and make any necessary changes in the weld parameters, gun angle, or weave pattern to correct for discrepancies.

6. Repeat the weld until the bead can be made straight, uniform, free of any visual defects, and with good penetration.

7. Turn off the welding machine and shielding gas and clean up your work area when finished with this practice.

INSTRUCTOR'S COMMENTS _____

_____

_____

■ PRACTICE 11-22

Name _____   Date _____

Class _____   Instructor _____   Grade _____

OBJECTIVE: After completing this practice, you should be able to make a butt joint in the flat (1G) position using pulsed-arc globular metal transfer.

EQUIPMENT AND MATERIALS NEEDED FOR THIS PRACTICE

1. A properly set-up and adjusted GMA welding machine.

2. Proper safety protection (welding hood, safety glasses, wire brush, chipping hammer, leather gloves, pliers, long-sleeved shirt, long pants, and leather boots or shoes). Refer to Chapter 2 in the text for more specific safety information.

3. 0.035-in. and/or 0.045-in. (0.9-mm and/or 1.2-mm) diameter welding wire.

4. Two or more pieces of mild steel plate, 12-in. (305-mm) long by 3-in. (76-mm) wide by 1/8-in. (6-mm) thick. Prepare the long edges by grinding or burning a 45° bevel on the plates.

INSTRUCTIONS

1. Have your instructor show you how to correctly set the controls on the welding machine so that you will be welding with a pulsed-arc and globular transfer. The current and wire-feed speed as well as the frequency, amplitude, and pulse width must be correct for the metal being welded.

2. Tack the plates together. Leave a 1/8 in. (3 mm) root opening between the plates.

3. Place the metal flat on the weld table. Start at one end using either a push or dragging technique to make a single pass weld along the entire 12 in. (305 mm) of the plate.

4. Use a weave pattern that follows the contour of the groove.

5. Inspect the appearance of the bead and make any necessary changes in the weld parameters, gun angle, or weave pattern to correct for discrepancies.

6. Repeat the weld until the bead can be made straight, uniform, free of any visual defects, and with good penetration.

7. Turn off the welding machine and shielding gas, and clean up your work area when finished with this practice.

INSTRUCTOR'S COMMENTS _____

_____

_____

■ PRACTICE 11-23

Name _____ Date _____

Class _____ Instructor _____ Grade _____

OBJECTIVE: After completing this practice, you should be able to make a butt joint in the flat (1G) position using pulsed-arc globular metal transfer. The weld must have 100% penetration.

EQUIPMENT AND MATERIALS NEEDED FOR THIS PRACTICE

1. A properly set-up and adjusted GMA welding machine.

2. Proper safety protection (welding hood, safety glasses, wire brush, chipping hammer, leather gloves, pliers, long-sleeved shirt, long pants, and leather boots or shoes). Refer to Chapter 2 in the text for more specific safety information.

3. 0.035-in. and/or 0.045-in. (0.9-mm and/or 1.2-mm) diameter welding wire.

4. Two or more pieces of mild steel plate, 12-in. (305-mm) long by 3-in. (76-mm) wide by 1/8-in. (6-mm) thick. Prepare the long edges by grinding or burning a 45° bevel on the plates.

INSTRUCTIONS

1. Have your instructor show you how to correctly set the controls on the welding machine so that you will be welding with a pulsed-arc and globular transfer. The current and wire-feed speed as well as the frequency, amplitude, and pulse width must be correct for the metal being welded.

2. Tack the plates together. Leave a 1/8 in. (3 mm) root opening between the plates.

3. Place the metal flat on the weld table. Start at one end and regulate your weld parameters, travel speed, gun angle, and weave pattern so that 100% penetration is achieved.

4. If the weld burns through, appears to sink or will not fill up, increase the travel speed.

5. Inspect the appearance of the bead and make any necessary changes in the weld parameters, gun angle, or weave pattern to correct for discrepancies.

6. Repeat the weld until the bead can be made straight, uniform, free of any visual defects, and with 100% penetration.

7. Turn off the welding machine and shielding gas and clean up your work area when finished with this practice.

INSTRUCTOR'S COMMENTS _____
_____
_____

■ PRACTICE 11-24

Name _____   Date _____

Class _____   Instructor _____   Grade _____

OBJECTIVE: After completing this practice, you should be able to make a tee joint and a lap joint in the flat (1F) position using pulsed-arc globular metal transfer.

EQUIPMENT AND MATERIALS NEEDED FOR THIS PRACTICE

1. A properly set-up and adjusted GMA welding machine.

2. Proper safety protection (welding hood, safety glasses, wire brush, chipping hammer, leather gloves, pliers, long-sleeved shirt, long pants, and leather boots or shoes). Refer to Chapter 2 in the text for more specific safety information.

3. 0.035-in. and/or 0.045-in. (0.9-mm and/or 1.2-mm) diameter welding wire.

4. Two or more pieces of mild steel plate, 12-in. (305-mm) long by 3-in. (76-mm) wide by 1/8-in. (6-mm) thick. Edge preparation is not required.

INSTRUCTIONS

1. Have your instructor show you how to correctly set the controls on the welding machine so that you will be welding with a pulsed-arc and globular transfer. The current and wire-feed speed as well as the frequency, amplitude, and pulse width must be correct for the metal being welded.

2. Tack the plates together in the tee and lap configurations.

3. Place metal flat on the weld table. Start at one end using either a push or dragging technique to make a single pass weld along the entire 12 in. (305 mm) of the plate.

4. Use a weave pattern that follows the contour of the joint.

5. Inspect the appearance of the bead and make any necessary changes in the weld parameters, gun angle, or weave pattern to correct for discrepancies.

6. Repeat the weld until the bead can be made straight, uniform, free of any visual defects, and with good penetration.

7. Turn off the welding machine and shielding gas and clean up your work area when finished with this practice.

INSTRUCTOR'S COMMENTS _____

_____

_____

■ PRACTICE 11-25

Name _____ Date _____

Class _____ Instructor _____ Grade _____

OBJECTIVE: After completing this practice, you should be able to make a tee joint and a lap joint in the horizontal (2F) position using pulsed-arc globular metal transfer.

EQUIPMENT AND MATERIALS NEEDED FOR THIS PRACTICE

1. A properly set-up and adjusted GMA welding machine.

2. Proper safety protection (welding hood, safety glasses, wire brush, chipping hammer, leather gloves, pliers, long-sleeved shirt, long pants, and leather boots or shoes). Refer to Chapter 2 in the text for more specific safety information.

3. 0.035-in. and/or 0.045-in. (0.9-mm and/or 1.2-mm) diameter welding wire.

4. Two or more pieces of mild steel plate, 12-in. (305-mm) long by 3-in. (76-mm) wide by 1/8-in. (6-mm) thick. Edge preparation is not required.

INSTRUCTIONS

1. Have your instructor show you how to correctly set the controls on the welding machine so that you will be welding with a pulsed-arc globular metal transfer. The current and wire-feed speed as well as the frequency, amplitude, and pulse width must be correct for the metal being welded.

2. Tack the plates together in the tee and lap configurations.

3. Secure the pieces in the horizontal position on the weld table. Start at one end using either a push or dragging technique to make a single pass weld along the entire 12 in. (305 mm) of the plate.

4. The weave pattern must follow the plate surfaces and establish a shelf to support the weld.

5. To prevent undercutting, the beads must be small and quickly made.

6. Inspect the appearance of the bead and make any necessary changes in the weld parameters, gun angle, or weave pattern to correct for discrepancies.

7. Repeat the weld until the bead can be made straight, uniform, and free of any visual defects.

8. Turn off the welding machine and shielding gas and clean up your work area when finished with this practice.

INSTRUCTOR'S COMMENTS _____

_____

_____

■ PRACTICE 11-26

Name _____ Date _____

Class _____ Instructor _____ Grade _____

OBJECTIVE: After completing this practice, you should be able to make a stringer bead in the flat (1G) position using axial spray metal transfer.

EQUIPMENT AND MATERIALS NEEDED FOR THIS PRACTICE

1. A properly set-up and adjusted GMA welding machine.

2. Proper safety protection (welding hood, safety glasses, wire brush, chipping hammer, leather gloves, pliers, long-sleeved shirt, long pants, and leather boots or shoes). Refer to Chapter 2 in the text for more specific safety information.

3. 0.035-in. and/or 0.045-in. (0.9-mm and/or 1.2-mm) diameter welding wire.

4. One or more pieces of mild steel plate, 12-in. (305-mm) long by 3-in. (76-mm) wide by 1/8-in. (6-mm) thick.

INSTRUCTIONS

1. Have your instructor show you how to correctly set the controls on the welding machine so that you will be welding with axial spray transfer. The current and wire-feed speed must be correct for the metal being welded.

2. Be sure that the shielding gas you are using is either pure argon or has a high percentage of argon. Spray transfer only works correctly if the shielding gas being used is pure or nearly pure argon.

3. Place the metal flat on the weld table and start at one end using either a push or dragging technique to make a bead along the entire 12 in. (305 mm) of the plate.

4. Inspect the appearance of the bead and make any necessary changes in the weld parameters or gun angle to correct for discrepancies.

5. Repeat the weld until the bead can be made straight, uniform, free of any visual defects, and with good penetration.

6. Turn off the welding machine and shielding gas and clean up your work area when finished with this practice.

INSTRUCTOR'S COMMENTS _____

_____

_____

■ PRACTICE 11-27

Name _____ Date _____

Class _____ Instructor _____ Grade _____

OBJECTIVE: After completing this practice, you should be able to make a butt, lap, and tee joint in the flat position using axial spray metal transfer.

EQUIPMENT AND MATERIALS NEEDED FOR THIS PRACTICE

1. A properly set-up and adjusted GMA welding machine.

2. Proper safety protection (welding hood, safety glasses, wire brush, chipping hammer, leather gloves, pliers, long-sleeved shirt, long pants, and leather boots or shoes). Refer to Chapter 2 in the text for more specific safety information.

3. 0.035-in. and/or 0.045-in. (0.9-mm and/or 1.2-mm) diameter welding wire.

4. Two or more pieces of mild steel plate, 12-in. (305-mm) long by 3-in. (76-mm) wide by 1/8-in. (6-mm) thick.

INSTRUCTIONS

1. Have your instructor show you how to correctly set the controls on the welding machine so that you will be welding with axial spray transfer. The current and wire-feed speed must be correct for the metal being welded.

2. Be sure that the shielding gas you are using is either pure argon or has a high percentage of argon. Spray transfer only works correctly if the shielding gas being used is pure or nearly pure argon.

3. Tack the pieces together in the butt, lap, and tee configurations.

4. Place the metal flat on the weld table and start at one end using either a push or dragging technique to make a bead along the entire 12 in. (305 mm) of the plate.

5. Inspect the appearance of the bead and make any necessary changes in the weld parameters or gun angle to correct for discrepancies.

6. Repeat the weld until the bead can be made straight, uniform, free of any visual defects, and with good penetration.

7. Turn off the welding machine and shielding gas and clean up your work area when finished with this practice.

INSTRUCTOR'S COMMENTS _____

_____

_____

■ PRACTICE 11-28

Name _____   Date _____

Class _____   Instructor _____   Grade _____

OBJECTIVE: After completing this practice, you should be able to make a butt joint and tee joint in the flat position using axial spray metal transfer. All welds must pass the bend test.

EQUIPMENT AND MATERIALS NEEDED FOR THIS PRACTICE

1. A properly set-up and adjusted GMA welding machine.

2. Proper safety protection (welding hood, safety glasses, wire brush, chipping hammer, leather gloves, pliers, long-sleeved shirt, long pants, and leather boots or shoes). Refer to Chapter 2 in the text for more specific safety information.

3. 0.035-in. and/or 0.045-in. (0.9-mm and/or 1.2-mm) diameter welding wire.

4. Two or more pieces of mild steel plate, 12-in. (305-mm) long by 3-in. (76-mm) wide by 1/8-in. (6-mm) thick.

INSTRUCTIONS

1. Have your instructor show you how to correctly set the controls on the welding machine so that you will be welding with axial spray transfer. The current and wire-feed speed must be correct for the metal being welded.

2. Be sure that the shielding gas you are using is either pure argon or has a high percentage of argon. Spray transfer only works correctly if the shielding gas being used is pure or nearly pure argon.

3. Tack the pieces together in the butt and tee configurations.

4. Place the metal flat on the weld table and start at one end using either a push or dragging technique to make a bead along the entire 12 in. (305 mm) of the plate.

5. Cut 1-in. (25-mm) sections from your welded piece and subject them to the bend test.

6. After bending, inspect the specimens for defects.

7. Repeat the weld until the bead can consistently pass the bend test.

8. Turn off the welding machine and shielding gas and clean up your work area when finished with this practice.

INSTRUCTOR'S COMMENTS _____

_____

_____

CHAPTER 11: QUIZ 1

Name _____   Date _____

Class _____   Instructor _____   Grade _____

INSTRUCTIONS
Carefully read Chapter 11 in the textbook and answer each question.

MATCHING
In the space provided to the left of Column A, write the letter from Column B that best answers or completes the statement in Column A.

Column A

Column B

_____   1. The very best welding conditions are those that will allow a welder to produce the largest _____ of successful welds in the shortest period of time with the highest productivity.

a. set up

_____   2. The same _____ may be used for semiautomatic GMAW, FCAW, and SAW.

b. lower

_____   3. If the shielding gas supply is a cylinder, then it must be _____ securely in place before the valve protection cap is removed.

c. cast

_____   4. _____ of the electrode wire results when the feed roller pushes the wire into a tangled ball because the wire would not go through the outfeed side conduit and appears to look like a bird's nest.

d. little

_____   5. The electrode wire has a natural curve that is known as its _____.

e. wire-feed speed

_____   6. The cast helps the wire make a good electrical _____ as it passes through the contact tube.

f. quantity

_____   7. Because changes in the wire-feed speed automatically change the _____, it is possible to set the amperage by using a chart and measuring the length of wire fed per minute.

g. welding gun angle

_____   8. The _____ the density of a gas, the higher will be the flow rate required for equal arc protection.

h. higher

_____   9. The correct flow rate can be set by checking welding guides that are available from the welding equipment and filler metal _____.

i. equipment

_____   10. The voltage is set on the welder, and the amperage is set by changing the _____.

j. bright metal

_____   11. Because of the constant-potential (CP) power supply, the welding _____ will change as the distance between the contact tube and the work changes.

k. mill scale

_____   12. The term _____ refers to the angle between the GMA welding gun and the work as it relates to the direction of travel.

_____   13. Backhand welding, or dragging angle, produces a weld with deep penetration and _____ buildup.

_____   14. Forehand welding, or pushing angle, produces a weld with shallow penetration and _____ buildup.

_____   15. Shielding gases in the gas metal arc process are used primarily to protect the molten metal from oxidation and _____.

_____   16. To make acceptable GMA welds consistently, the major skill required is the ability to _____ the equipment and weldment.

_____   17. All hot-rolled steel has an oxide layer, which is formed during the rolling process, called _____.

_____   18. The porosity that mill scale causes is most often confined within the weld groove and the surrounding surfaces within 1 in. (25 mm) must be cleaned to _____.

_____   19. Vertical down welds are often used on thin sheet metals or in the _____ pass in grooved joints.

_____   20. An advantage of short-circuiting arc metal transfer in the overhead position is that the _____ size of the molten weld pool allows surface tension to hold it in place.

l. contamination

m. contact

n. smaller

o. chained

p. root

q. amperage

r. manufacturers

s. Bird-nesting

t. current

CHAPTER 11: QUIZ 2

Name _____   Date _____

Class _____   Instructor _____   Grade _____

INSTRUCTIONS

Carefully read Chapter 11 in the textbook and answer each question.

IDENTIFICATION

Identify the numbered items on the drawing by writing the letter next to the identifying term in the space provided.

1. _____   FEED ROLLERS

2. _____   ALIGN SIDE TO SIDE

3. _____   NOTE: DO NOT TOUCH

4. _____   CONTACT TUBE

5. _____   CONDUIT LINER

6. _____   HELIX CAUSES TWISTING FOR IMPROVED WEAR

7. _____   FRONT VIEW

8. _____   ELECTRODE

9. _____   ALLEN WRENCH

10. _____   ALIGN TOP TO BOTTOM

11. _____   CONDUIT

12. _____   CAST

13. _____   TOP VIEW

14. _____   GAS DIFFUSER

15. _____   CAST CAUSES IMPROVED ELECTRICAL CONTACT

16. _____   HELIX

17. _____   THE WIRE PICKS UP THE WELDING CURRENT IN THIS AREA.

18. _____   LINER SETSCREW

CHAPTER 11: QUIZ 3

Name _____ Date _____

Class _____ Instructor _____ Grade _____

INSTRUCTIONS

Carefully read Chapter 11 in the textbook and answer each question.

IDENTIFICATION

Identify the numbered items on the drawing by writing the letter next to the identifying term in the space provided.

1. _____ SOLIDIFIED SLAG

2. _____ BASE METAL

3. _____ SOLIDIFIED WELD METAL

4. _____ DEEP PENETRATION — NARROW AND HIGH BEAD CONTOUR

5. _____ SHALLOW PENETRATION — WIDE AND LOW BEAD CONTOUR

6. _____ FOREHAND DIRECTION OF TRAVEL

7. _____ BACKHAND DIRECTION OF TRAVEL

8. _____ DEEP PENETRATION – NARROW AND HIGH BEAD CONTOUR

9. _____ PERPENDICULAR DIRECTION OF TRAVEL

10. _____ SHIELDING GAS

11. _____ MOLTEN WELD METAL

CHAPTER 11: QUIZ 4

Name _____ Date _____

Class _____ Instructor _____ Grade _____

INSTRUCTIONS

Carefully read Chapter 11 in the textbook and answer each question.

IDENTIFICATION

Identify the numbered items on the drawing by writing the letter next to the identifying term in the space provided.

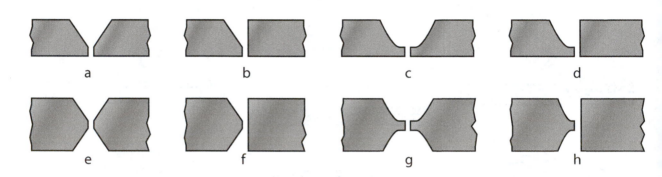

1. _____ DOUBLE J

2. _____ DOUBLE U

3. _____ DOUBLE V

4. _____ SINGLE BEVEL

5. _____ SINGLE V

6. _____ DOUBLE BEVEL

7. _____ SINGLE J

8. _____ SINGLE U

CHAPTER 12

Flux Cored Arc Welding Equipment, Setup, and Operation

CHAPTER 12: QUIZ 1

Name _____ Date _____

Class _____ Instructor _____ Grade _____

INSTRUCTIONS

Carefully read Chapter 12 in the textbook and answer each question.

MATCHING

In the space provided to the left of Column A, write the letter from Column B that best answers or completes the statement in Column A.

Column A	Column B

_____ 1. Flux cored arc welding (FCAW) is a _____ welding process in which weld heating is produced from an arc between the work and a continuously fed filler metal electrode. **a. air**

_____ 2. The CP welding machine's amperage naturally increases to whatever level is needed to maintain the same _____. **b. two sets**

_____ 3. The flux core additives that serve as deoxidizers, gas formers, and slag formers either protect the _____ or help to remove impurities from the base metal. **c. water-cooled**

_____ 4. Some elements in the flux that rapidly expand, called gas formers, push the surrounding _____ away from the molten weld pool. **d. fusion**

_____ 5. Slag helps the weld by protecting the hot metal from the effects of the atmosphere, controlling the bead shape by serving as a dam or mold, and serving as a blanket to slow the weld's _____, which improves its physical properties. **e. slag**

_____ 6. The FCA welding power supply is the same type that is required for GMAW, called _____. **f. fumes**

_____ 7. Although most of the FCA welding guns that you will find in schools are air-cooled, our industry often needs _____ guns because of the higher heat caused by longer welds made at higher currents. **g. productivity**

261

_____ 8. Because of the large quantity of _____ that can be generated during FCA welding, fume removal guns have been designed.

_____ 9. The major difference between electrode feed systems used for GMAW and FCAW is that larger FCAW machines that can use large diameter wire most often have _____ of feed rollers.

_____ 10. FCA welding has no _____, so nearly 100% of the FCAW electrode purchased is used.

_____ 11. Because of the deep _____ characteristic, no edge-beveling preparation is required on some joints in metal up to 1/2 in. (13 mm) in thickness.

_____ 12. The equipment and electrodes used for the FCAW process are more expensive; however, the cost is quickly recoverable through higher _____.

_____ 13. A limitation of flux cored arc welding is that the removal of postweld _____ requires another production step.

_____ 14. FCA welding electrodes are available as _____ electrodes.

_____ 15. Electrodes are available in sizes from 0.023 in. to 5/32 in. (0.58 mm to 3.9 mm) in _____.

_____ 16. The finished FCA filler metal is _____ on spools, reels, and coils made from metal, wood, pressed fiber, or plastic.

_____ 17. The _____ cause the wire to rub on the inside of the contact tube.

_____ 18. Deoxidizers are added to the flux to produce a weld with better mechanical properties with less _____.

_____ 19. In the molten state slag moves through the molten weld pool and acts as a magnet or sponge to chemically combine with _____ in the metal and remove them.

_____ 20. Fluxing agents make the weld more _____ and allow it to flow outward, filling the undercut.

h. seamed-type and seamless-type

i. porosity

j. arc length

k. cast and helix

l. wound (rolled)

m. stub loss

n. fluid

o. impurities

p. cooling rate

q. penetration

r. diameter

s. molten weld pool

t. constant-potential, constant-voltage (CP, CV)

CHAPTER 12: QUIZ 2

Name _____ Date _____

Class _____ Instructor _____ Grade _____

INSTRUCTIONS
Carefully read Chapter 12 in the textbook and answer each question.

MATCHING
In the space provided to the left of Column A, write the letter from Column B that best answers or completes the statement in Column A.

Column A

_____ 1. All FCAW fluxes are divided into two groups based on the acid or basic chemical reactivity of the _____.

_____ 2. T-1 fluxes are _____ fluxes and they are acidic.

_____ 3. T-5 fluxes are _____ fluxes and they are basic.

_____ 4. In the electrode number *E70T-10*, the _____ indicates that the electrode is tubular (flux cored).

_____ 5. The AWS classification for stainless steel for FCAW electrodes starts with the letter _____ as its prefix.

_____ 6. The designation for metal cored steel electrodes was changed from the letter T for tubular to the letter _____ for core.

_____ 7. Some wire electrodes require storage in an electric rod oven to prevent contamination from excessive _____.

_____ 8. When the electrode provides all of the shielding, it is called _____.

_____ 9. Changes in the electrode travel angle will affect the weld bead shape and _____.

_____ 10. The forehand technique is sometimes referred to as pushing the weld bead, and _____ may be referred to as pulling or dragging the weld bead.

_____ 11. An advantage of using the _____ welding technique is that you can easily see the joint where the bead will be deposited.

_____ 12. With the forehand technique, because less weld metal is being applied, the rate of travel along the joint can be _____, which may make it harder to create a uniform weld.

_____ 13. The _____ gun angle is used on automated welding because there is no need to change the gun angle when the weld changes direction.

Column B

a. moisture

b. contact tube

c. slag

d. faster

e. bead shape

f. perpendicular

g. spray transfer

h. globular transfer

i. rutile-based

j. E

k. T

l. C

m. lime-based

———   14. A disadvantage of the perpendicular technique is limited _____ of the weld unless you lean your head way over to the side.

———   15. An advantage of the backhand technique is it is easy to see the back of the molten weld pool as you are welding, which makes it easier to control the _____.

———   16. The American Welding Society defines _____ as the linear rate at which the arc is moved along the weld joint.

———   17. The _____ mode of metal transfer is the most common process used with gas shielded FCAW.

———   18. With _____, the electrode forms a molten ball at its end that grows in size to approximately two times to three times the original electrode diameter.

———   19. The electrode extension is measured from the end of the electrode _____ to the point the arc begins at the end of the electrode.

———   20. Porosity can be caused by moisture in the flux, improper gun manipulation, or _____.

n. travel speed

o. penetration

p. visibility

q. surface contamination

r. forehand

s. backhand

t. self-shielding

CHAPTER 12: QUIZ 3

Name _____ Date _____

Class _____ Instructor _____ Grade _____

INSTRUCTIONS
Carefully read Chapter 12 in the textbook and answer each question.

IDENTIFICATION
Identify the numbered items on the drawing by writing the letter next to the identifying term in the space provided.

1. _____ WIRE GUIDE AND CONTACT TUBE
2. _____ SELF-SHIELDED FLUX CORED ARC WELDING (FCAW-S)
3. _____ ELECTRODE EXTENSION
4. _____ CONTACT TUBE
5. _____ SHIELDING GAS
6. _____ BASE METAL
7. _____ FLUX FILLED TUBULAR WIRE ELECTRODE
8. _____ FLUX CORED ELECTRODE
9. _____ MOLTEN SLAG
10. _____ POWDERED METAL FLUX AND SLAG FORMING MATERIALS
11. _____ GAS NOZZLE
12. _____ WELD METAL
13. _____ GAS-SHIELDED FLUX CORED ARC WELDING (FCAW-G)
14. _____ INSULATOR
15. _____ MOLTEN WELD POOL
16. _____ ARC SHIELDING COMPOSED OF VAPORIZED COMPOUNDS
17. _____ VISIBLE EXTENSION
18. _____ SLAG
19. _____ ARC AND METAL TRANSFER

CHAPTER 12: QUIZ 4

Name _____ Date _____

Class _____ Instructor _____ Grade _____

INSTRUCTIONS

Carefully read Chapter 12 in the textbook and answer each question.

IDENTIFICATION

Identify the numbered items on the drawing by writing the letter next to the identifying term in the space provided.

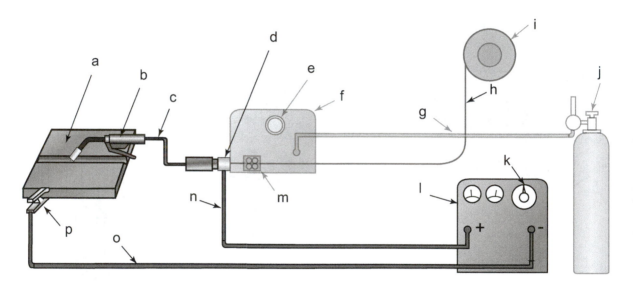

1. _____ WIRE REEL

2. _____ WORK CLAMP

3. _____ WELDING CABLE

4. _____ WIRE FEED UNIT

5. _____ SHIELDING GAS SOURCE

6. _____ WIRE SPEED FEED CONTROL

7. _____ WORK CABLE

8. _____ WELDING MACHINE

9. _____ POWER TERMINAL

10. _____ VOLTAGE CONTROL

11. _____ WELDING GUN CABLE

12. _____ WORK

13. _____ WIRE

14. _____ GAS HOSE

15. _____ WIRE FEED DRIVE ROLLERS

16. _____ GUN

CHAPTER 12: QUIZ 5

Name _____ Date _____

Class _____ Instructor _____ Grade _____

INSTRUCTIONS
Carefully read Chapter 12 in the textbook and answer each question.

IDENTIFICATION
Identify the numbered items on the drawing by writing the letter next to the identifying term in the space provided.

1. _____ CLOSING ROLLERS

2. _____ SHAPE OF METAL AFTER PASSING THROUGH ROLLERS

3. _____ ROLLERS OR DIES

4. _____ SEAMLESS TUBING

5. _____ DRAW DIE

6. _____ SEAMED WIRE WITH FLUX CORE

7. _____ TAKE-UP REEL

8. _____ SEALED END

9. _____ STRIP REEL

10. _____ FORMING ROLLERS

11. _____ FLUX INPUT

12. _____ FLUX INSIDE TUBING

13. _____ THIN SHEET METAL

14. _____ SEAMLESS WIRE WITH FLUX CORE

15. _____ VIBRATING TABLE

CHAPTER 12: QUIZ 6

Name _____ Date _____

Class _____ Instructor _____ Grade _____

INSTRUCTIONS

Carefully read Chapter 12 in the textbook and answer each question.

IDENTIFICATION

Identify the numbered items on the drawing by writing the letter next to the identifying term in the space provided.

Mandatory Classification Designators for FCAW Electrodes

```
                    ┌──────────────────────── a
                    │ ┌────────────────────── b
                    │ │ ┌──────────────────── c
                    │ │ │ ┌────────────────── d
                    │ │ │ │ ┌──────────────── e
                    │ │ │ │ │ ┌────────────── f
                    │ │ │ │ │ │ ┌──────────── g
                    │ │ │ │ │ │ │ ┌───────── h
        E X X T X - X X X - X - X H X        i
                                  │ └──────── j
                                  └────────── k
```

1. _____ USABILITY DESIGNATOR

2. _____ TENSILE STRENGTH DESIGNATOR

3. _____ FOR FLUX CORED ELECTRODES

4. _____ DESIGNATES AN ELECTRODE

5. _____ OPTIONAL, SUPPLEMENTAL DIUSIBLE HYDROGEN DESIGNATOR

6. _____ POSITION DESIGNATOR

7. _____ DEPOSIT COMPOSITION DESIGNATOR

8. _____ DESIGNATES THE CONDITION OF HEAT TREATMENT

9. _____ SHIELDING GAS DESIGNATOR

10. _____ IMPACT DESIGNATOR

11. _____ OPTIONAL SUPPLEMENTAL DESIGNATORS

Flux Cored Arc Welding

ELECTRODE		WELDING POWER			SHIELDING GAS		BASE METAL	
Type	Size	Amps	Wire–Feed Speed IPM (cm/min)	Volts	Type	Flow	Type	Thickness
E70T–1	0.035 in. (0.9 mm)	130 to 150	288 to 380 (732 to 975)	22 to 25	none	n/a	Low carbon steel	1/4 in. to 1/2 in. (6 mm to 13 mm)
E70T–1	0.045 in. (1.2 mm)	150 to 210	200 to 300 (508 to 762)	28 to 29	none	n/a	Low carbon steel	1/4 in. to 1/2 in. (6 mm to 13 mm)
E70T–5	0.035 in. (0.9 mm)	130 to 200	288 to 576 (732 to 1463)	20 to 28	75% Argon 25% CO_2	30 cfh	Low carbon steel	1/4 in. to 1/2 in. (6 mm to 13 mm)
E70T–5	0.045 in. (1.2 mm)	150 to 250	200 to 400 (508 to 1016)	23 to 29	75% Argon 25% CO_2	35 cfh	Low carbon steel	1/4 in. to 1/2 in. (6 mm to 13 mm)

Table 13-1 FCA welding parameters that can be used if specific settings are not available from the electrode manufacturer.

Electrode			Welding Power			Shielding Gas		Base Metal	
Type	Size	Amps	Wire-Feed Speed IPM (cm/min)	Volts		Type	Flow	Type	Thickness
E70T-1	0.035 in. (0.9 mm)	130 to 150	288 to 380 (732 to 975)	22 to 25		None	n/a	Low carbon steel	1/4 in. to 3/4 in. (6 mm to 19 mm)
E70T-1	0.045 in. (1.2 mm)	150 to 210	200 to 300 (508 to 762)	28 to 29		None	n/a	Low carbon steel	1/4 in. to 3/4 in. (6 mm to 19 mm)
E70T-1	0.052 in. (1.4 mm)	150 to 300	150 to 350 (381 to 889)	25 to 33		None	n/a	Low carbon steel	1/4 in. to 3/4 in. (6 mm to 19 mm)
E70T-1	1/16 in. (1.6 mm)	200 to 400	150 to 300 (381 to 762)	27 to 33		None	n/a	Low carbon steel	1/4 in. to 3/4 in. (6 mm to 19 mm)
E70T-5	0.035 in. (0.9 mm)	130 to 200	288 to 576 (732 to 1463)	20 to 28		75% argon 25% CO_2	30 cfh	Low carbon steel	1/4 in. to 3/4 in. (6 mm to 19 mm)
E70T-5	0.045 in. (1.2 mm)	150 to 250	200 to 400 (508 to 1016)	23 to 29		75% argon 25% CO_2	35 cfh	Low carbon steel	1/4 in. to 3/4 in. (6 mm to 19 mm)
E70T-5	0.052 in. (1.4 mm)	150 to 300	150 to 350 (381 to 889)	21 to 32		75% argon 25% CO_2	35 cfh	Low carbon steel	1/4 in. to 3/4 in. (6 mm to 19 mm)
E70T-5	1/16 in. (1.6 mm)	180 to 400	145 to 350 (368 to 889)	21 to 34		75% argon 25% CO_2	40 cfh	Low carbon steel	1/4 in. to 3/4 in. (6 mm to 19 mm)

Table 13-2 FCA welding parameters that can be used if specific settings are not available from the electrode manufacturer.

Electrode			Welding Power			Shielding Gas		Base Metal	
Type	Size	Amps	Wire-Feed Speed IPM (cm/min)	Volts		Type	Flow	Type	Thickness
E70T-1	0.030 in. (0.8 mm)	40 to 145	90 to 340 (228 to 864)	20 to 27		None	n/a	Low carbon steel	16 gauge to 18 gauge
E70T-1	0.035 in. (0.9 mm)	130 to 200	288 to 576 (732 to 1463)	20 to 28		None	n/a	Low carbon steel	16 gauge to 18 gauge
E70T-5	0.035 in. (0.9 mm)	90 to 200	190 to 576 (483 to 1463)	16 to 29		75% argon 25% CO_2	35 cfh	Low carbon steel	16 gauge to 18 gauge

Table 13-3 FCA welding parameters that can be used if specific settings are not available from the electrode manufacturer.

■ PRACTICE 13-1

Name _____   Date _____

Class _____   Instructor _____   Grade _____

OBJECTIVE: After completing this practice, you should be able to safely set up FCAW equipment.

EQUIPMENT AND MATERIALS NEEDED FOR THIS PRACTICE

1. Semiautomatic welding power source approved for FCA welding.

2. Welding gun.

3. Electrode feed unit, electrode supply.

4. Shielding gas supply, shielding gas flowmeter regulator.

5. Electrode conduit, power and work leads.

6. Shielding gas hoses (if required).

7. Assorted hand tools, spare parts, and any other required materials.

INSTRUCTIONS

> **NOTE: Some manufacturers include detailed setup instructions with their equipment. If such instructions are available for your equipment, follow them. However, if no instructions are available, then follow the instructions listed below.**

1. If the shielding gas is to be used and it comes from a cylinder, the cylinder must be chained securely in place before the valve protection cap is removed. Stand to one side of the valve opening and quickly crack the valve to blow out any dirt in the valve before the flowmeter regulator is attached. Attach the correct hose from the regulator to the "gas-in" connection on the electrode feed unit or machine.

2. Install the reel of electrode (welding wire) on the holder and secure it. Check the feed roller size to ensure that it matches the wire size. The conduit liner size should be checked to be sure that it is compatible with the wire size. Connect the conduit to the feed unit. The conduit or an extension should be aligned with the groove in the roller and set as close to the roller as possible without touching. Misalignment at this point can contribute to a bird's nest.

3. Be sure that the power is off before attaching the welding cables. The electrode and work leads should be attached to the proper terminals. The electrode lead should be attached to electrode or positive (+) terminal. If necessary, it is also attached to the power cable part of the gun lead. The work lead should be attached to work or negative (−) terminal.

4. The shielding "gas-out" side of the solenoid is then also attached to the gun lead. If a separate splice is required from the gun switch circuit to the feed unit, it should be connected at this time.

Check to see that the welding contactor circuit is connected from the wire-feed unit to the power source.

5. The welding cable liner or wire conduit must be securely attached to the gas diffuser and contact tube (tip). The contact tube must be the correct size to match the electrode wire size being used. If a shielding gas is to be used, a gas nozzle would be attached to complete the assembly. If a gas nozzle is not needed for a shielding gas, it may still be installed. Because it is easy for a student to touch the work with the contact tube during welding, an electrical short may occur. This short out of the contact tube will immediately destroy the tube. Although the gas nozzle may interfere with some visibility, it may be worth the trouble for a welder in training.

INSTRUCTOR'S COMMENTS _____

_____

_____

■ PRACTICE 13-2

Name _____ Date _____

Class _____ Instructor _____ Grade _____

OBJECTIVE: After completing this practice, you should be able to thread FCAW wire.

EQUIPMENT AND MATERIALS NEEDED FOR THIS PRACTICE

1. Semiautomatic welding power source approved for FCA welding.

2. Welding gun.

3. Electrode feed unit, electrode supply.

4. Shielding gas supply, shielding gas flowmeter regulator.

5. Electrode conduit, power and work leads.

6. Shielding gas hoses (if required).

7. Assorted hand tools, spare parts, and any other required materials.

INSTRUCTIONS

> **NOTE: Some manufacturers include detailed setup instructions with their equipment. If such instructions are available for your equipment, follow them. However, if no instructions are available, then follow the instructions listed below.**

1. Check to see that the unit is assembled correctly according to the manufacturer's specifications. Switch on the power and check the gun switch circuit by depressing the switch. The power source relays, feed relays, gas solenoid, and feed motor should all activate.

2. Cut off the end of the electrode wire if it is bent. When you are working with the wire, be sure to hold it tightly. The wire will become tangled if it is released. The wire has a natural curl called cast. Straighten out about 12 in. (300 mm) of the cast to make threading easier.

3. Separate the wire-feed rollers and push the wire first through the guides, then between the rollers, and finally into the conduit liner. Reset the rollers so that there is a slight amount of compression on the wire. Set the wire-feed speed control to a slow speed. Hold the welding gun so that the electrode conduit and cable are as straight as possible.

4. Remove the contact tube on the gun.

5. Press the gun switch. The wire should start feeding into the liner. Watch to make certain that the wire feeds smoothly. Release the gun switch as soon as the end comes through the gun. If the wire stops feeding before it reaches the end of the conduit, stop and check the system. If no obvious problem can be found, mark the wire with tape and remove it from the gun. It then can be held next to the electrode conduit and cable system to determine the location of the problem.

6. Replace the contact tube on the gun.

7. With the wire-feed running, adjust the feed roller compression so that the wire reel can be stopped easily by a slight pressure. Too light a roller pressure will cause the wire to feed erratically. Too high a pressure can crush some wires, causing some flux to be dropped inside the wire liner. If this happens, you will have a continual problem with the wire not feeding smoothly or jamming.

8. With the wire-feed running, adjust the spool drag so that the reel stops when the feed stops. The reel should not coast to a stop because the wire can be snagged easily. Also, when the feed restarts, a jolt occurs when the slack in the wire is taken up. This jolt can be enough to momentarily stop the wire, possibly causing a discontinuity in the weld or may cause the wire to break.

9. When the test runs are completed, the wire can either be rewound or cut off. Some wire-feed units have a retract button. This allows the feed driver to reverse and retract the wire automatically. To rewind the wire on units without this retract feature, release the rollers and turn them backward by hand. If the machine will not allow the feed rollers to be released without upsetting the tension, you must cut the wire. Some wire reels have covers to prevent the collection of dust, dirt, and metal filings on the wire.

INSTRUCTOR'S COMMENTS _____

_____

_____

■ PRACTICE 13-3

Name _____ Date _____

Class _____ Instructor _____ Grade _____

OBJECTIVE: After completing this practice, you should be able to make a stringer bead weld in the flat position.

EQUIPMENT AND MATERIALS NEEDED FOR THIS PRACTICE

1. A properly set-up and adjusted FCA welding machine.

2. Table 13-1 or specific settings from the electrode manufacturer.

3. Proper safety protection.

4. E70T-1 and/or E70T-5 electrodes with 0.035-in. and/or 0.045-in. (0.9-mm and/or 1.2-mm) diameter.

5. One or more pieces of mild steel plate, 12-in. (305-mm) long by 3-in. (76-mm) wide and 1/4-in. (6-mm) thick.

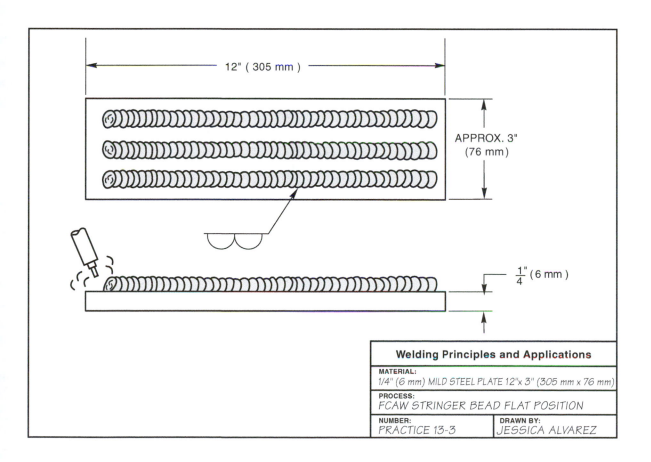

12" (305 mm)

APPROX. 3" (76 mm)

$\frac{1}{4}$" (6 mm)

Welding Principles and Applications

MATERIAL: 1/4" (6 mm) MILD STEEL PLATE 12"x 3" (305 mm x 76 mm)	
PROCESS: FCAW STRINGER BEAD FLAT POSITION	
NUMBER: PRACTICE 13-3	**DRAWN BY:** JESSICA ALVAREZ

INSTRUCTIONS

1. Starting at one end of the plate, and using a dragging technique, make a weld bead along the entire 12 in. (305 mm) length of the metal. Use 1 in. (25 mm) of stickout.

2. After the weld is complete, check its appearance. Make any needed changes to correct the weld.

3. Repeat the weld and make additional adjustments.

4. After the machine is set properly, start to work on improving the straightness and uniformity of the weld.

5. Use weave patterns of different widths and straight stringers without weaving.

6. Repeat with both classifications of electrodes as needed until beads can be made straight, uniform, and free from any visual defects.

7. Turn off the welding machine and shielding gas and clean up your work area when you are finished welding.

INSTRUCTOR'S COMMENTS _____

_____

_____

■ PRACTICE 13-4

Name _____   Date _____

Class _____   Instructor _____   Grade _____

OBJECTIVE: After completing this practice, you should be able to make a square groove weld in the flat position.

EQUIPMENT AND MATERIALS NEEDED FOR THIS PRACTICE

1. A properly set-up and adjusted FCA welding machine.

2. Table 13-1 or specific settings from the electrode manufacturer.

3. Proper safety protection.

4. E70T-1 and/or E70T-5 electrodes with 0.035-in. and/or 0.045-in. (0.9-mm and/or 1.2-mm) diameter.

5. Two or more pieces of mild steel plate, 12-in. (305-mm) long by 3-in. (76-mm) wide and 1/4-in. (6-mm) thick.

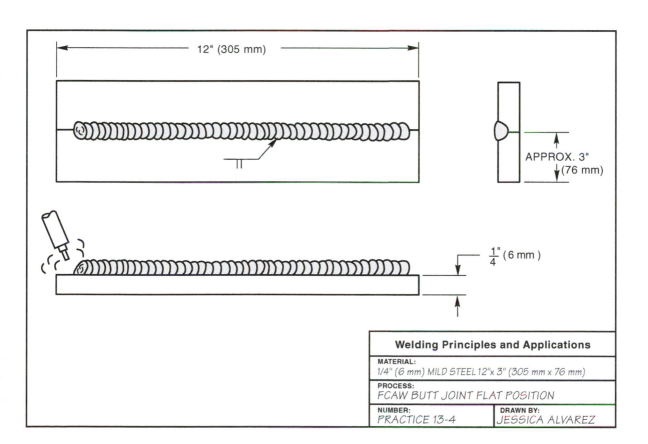

Welding Principles and Applications

MATERIAL:
1/4" (6 mm) MILD STEEL 12"x 3" (305 mm x 76 mm)

PROCESS:
FCAW BUTT JOINT FLAT POSITION

NUMBER:	DRAWN BY:
PRACTICE 13-4	JESSICA ALVAREZ

INSTRUCTIONS

1. Tack weld the plates together and place them in the flat position to be welded.

2. Starting at one end, run a bead along the joint. Watch the molten weld pool and bead for signs that a change in technique may be required.

3. In order to produce a uniform weld, make any needed changes as the weld progresses.

4. Repeat with both classifications of electrodes as needed until consistently defect-free welds can be made in the 1/4-in. (6-mm) thick plate.

5. Turn off the welding machine and shielding gas and clean up your work area when you are finished welding.

INSTRUCTOR'S COMMENTS _____

_____

_____

■ PRACTICE 13-5

Name _____ Date _____

Class _____ Instructor _____ Grade _____

OBJECTIVE: After completing this practice, you should be able to make a V-groove butt weld in the flat position.

EQUIPMENT AND MATERIALS NEEDED FOR THIS PRACTICE

1. A properly set-up and adjusted FCA welding machine.

2. Table 13-2 or specific settings from the electrode manufacturer.

3. Proper safety protection.

4. E70T-1 and/or E70T-5 electrodes with 0.035-in. and/or through 1/16-in. (0.9-mm and/or through 1.6-mm) diameter.

5. Two pieces of mild steel plate, 12-in. (305-mm) long by 3-in. (76-mm) wide and 3/8-in. (9.5-mm) thick or thicker beveled plate.

6. One piece of mild steel plate, 14-in. (355-mm) long, 1-in. (25-mm) wide, and 1/4-in. (6-mm) thick for a backing strip.

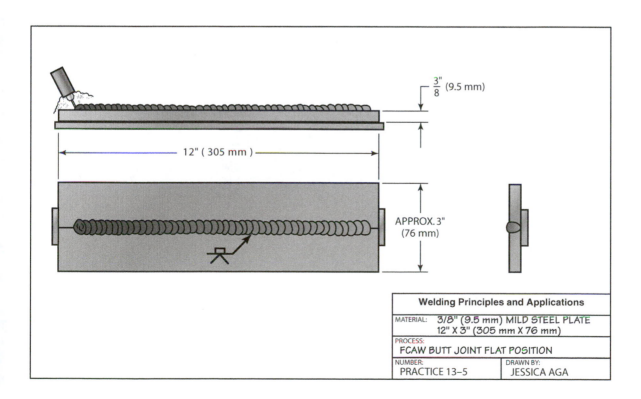

$\frac{3"}{8}$ (9.5 mm)

12" (305 mm)

APPROX. 3"
(76 mm)

Welding Principles and Applications

MATERIAL:	3/8" (9.5 mm) MILD STEEL PLATE 12" X 3" (305 mm X 76 mm)
PROCESS:	FCAW BUTT JOINT FLAT POSITION
NUMBER: PRACTICE 13–5	DRAWN BY: JESSICA AGA

INSTRUCTIONS

1. Tack weld the backing strip to the plates. There should be approximately 1/8 in. (3 mm) root gap between the plates. The beveled surface can be made with or without a root face.

2. Place the test plates in position at a comfortable height and location. Be sure that you will have complete and free movement along the full length of the weld joint. Make a practice pass along the joint with the welding gun without power to make sure nothing will interfere with your making the weld. Be sure the welding cable is free and will not get caught on anything during the welding.

3. Start the weld outside of the groove on the backing strip tab. This is done so that the arc is smooth and the molten weld pool size is established at the beginning of the groove. Continue the weld out on to the tab at the other end of the groove. This is to insure that the end of the groove is completely filled with weld.

4. Repeat with both classifications of electrodes as needed until consistently defect-free welds can be made.

5. Turn off the welding machine and shielding gas and clean up your work area when you are finished welding.

INSTRUCTOR'S COMMENTS _____

_____

_____

■ PRACTICE 13-6

Name _____ Date _____

Class _____ Instructor _____ Grade _____

OBJECTIVE: After completing this practice, you should be able to make fillet welds in both lap and tee joints in the flat position.

EQUIPMENT AND MATERIALS NEEDED FOR THIS PRACTICE

1. A properly set-up and adjusted FCA welding machine.

2. Table 13-1 or specific settings from the electrode manufacturer.

3. Proper safety protection.

4. E70T-1 and/or E70T-5 electrodes with 0.035-in. and/or 0.045-in. (0.9-mm and/or 1.2-mm) diameter.

5. Two or more pieces of mild steel plate, 12-in. (305-mm) long by 3-in. (76-mm) wide and 3/8-in. (9.5-mm) thick.

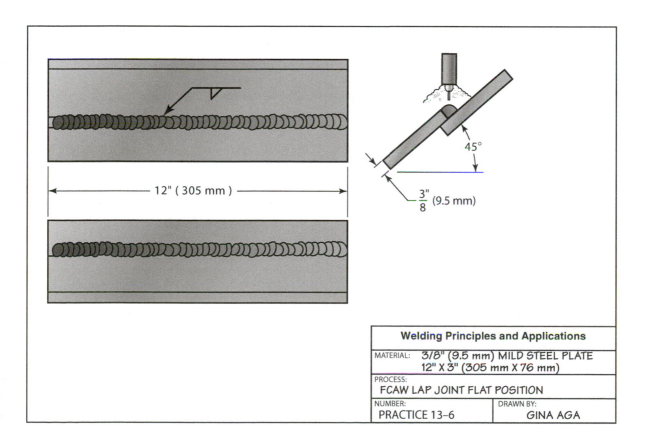

Welding Principles and Applications

MATERIAL: 3/8" (9.5 mm) MILD STEEL PLATE
12" X 3" (305 mm X 76 mm)

PROCESS:
FCAW LAP JOINT FLAT POSITION

NUMBER: PRACTICE 13–6   DRAWN BY: GINA AGA

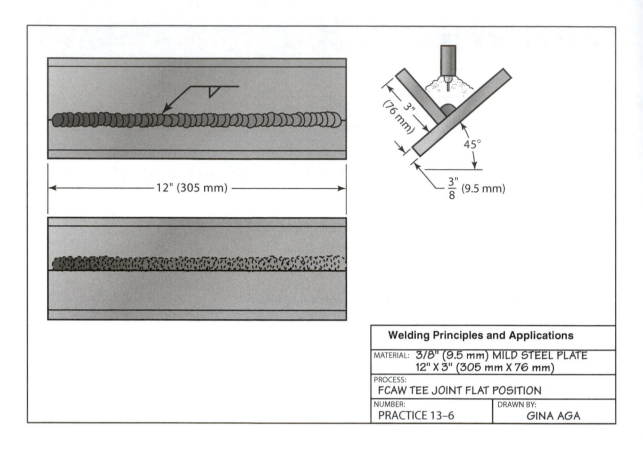

Welding Principles and Applications

MATERIAL:	3/8" (9.5 mm) MILD STEEL PLATE 12" X 3" (305 mm X 76 mm)
PROCESS:	FCAW TEE JOINT FLAT POSITION

NUMBER:	DRAWN BY:
PRACTICE 13-6	GINA AGA

INSTRUCTIONS

1. Tack weld the plates together and place them in position to be welded. When making the lap or tee joints in the flat position, the plates must be at a 45° angle so that the surface of the weld will be flat.

2. Starting at one end, run a bead along the joint. Watch the molten weld pool and bead for signs that a change in technique may be required.

3. Make any needed changes as the weld progresses in order to produce a weld of uniform height and width.

4. Repeat with both classifications of electrodes as needed until consistently defect-free welds can be made.

5. Turn off the welding machine and shielding gas and clean up your work area when you are finished welding.

INSTRUCTOR'S COMMENTS _____

_____

_____

■ PRACTICE 13-7

Name _____   Date _____

Class _____   Instructor _____   Grade _____

OBJECTIVE: After completing this practice, you should be able to make V-groove joint welds in the vertical up position.

EQUIPMENT AND MATERIALS NEEDED FOR THIS PRACTICE

1. A properly set-up and adjusted FCA welding machine.

2. Table 13-1 or specific settings from the electrode manufacturer.

3. Proper safety protection.

4. E70T-1 and/or E70T-5 electrodes with 0.035-in. and/or 0.045-in. (0.9-mm and/or 1.2-mm) diameter.

5. Two or more pieces of mild steel plate, 12-in. (305-mm) long by 3-in. (76-mm) wide by 1/4-in. (6-mm) thick (or thinner) plate.

INSTRUCTIONS

1. Tack weld the plates together in the groove joint configuration.

2. Start with the plate at a 45° angle. When making these joints, you will increase the plate angle gradually as you develop skill until you are making satisfactory welds in the full vertical up position.

3. Starting at the bottom, run a bead up the joint. Watch the molten weld pool and bead for signs that a change in technique may be required.

4. Repeat with both classifications of electrodes as needed until defect-free welds can be made.

5. Turn off the welding machine and shielding gas and clean up your work area when finished with this practice.

INSTRUCTOR'S COMMENTS _____

_____

_____

■ PRACTICE 13-8

Name _____   Date _____

Class _____   Instructor _____   Grade _____

OBJECTIVE: After completing this practice, you should be able to make a square groove butt joint welds at a full vertical up angle.

EQUIPMENT AND MATERIALS NEEDED FOR THIS PRACTICE

1. A properly set-up and adjusted FCA welding machine.

2. Table 13-1 or specific settings from the electrode manufacturer.

3. Proper safety protection.

4. E70T-1 and/or E70T-5 electrodes with 0.035-in. and/or 0.045-in. (0.9-mm and/or 1.2-mm) diameter.

5. Two or more pieces of mild steel plate, 12-in. (305-mm) long by 6-in. (152-mm) wide by 1/4-in. (6-mm) thick (or thinner) plate.

INSTRUCTIONS

1. Tack weld the plates together in the butt configuration.

2. Start with the plate at a 45° angle. When making these joints, you will increase the plate angle gradually as you develop skill until you are making satisfactory welds in the full vertical up position.

3. Starting at the bottom, run a bead up the joint. Watch the molten weld pool and bead for signs that a change in technique may be required.

4. Repeat with both classifications of electrodes as needed until defect-free welds can be made.

5. Turn off the welding machine and shielding gas and clean up your work area when finished with this practice.

INSTRUCTOR'S COMMENTS _____

_____

_____

■ PRACTICE 13-9

Name _____ Date _____

Class _____ Instructor _____ Grade _____

OBJECTIVE: After completing this practice, you should be able to make V-groove joint welds in the vertical up position with a backing strip.

EQUIPMENT AND MATERIALS NEEDED FOR THIS PRACTICE

1. A properly set-up and adjusted FCA welding machine.

2. Table 13-1 or specific settings from the electrode manufacturer.

3. Proper safety protection.

4. E70T-1 and/or E70T-5 electrodes with 0.035-in. and/or 0.045-in. (0.9-mm and/or 1.2 mm) diameter.

5. Two or more pieces of mild steel plate, 12-in. (305-mm) long by 3-in. (76-mm) wide by 3/8-in. (9.5-mm) thick beveled plate and a 14-in. (355-mm) long by 1-in. (25-mm) wide by 1/4-in. (6-mm) thick backing strip.

INSTRUCTIONS

1. Tack weld the plates in a groove joint configuration with the backing strip in place.

2. Start with the plate at a 45° angle. When making these joints, you will increase the angle until satisfactory welds can be made in a full vertical up angle.

3. Starting at the bottom, run a bead up the joint. Watch the molten weld pool and bead for signs that a change in technique may be required.

4. Repeat with both classifications of electrodes as needed until defect-free welds can be made.

5. Turn off the welding machine and shielding gas and clean up your work area when finished with this practice.

INSTRUCTOR'S COMMENTS _____

_____

_____

■ PRACTICE 13-10

Name _____ Date _____ Electrode Used _____

Class _____ Instructor _____ Grade _____

OBJECTIVE: After completing this practice, you should be able to make butt joint welds at a 45° vertical up position.

EQUIPMENT AND MATERIALS NEEDED FOR THIS PRACTICE

1. A properly set-up and adjusted FCA welding machine.

2. Table 13-2 or specific settings from the electrode manufacturer.

3. Proper safety protection.

4. E70T-1 and/or E70T-5 electrodes with 0.035-in. and/or through 0.045-in. (0.9-mm and/or through 1.2-mm) diameter.

5. Two or more pieces of mild steel plate, 12-in. (305-mm) long by 3-in. (76-mm) wide by 3/8-in. (9.5-mm) thick (or thicker) beveled plate and a 9-in. (230-mm) long by 1-in. (25-mm) wide by 1/4-in. (6-mm) thick backing strip.

INSTRUCTIONS

1. Tack weld the plates together in the butt joint configuration with the backing strip in place.

2. Start with the plate at a 45° angle. When making these joints, you will increase the plate angle gradually as you develop skill until you are making satisfactory welds in the full vertical up position.

3. Starting at the bottom, run a bead up the joint. Watch the molten weld pool and bead for signs that a change in technique may be required.

4. Repeat with both classifications of electrodes as needed until defect-free welds can be made.

5. Turn off the welding machine and shielding gas and clean up your work area when finished with this practice.

INSTRUCTOR'S COMMENTS _____

_____

_____

■ PRACTICE 13-11

Name _____ Date _____

Class _____ Instructor _____ Grade _____

OBJECTIVE: After completing this practice, you should be able to make a lap weld joint and a tee joint in the vertical up position.

EQUIPMENT AND MATERIALS NEEDED FOR THIS PRACTICE

1. A properly set-up and adjusted FCA welding machine.

2. Table 13-1 or specific settings from the electrode manufacturer.

3. Proper safety protection.

4. E70T-1 and/or E70T-5 electrodes with 0.035-in. and/or through 0.045-in. (0.9-mm and/or through 1.2-mm) diameter.

5. Two or more pieces of mild steel plate, 12-in. (305-mm) long by 3-in. (76-mm) wide by 3/8-in. (9.5-mm) thick.

INSTRUCTIONS

1. Tack weld the plates together in lap joint and tee joint configurations.

2. Start with the plate at a 45° angle to the table. You will gradually increase this angle as your skill develops until you are making satisfactory welds in the full vertical up position.

3. Starting at the bottom, run a bead up the joint. Watch the molten weld pool and bead for signs that a change in technique may be required.

4. Repeat with both joint types and with both classifications of electrodes as needed until defect-free welds can be made.

5. Turn off the welding machine and shielding gas and clean up your work area when finished with this practice.

INSTRUCTOR'S COMMENTS _____
_____
_____

■ PRACTICE 13-12

Name _____ Date _____

Class _____ Instructor _____ Grade _____

OBJECTIVE: After completing this practice, you should be able to make a lap joint and a tee joint in the horizontal position.

EQUIPMENT AND MATERIALS NEEDED FOR THIS PRACTICE

1. A properly set-up and adjusted FCA welding machine.

2. Table 13-1 or specific settings from the electrode manufacturer.

3. Proper safety protection.

4. E70T-1 and/or E70T-5 electrodes with 0.035-in. and/or 0.045-in. (0.9-mm and/or 1.2-mm) diameter.

5. Two or more pieces of mild steel plate, 12-in. (305-mm) long by 3-in. (76-mm) wide by 3/8-in. (9.5-mm) thick beveled plate.

INSTRUCTIONS

1. Tack weld the plates together in the lap and tee joint configurations.

2. Place the plate in the horizontal 2F position.

3. Make the root weld small so that you do not trap slag under the overlap on the lower edge of the weld.

4. Clean and wire brush each pass before the next pass is added.

5. Make all the succeeding welds small to help control the contour.

6. Repeat with both classifications of electrodes and with both joint types as needed until defect-free welds can be made.

7. Turn off the welding machine and shielding gas and clean up your work area when finished with this practice.

INSTRUCTOR'S COMMENTS _____

_____

_____

■ PRACTICE 13-13

Name _____ Date _____

Class _____ Instructor _____ Grade _____

OBJECTIVE: After completing this practice, you should be able to make a multiple pass lap joint and a tee joint in the horizontal position.

EQUIPMENT AND MATERIALS NEEDED FOR THIS PRACTICE

1. A properly set-up and adjusted FCA welding machine.

2. Table 13-1 or specific settings from the electrode manufacturer.

3. Proper safety protection.

4. E70T-1 and/or E70T-5 electrodes with 0.035-in. and/or 0.045-in. (0.9-mm and/or 1.2-mm) diameter.

5. Two or more pieces of mild steel plate, 7-in. (178-mm) long by 3-in. (76-mm) wide by 3/4-in. (19-mm) thick beveled plate.

INSTRUCTIONS

1. Tack weld the plates together in the lap and tee joint configurations.

2. Place the plate in the horizontal 2F position.

3. Make the root weld small so that you do not trap slag under the overlap on the lower edge of the weld, Figure 13-43.

4. Clean and wire brush each pass before the next pass is added.

5. Make all the succeeding welds small to help control the contour, Figure 13-44.

6. Repeat with both classifications of electrodes and with both joint types as needed until defect-free welds can be made.

7. Turn off the welding machine and shielding gas and clean up your work area when finished with this practice.

INSTRUCTOR'S COMMENTS _____

_____

_____

■ PRACTICE 13-14

Name _____ Date _____

Class _____ Instructor _____ Grade _____

OBJECTIVE: After completing this practice, you should be able to make a stringer bead at a 45° angle.

EQUIPMENT AND MATERIALS NEEDED FOR THIS PRACTICE

1. A properly set-up and adjusted FCA welding machine.

2. Table 13-1 or specific settings from the electrode manufacturer.

3. Proper safety protection.

4. E70T-1 and/or E70T-5 electrodes with 0.035-in. and/or 0.045-in. (0.9-mm and/or 1.2-mm) diameter.

5. One or more pieces of mild steel plate, 12-in. (305-mm) long by 3-in. (76-mm) wide by 1/4-in. (6-mm) thick or thicker.

INSTRUCTIONS

1. Start with the plate at a 45° angle to the table. You will gradually increase this angle as your skill develops until you are making satisfactory welds on the vertical face of the plate in the horizontal position.

2. Repeat with both classifications of electrodes as needed until defect-free welds can be made.

3. Turn off the welding machine and shielding gas and clean up your work area when finished with this practice.

INSTRUCTOR'S COMMENTS _____

_____

_____

■ PRACTICE 13-15

Name _____ Date _____

Class _____ Instructor _____ Grade _____

OBJECTIVE: After completing this practice, you should be able to make a butt (groove) joint in the horizontal position.

EQUIPMENT AND MATERIALS NEEDED FOR THIS PRACTICE

1. A properly set-up and adjusted FCA welding machine.

2. Table 13-1 or specific settings from the electrode manufacturer.

3. Proper safety protection.

4. E70T-1 and/or E70T-5 electrodes with 0.035-in. and/or 0.045-in. (0.9-mm and/or 1.2-mm) diameter.

5. Two or more pieces of mild steel plate, 12-in. (305-mm) long by 3-in. (76-mm) wide by 1/4-in. (6-mm) thick or thicker.

INSTRUCTIONS

1. Tack weld the plates together leaving a slight space between them of approximately one-half the thickness of the metal.

2. Position the plate horizontally.

3. Using a slightly upward gun angle begin to weld at one end of the groove. Travel at a relatively fast speed to prevent the bead from sagging down. If sagging occurs, travel faster.

4. Repeat with both classifications of electrodes as needed until defect-free welds can be made.

5. Turn off the welding machine and shielding gas and clean up your work area when finished with this practice.

INSTRUCTOR'S COMMENTS _____
_____

■ PRACTICE 13-16

Name _____ Date _____

Class _____ Instructor _____ Grade _____

OBJECTIVE: After completing this practice, you should be able to make a beveled butt joint in the horizontal position using a backing strip.

EQUIPMENT AND MATERIALS NEEDED FOR THIS PRACTICE

1. A properly set-up and adjusted FCA welding machine.

2. Table 13-2 or specific settings from the electrode manufacturer.

3. Proper safety protection.

4. E70T-1 and/or E70T-5 electrodes with 0.035-in. and/or through 0.045-in. (0.9-mm and/or through 0.9-mm) diameter.

5. Two or more pieces of mild steel plate, 7-in. (178-mm) long by 3-in. (76-mm) wide by 3/4-in. (19-mm) thick or thicker beveled plates and a 9-in. (230-mm) long by 1-in. (25-mm) wide by 1/4-in. (6-mm) thick backing strip.

INSTRUCTIONS

1. Tack weld the plates and the backing strip together. Leave a slight space between the plates and allow the backing strip to extend 1 in. (25-mm) beyond the plates at each end.

2. Position the plate horizontally.

3. Using a slightly upward gun angle begin to weld at one end of the groove. Travel at a relatively fast speed to prevent the bead from sagging down. Multiple passes will be required. Clean and wire brush the beads between passes.

4. Repeat with both classifications of electrodes as needed until defect-free welds can be made.

5. Turn off the welding machine and shielding gas and clean up your work area when finished with this practice.

INSTRUCTOR'S COMMENTS _____

_____

_____

■ PRACTICE 13-17

Name _____ Date _____

Class _____ Instructor _____ Grade _____

OBJECTIVE: After completing this practice, you should be able to make a butt (square groove) joint in the overhead 4G position.

EQUIPMENT AND MATERIALS NEEDED FOR THIS PRACTICE

1. A properly set-up and adjusted FCA welding machine.

2. Table 13-1 or specific settings from the electrode manufacturer.

3. Proper safety protection. (Extra protection is needed for this practice. Wear a leather jacket and cap.)

4. E70T-1 and/or E70T-5 electrodes with 0.035-in. and/or 0.045-in. (0.9-mm and/or 1.2-mm) diameter.

5. Two or more pieces of mild steel plate, 12-in. (305-mm) long by 3-in. (76-mm) wide by 1/4-in. (6-mm) thick.

INSTRUCTIONS

1. Tack weld the plates together leaving a slight space between the plates of approximately one half the thickness of the metal.

2. Position the plate in the overhead position. Get yourself into a comfortable position.

3. Use a lower current setting. Maintain a small weld pool and travel quickly to prevent dripping of the molten metal.

4. Repeat with both classifications of electrodes as needed until defect-free welds can be made.

5. Turn off the welding machine and shielding gas and clean up your work area when finished with this practice.

INSTRUCTOR'S COMMENTS _____

_____

_____

■ PRACTICE 13-18

Name _____ Date _____

Class _____ Instructor _____ Grade _____

OBJECTIVE: After completing this practice, you should be able to make a V-groove butt joint in the overhead 4G position with a backing strip.

EQUIPMENT AND MATERIALS NEEDED FOR THIS PRACTICE

1. A properly set-up and adjusted FCA welding machine.

2. Table 13-1 or specific settings from the electrode manufacturer.

3. Proper safety protection. (Extra protection is needed for this practice. Wear a leather jacket and cap.)

4. E70T-1 and/or E70T-5 electrodes with 0.035-in. and/or 0.045-in. (0.9-mm and/or 1.2-mm) diameter.

5. Two or more pieces of mild steel plate, 12-in. (305-mm) long by 3-in. (76-mm) wide by 3/8-in. (9.5 mm) thick beveled plate and a 14-in. (355-mm) long by 1-in. (25-mm) wide by 1/4-in (6-mm) thick backing strip.

INSTRUCTIONS

1. Tack weld the plates and the backing strip together leaving a slight space between the plates and allow the backing strip to extend 1 in. (25 mm) beyond the end of the plates.

2. Position the plate in the overhead position. Get yourself into a comfortable position.

3. Use a lower current setting. Maintain a small weld pool and travel quickly to prevent dripping of the molten metal.

4. Multiple passes will be required. Clean and wire brush the beads between passes.

5. Repeat with both classifications of electrodes as needed until defect-free welds can be made.

6. Turn off the welding machine and shielding gas and clean up your work area when finished with this practice.

INSTRUCTOR'S COMMENTS _____

_____

_____

■ PRACTICE 13-19

Name _____ Date _____

Class _____ Instructor _____ Grade _____

OBJECTIVE: After completing this practice, you should be able to make lap and tee joints in the overhead 4F position.

EQUIPMENT AND MATERIALS NEEDED FOR THIS PRACTICE

1. A properly set-up and adjusted FCA welding machine.

2. Table 13-1 or specific settings from the electrode manufacturer.

3. Proper safety protection. (Extra protection is needed for this practice. Wear a leather jacket and cap.)

4. E70T-1 and/or E70T-5 electrodes with 0.035-in. and/or 0.045-in. (0.9-mm and/or 1.2-mm) diameter.

5. Two or more pieces of mild steel plate, 12-in. (305-mm) long by 3-in. (76-mm) wide by 3/8-in. (9.5-mm) thick plate for the lap joint and two or more pieces of mild steel plate 12-in. (305-mm) long 3-in. (76-mm) wide and 3/4-in. (19-mm) thick plate for the tee joint.

INSTRUCTIONS

1. Tack weld the plates together in the lap and tee joint configuration.

2. Position the plate in the overhead position. Get yourself into a comfortable position.

3. Use a lower current setting. Maintain a small weld pool and travel quickly to prevent dripping.

4. Multiple passes will be required. Clean and wire brush the beads between passes.

5. Repeat with both classifications of electrodes as needed until defect-free welds can be made.

6. Turn off the welding machine and shielding gas and clean up your work area when finished with this practice.

INSTRUCTOR'S COMMENTS _____

_____

_____

■ PRACTICE 13-20

Name _____   Date _____

Class _____   Instructor _____   Grade _____

OBJECTIVE: After completing this practice, you should be able to make a butt joint in the flat 1G position.

EQUIPMENT AND MATERIALS NEEDED FOR THIS PRACTICE

1. A properly set-up and adjusted FCA welding machine.

2. Table 13-3 or specific settings from the electrode manufacturer.

3. Proper safety protection.

4. E70T-1 and/or E70T-5 electrodes with 0.030-in. and/or 0.035-in. (0.8-mm and/or 0.9-mm) diameter.

5. Two or more pieces of mild steel sheet, 12-in. (305-mm) long by 3-in. (76-mm) wide by 16-gauge to 18-gauge thick.

INSTRUCTIONS

1. Tack weld the plates together in the butt joint configuration without a root gap.

2. Position the plate in the flat position.

3. Use a low current setting and beware of burnthrough. If burnthrough occurs, the welder can be pulsed on and off to fill the hole.

4. Repeat with both classifications of electrodes as needed until defect-free welds can be made.

5. Turn off the welding machine and shielding gas and clean up your work area when finished with this practice.

INSTRUCTOR'S COMMENTS _____
_____
_____

■ PRACTICE 13-21

Name _____ Date _____

Class _____ Instructor _____ Grade _____

OBJECTIVE: After completing this practice, you should be able to make lap and tee joints in the flat 1F position.

EQUIPMENT AND MATERIALS NEEDED FOR THIS PRACTICE

1. A properly set-up and adjusted FCA welding machine.

2. Table 13-3 or specific settings from the electrode manufacturer.

3. Proper safety protection.

4. E70T-1 and/or E70T-5 electrodes with 0.030-in. and/or 0.035-in. (0.8-mm and/or 0.9-mm) diameter.

5. Two or more pieces of mild steel sheet, 12-in. (305-mm) long by 3-in. (76-mm) wide by 16-gauge to 18-gauge thick.

INSTRUCTIONS

1. Tack weld the plates together in the lap and tee joint configurations.

2. Position the plate in the flat position.

3. Use a low current setting and beware of burnthrough. If burnthrough occurs, the welder can be pulsed on and off to fill the hole.

4. Repeat with both classifications of electrodes as needed until defect-free welds can be made.

5. Turn off the welding machine and shielding gas and clean up your work area when finished with this practice.

INSTRUCTOR'S COMMENTS _____

_____

_____

■ PRACTICE 13-22

Name _____ Date _____

Class _____ Instructor _____ Grade _____

OBJECTIVE: After completing this practice, you should be able to make a butt joint in the vertical down 3G position.

EQUIPMENT AND MATERIALS NEEDED FOR THIS PRACTICE

1. A properly set-up and adjusted FCA welding machine.

2. Table 13-3 or specific settings from the electrode manufacturer.

3. Proper safety protection.

4. E70T-1 and/or E70T-5 electrodes with 0.030-in. and/or 0.035-in. (0.8-mm and/or 0.9-mm) diameter.

5. Two or more pieces of mild steel sheet, 12-in. (305-mm) long by 3-in. (76-mm) wide by 16-gauge to 18-gauge thick.

INSTRUCTIONS

1. Tack weld the plates together in the butt joint configuration without a root gap.

2. Position the plate in the vertical position.

3. Begin the bead at the top and weld downward. This works best on thin-gauge metal.

4. Use a low current setting and beware of burnthrough. If burnthrough occurs, the welder can be pulsed on and off to fill the hole.

5. Repeat with both classifications of electrodes as needed until defect-free welds can be made.

6. Turn off the welding machine and shielding gas and clean up your work area when finished with this practice.

INSTRUCTOR'S COMMENTS _____

_____

_____

■ PRACTICE 13-23

Name _____   Date _____

Class _____   Instructor _____   Grade _____

OBJECTIVE: After completing this practice, you should be able to make lap and tee joints in the vertical down 3F position.

EQUIPMENT AND MATERIALS NEEDED FOR THIS PRACTICE

1. A properly set-up and adjusted FCA welding machine.

2. Table 13-3 or specific settings from the electrode manufacturer.

3. Proper safety protection.

4. E70T-1 and/or E70T-5 electrodes with 0.030-in. and/or 0.035-in. (0.8-mm and/or 0.9-mm) diameter.

5. Two or more pieces of mild steel sheet, 12-in. (305-mm) long by 3-in. (76-mm) wide by 16-gauge to 18-gauge thick.

INSTRUCTIONS

1. Tack weld the plates together in the lap and tee joint configurations.

2. Position the plate in the vertical position.

3. Begin the bead at the top and weld downward. This works best on thin-gauge metal.

4. Use a low current setting and beware of burnthrough. If burnthrough occurs, the welder can be pulsed on and off to fill the hole.

5. Repeat with both classifications of electrodes as needed until defect-free welds can be made.

6. Turn off the welding machine and shielding gas and clean up your work area when finished with this practice.

INSTRUCTOR'S COMMENTS _____

_____

_____

■ PRACTICE 13-24

Name _____ Date _____

Class _____ Instructor _____ Grade _____

OBJECTIVE: After completing this practice, you should be able to make lap and tee joints in the horizontal 2F position.

EQUIPMENT AND MATERIALS NEEDED FOR THIS PRACTICE

1. A properly set-up and adjusted FCA welding machine.

2. Table 13-3 or specific settings from the electrode manufacturer.

3. Proper safety protection.

4. E70T-1 and/or E70T-5 electrodes with 0.030-in. and/or 0.035-in. (0.8-mm and/or 0.9-mm) diameter.

5. Two or more pieces of mild steel sheet, 12-in. (305-mm) long by 3-in. (76-mm) wide by 16-gauge to 18-gauge thick.

INSTRUCTIONS

1. Tack weld the plates together in the lap and tee joint configurations.

2. Position the plate in the horizontal position.

3. Angle the gun slightly upward and travel relatively fast to avoid sagging of the weld pool.

4. Use a low current setting and beware of burnthrough. If burnthrough occurs, the welder can be pulsed on and off to fill the hole.

5. Repeat with both classifications of electrodes as needed until defect-free welds can be made.

6. Turn off the welding machine and shielding gas and clean up your work area when finished with this practice.

INSTRUCTOR'S COMMENTS _____

_____

_____

■ PRACTICE 13-25

Name _____ Date _____

Class _____ Instructor _____ Grade _____

OBJECTIVE: After completing this practice, you should be able to make a butt joint in the horizontal 2G position.

EQUIPMENT AND MATERIALS NEEDED FOR THIS PRACTICE

1. A properly set-up and adjusted FCA welding machine.

2. Table 13-3 or specific settings from the electrode manufacturer.

3. Proper safety protection.

4. E70T-1 and/or E70T-5 electrodes with 0.030-in. and/or 0.035-in. (0.8-mm and/or 0.9-mm) diameter.

5. Two or more pieces of mild steel sheet, 12-in. (305-mm) long by 3-in. (76-mm) wide by 16-gauge to 18-gauge thick.

INSTRUCTIONS

1. Tack weld the plates together in the butt joint configuration with no root gap.

2. Position the plate in the horizontal position.

3. Angle the gun slightly upward and travel relatively fast to avoid sagging of the weld pool.

4. Use a low current setting and beware of burnthrough. If burnthrough occurs, the welder can be pulsed on and off to fill the hole.

5. Repeat with both classifications of electrodes as needed until defect-free welds can be made.

6. Turn off the welding machine and shielding gas and clean up your work area when finished with this practice.

INSTRUCTOR'S COMMENTS _____

_____

_____

■ PRACTICE 13-26

Name _____   Date _____

Class _____   Instructor _____   Grade _____

OBJECTIVE: After completing this practice, you should be able to make a butt joint in the overhead 4G position.

EQUIPMENT AND MATERIALS NEEDED FOR THIS PRACTICE

1. A properly set-up and adjusted FCA welding machine.

2. Table 13-3 or specific settings from the electrode manufacturer.

3. Proper safety protection. (Extra protection is required for this practice. Wear a leather jacket and cap.)

4. E70T-1 and/or E70T-5 electrodes with 0.030-in. and/or 0.035-in. (0.8-mm and/or 0.9-mm) diameter.

5. Two or more pieces of mild steel sheet, 12-in. (305-mm) long by 3-in. (76-mm) wide by 16-gauge to 18-gauge thick.

INSTRUCTIONS

1. Tack weld the plates together in the butt joint configuration with no root gap.

2. Position the plate in the overhead position. Get yourself into a comfortable position.

3. Travel relatively fast to avoid dripping of the weld pool.

4. Use a low current setting and beware of burnthrough. If burnthrough occurs, the welder can be pulsed on and off to fill the hole.

5. Repeat with both classifications of electrodes as needed until defect-free welds can be made.

6. Turn off the welding machine and shielding gas and clean up your work area when finished with this practice.

INSTRUCTOR'S COMMENTS _____

_____

_____

■ PRACTICE 13-27

Name _____ Date _____

Class _____ Instructor _____ Grade _____

OBJECTIVE: After completing this practice, you should be able to make a lap and tee joints in the overhead 4F position.

EQUIPMENT AND MATERIALS NEEDED FOR THIS PRACTICE

1. A properly set-up and adjusted FCA welding machine.

2. Table 13-3 or specific settings from the electrode manufacturer.

3. Proper safety protection. (Extra protection is required for this practice. Wear a leather jacket and cap.)

4. E70T-1 and/or E70T-5 electrodes with 0.030-in. and/or 0.035-in. (0.8-mm and/or 0.9-mm) diameter.

5. Two or more pieces of mild steel sheet, 12-in. (305-mm) long by 3-in. (76-mm) wide by 16-gauge to 18-gauge thick.

INSTRUCTIONS

1. Tack weld the plates together in the lap and tee joint configurations.

2. Position the plate in the overhead position. Get yourself into a comfortable position.

3. Travel relatively fast to avoid dripping of the weld pool.

4. Use a low current setting and beware of burnthrough. If burnthrough occurs, the welder can be pulsed on and off to fill the hole.

5. Repeat with both classifications of electrodes as needed until defect-free welds can be made.

6. Turn off the welding machine and shielding gas and clean up your work area when finished with this practice.

INSTRUCTOR'S COMMENTS _____

_____

_____

■ PRACTICE 13-28

Name _____ Date _____

Class _____ Instructor _____ Grade _____

OBJECTIVE: After completing this practice you should be able to make plug weld.

EQUIPMENT AND MATERIALS NEEDED FOR THIS PRACTICE

1. A properly set-up and adjusted FCA welding machine.

2. Table 13-1 or specific settings from the electrode manufacturer.

3. Proper safety protection.

4. E70T-1 and/or E70T-5 electrodes with 0.035-in. and/or 0.045-in. (0.9-mm and/or 1.2-mm) diameter.

5. Two or more pieces of mild steel plate, 6-in. (152-mm) square by 1/4-in. (6-mm) thick or thicker.

INSTRUCTIONS

1. Using an OFC or PAC torch and proper safety protection, lay out and cut a 1-in. (25-mm) diameter hole through the top plate.

2. Clean any slag off the backside of the cut, grinding it smooth if necessary so that the top plate will sit flat on the bottom plate.

3. Clamp the plates together using a C-clamp, **Figure 13-55**.

4. Be sure you have full range of movement so that you can keep the welding gun pointed at the root of the weld at approximately a 45° angle.

5. Start the weld at the bottom edge of the hole, **Figure 13-56**, and make a full circumference of the plug weld joint.

6. Pull the trigger and make the weld.

7. Try to make the weld all the way around the bottom edge of the hole, but if you cannot, then stop and chip the weld before starting the next weld.

8. Once the weld has been completed all the way around the base, stop and chip the slag.

9. If you do not chip the slag out of a deep plug weld, then the slag can build up excessively, making it impossible for you to maintain visibility of the weld and prevent slag inclusions.

10. The next weld pass will be mainly on the base plate, lapping halfway up on the previous weld, **Figure 13-57**.

11. Hold the welding gun perpendicular to the plate and make the weld all the way around.

12. Continue the circular pattern, expanding it all the way out to the outside, and make two complete passes.

13. Stop and chip the slag.

14. Finish the weld by starting in the center, and build it up until the weld bead is flush with the surface.

15. Turn off the welding machine and shielding gas and clean your work area when you are finished welding.

INSTRUCTOR'S COMMENTS _____
_____
_____

CHAPTER 13: QUIZ 1

Name _____   Date _____

Class _____   Instructor _____   Grade _____

INSTRUCTIONS

Carefully read Chapter 13 in the textbook and answer each question.

MATCHING

In the space provided to the left of Column A, write the letter from Column B that best answers or completes the statement in Column A.

Column A

Column B

_____ 1. _____ of the FCA weld station is the key to making quality welds.

a. cast

_____ 2. It is always best to follow the specific _____ recommendations regarding setup as provided in its equipment literature.

b. bevel

_____ 3. FCA welding produces a lot of _____ light, heat, sparks, slag, and welding fumes.

c. Setup

_____ 4. Forced ventilation and possibly a _____ must be used to prevent fume-related injuries.

d. root

_____ 5. Changes such as variations in material thickness, position, and type of joint require changes both in _____ and setup.

e. Bird-nesting

_____ 6. If the shielding gas is to be used and it comes from a cylinder, then the cylinder must be chained securely in place before the _____ is removed.

f. technique

_____ 7. _____ of the electrode wire results when the feed roller pushes the wire into a tangled ball because the wire would not go through the outfeed side conduit.

g. groove

_____ 8. Electrode wire has a natural curl known as _____.

h. weave

_____ 9. One advantage of FCA welding is the ability to make 100% joint penetrating welds without beveling the _____ of the plates.

i. square-jointed

_____ 10. More filler metal and welding time are required to fill a beveled joint than are required to make a _____ weld.

j. cover

_____ 11. Beveled joints have more heat from the thermal beveling and additional welding required to fill the _____.

k. ultraviolet

_____ 12. The lower heat input to the square joint means less _____.

l. filler

_____ 13. Any time the metal being welded is thicker than 1/4 in. (6 mm) and a 100% joint penetration weld is required, the edges of the plate must be prepared with a _____.

m. manufacturer's

_____ 14. The _____ pass fuses the two parts together and establishes the depth of weld metal penetration.

_____ 15. The _____ pass is made after the root pass is completed and used to fill the groove with weld metal.

_____ 16. The _____ pass is the last weld pass on a multipass weld.

_____ 17. The root may be either open or closed using a _____.

_____ 18. Filler passes are made with either stringer beads or _____ beads for flat or vertically positioned welds, but stringer beads work best for horizontal and overhead-positioned welds.

_____ 19. When multiple pass filler welds are required, each weld bead must _____ the others along the edges.

_____ 20. Each weld bead must be _____ before the next bead is started.

n. backing strip

o. overlap

p. respirator

q. cleaned

r. distortion

s. edges

t. valve protection cap

CHAPTER 13: QUIZ 2

Name _____ Date _____

Class _____ Instructor _____ Grade _____

INSTRUCTIONS

Carefully read Chapter 13 in the textbook and answer each question.

MATCHING

In the space provided to the left of Column A, write the letter from Column B that best answers or completes the statement in Column A.

Column A

Column B

_____ 1. The cover pass may or may not simply be a continuation of the weld beads used to make the _____ pass(es).

a. cracks

_____ 2. The _____ inspection checks to see that the weld is uniform in width and reinforcement.

b. plate

_____ 3. There should be no _____ on the plate other than those on the weld itself.

c. tee

_____ 4. The weld must be free of both incomplete fusion and _____.

d. flat

_____ 5. A fillet weld should be built up equal to the thickness of the _____.

e. melted

_____ 6. The lap joint is made by _____ the edges of the plates.

f. inclusions

_____ 7. The _____ joint is made by tack welding one piece of metal on another piece of metal at a right angle.

g. slag

_____ 8. A fillet welded lap or tee joint can be strong if it is welded on _____, even without having deep penetration.

h. personal protection

_____ 9. The root of fillet welds must be _____ to ensure a completely fused joint.

i. filler

_____ 10. When making the lap or tee joints in the flat position, the plates must be at a 45° angle so that the surface of the weld will be _____.

j. arc strikes

_____ 11. It is easier to make a quality weld in the vertical up position if both the amperage range and voltage range are set at the _____ end.

k. thicker

_____ 12. A root pass that is too large can trap _____ under overlap along the lower edge of the weld.

l. distortion

_____ 13. A small molten weld pool can be achieved by using lower current settings and _____ traveling speeds.

m. Fillet

_____ 14. The higher speed also reduces the amount of weld _____ by reducing the amount of time that heat is applied to a joint.

n. visual

_____ 15. When welding overhead, extra _____ is required to reduce the danger of burns.

o. spatter

_____ 16. Much of the _____ created during overhead welding falls into or on the nozzle and contact tube.

p. Plug

_____ 17. _____ welds are the easiest weld to make on thin stock.

q. overlapping

_____ 18. A common use for FCA welding on thin stock is to join it to a _____ member.

r. both sides

_____ 19. _____ welds are made by cutting or drilling a hole through the top plate and making a weld through that hole onto the plate that is directly behind the top plate.

s. faster

_____ 20. If you do not chip the slag out of a deep plug weld, then the slag can build up excessively, making it impossible for you to maintain visibility of the weld and prevent slag _____.

t. lower

CHAPTER 13: QUIZ 3

Name _____ Date _____

Class _____ Instructor _____ Grade _____

INSTRUCTIONS
Carefully read Chapter 13 in the textbook and answer each question.

IDENTIFICATION
Identify the numbered items on the drawing by writing the letter next to the identifying term in the space provided.

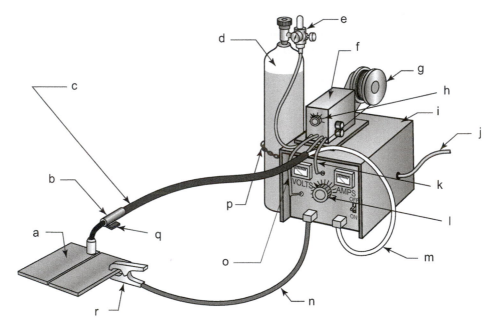

1. _____ WIRE FEED AND CONTROL UNIT

2. _____ WIRE SPOOL

3. _____ GUN START/STOP TRIGGER

4. _____ WORK

5. _____ COMBINATION REGULATOR AND FLOWMETER

6. _____ WORK CLAMP

7. _____ MAIN POWER SUPPLY CABLE

8. _____ WIRE SPEED ADJUSTMENT

9. _____ WELDING VOLTAGE ADJUSTMENT

10. _____ WELDING MACHINE

11. _____ POWER SUPPLY CONTACTOR CONNECTION

12. _____ WELDING POWER CABLE

13. _____ CYLINDER SAFETY CHAIN

14. _____ WELDING CABLE ASSEMBLY

15. _____ SHIELDING GAS CYLINDER

16. _____ WORK CABLE

17. _____ WIRE FEEDER POWER CABLE

18. _____ WELDING GUN

CHAPTER 13: QUIZ 4

Name _____ Date _____

Class _____ Instructor _____ Grade _____

INSTRUCTIONS
Carefully read Chapter 13 in the textbook and answer each question.

IDENTIFICATION
Identify the numbered items on the drawing by writing the letter next to the identifying term in the space provided.

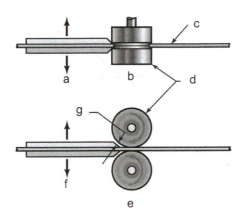
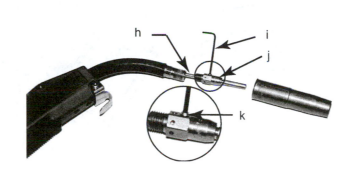

1. _____ FEED ROLLERS

2. _____ ELECTRODE

3. _____ CONDUIT

4. _____ SETSCREW

5. _____ FRONT VIEW

6. _____ NOTE: DO NOT TOUCH

7. _____ ALLEN WRENCH

8. _____ ALIGN TOP TO BOTTOM

9. _____ TOP VIEW

10. _____ GAS DIFFUSER

11. _____ ALIGN SIDE TO SIDE

CHAPTER 13: QUIZ 5

Name _____ Date _____

Class _____ Instructor _____ Grade _____

INSTRUCTIONS

Carefully read Chapter 13 in the textbook and answer each question.

IDENTIFICATION

Identify the numbered items on the drawing by writing the letter next to the identifying term in the space provided.

1. _____ BACKING STRIP

2. _____ LEG

3. _____ SMOOTH WELD TRANSITION

4. _____ ROOT FACE

5. _____ THICKNESS

6. _____ TACK WELDS

7. _____ TEST PLATES

8. _____ OPEN ROOT

9. _____ ROOT PASS

10. _____ ROOT EDGE

11. _____ WELD SIZE

CHAPTER 13: QUIZ 6

Name _____   Date _____

Class _____   Instructor _____   Grade _____

INSTRUCTIONS

Carefully read Chapter 13 in the textbook and answer each question.

IDENTIFICATION

Identify the numbered items on the drawing by writing the letter next to the identifying term in the space provided.

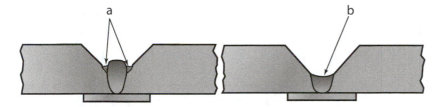

1. _____   EXCESSIVE PENETRATION

2. _____   EXCESSIVE REINFORCEMENT

3. _____   CONCAVE FACE NO SLAG TRAPPED

4. _____   SLAG TRAPPED

5. _____   SLAG INCLUSION

6. _____   UNDERCUT

7. _____   OVERLAP

8. _____   ARC STRIKE

9. _____   POROSITY

10. _____   EXCESSIVE WIDTH

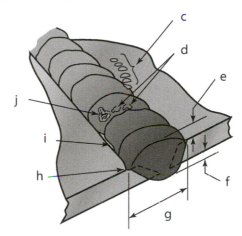

CHAPTER 13: QUIZ 7

Name _____ Date _____

Class _____ Instructor _____ Grade _____

INSTRUCTIONS

Carefully read Chapter 13 in the textbook and answer each question.

IDENTIFICATION

Identify the numbered items on the drawing by writing the letter next to the identifying term in the space provided.

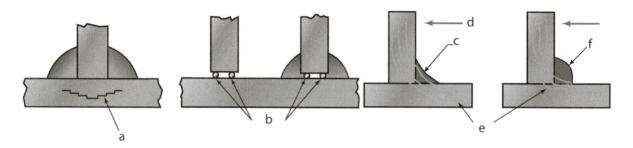

1. _____ FLAT TO CONCAVE CONTOUR

2. _____ STRESS LINES

3. _____ SPACER RODS

4. _____ UNDERBEAD CRACKING OR LAMELLAR TEARING

5. _____ CONVEX CONTOUR

6. _____ FORCE

Gas Metal Arc and Flux Cored Arc Welding of Pipe

■ PRACTICE 14-1

Name _____ Date _____

Class _____ Instructor _____ Grade _____

OBJECTIVE: After completing this practice, you will make a pipe-to-plate fillet welded joint in the 1F flat rolled position using each of the three wire welding processes.

EQUIPMENT AND MATERIALS NEEDED FOR THIS PRACTICE

1. Semiautomatic welding power system approved for GMA and FCA welding.

2. Assorted hand tools, spare parts, and any other required materials.

3. Proper PPE.

4. E70T-1 and/or E70T-5 and ER70S-2 and/or ER70S-3 electrodes with 0.035-in. and/or 0.045-in. (0.9-mm and/or 1.2-mm) diameter.

5. One piece of schedule 40 mild steel pipe 3 in. to 6 in. (76 mm to 150 mm) in diameter.

6. One 4 in. or 7 in. (100 mm or 170 mm) by 1/4 in. (6 mm) mild steel plate.

INSTRUCTIONS

1. Mark a straight line around the pipe using a pipe wrap-around or other pipe layout tool. Using all of the appropriate PPE for a flame cutting torch or plasma cutting torch and following all of the equipment manufacturer's safety and operating guidelines, make a cut along the line. Clean any slag off, and grind the end if necessary to have the pipe stand vertically on the plate.

2. Tack weld the plate to the pipe with four equally spaced approximately 1 in. (25 mm) long welds. Grind the ends of the weld to a featheredge using a thin disk-grinding wheel.

3. The welds on this practice will have three passes. The first pass will be made with the short-circuiting metal transfer process, and the last two passes will be made using the spray metal transfer process.

4. Mount the pipe and plate at a 45° angle so that it can be turned between welds and so that the surface of the fillet weld will remain in the flat position. Position the pipe so that the tack welds are at approximately the 1:30 and 10:30 o'clock positions. Start the first weld on the 1:30 o'clock position and weld toward the 10:30 o'clock position. The gun should be a slight forehand angle; transition to a perpendicular angle around the 12:00 o'clock position, then onto a slight backhand position as it moves toward the 10:30 o'clock position. The weld bead should be approximately 3/16 in. (4.8 mm).

5. Visually inspect the root surface to see if you have complete fusion and no overlap. Discuss with your instructor what you need to do to improve your root weld.

6. Rotate the pipe so you can make the next root weld. Make the necessary changes in the machine settings before starting the next weld. Repeat this process until you have made the root pass all the way around the pipe.

7. Chip and wire brush the root and grind any areas that need additional cleanup.

8. The next two weld passes will be made using the spray metal transfer method. Using the same gun angle as you used for the root pass, make a filler pass over the root weld. Use a slight weaving pattern to make the weld bead wide enough to cover the root weld. To reduce the need for grinding, speed up your travel rate slightly just before stopping the weld. This will result in a slight tapering down of the weld bead and can eliminate the need to grind the end of the weld. However, you should grind the starting point of this first weld so that when you get all the way around, you can tie the ending of the weld smoothly into the starting point of the weld pass.

9. Visually inspect the root surface to see if you have complete fusion and no overlap. Discuss with your instructor what you need to do to improve your root weld.

10. Rotate the pipe so you can make the next root weld. Make the necessary changes in the machine settings before starting the next weld. Repeat this process until you have made the filler pass all the way around the pipe.

11. Chip and wire brush the root and grind any areas that need additional cleanup.

12. The cover pass will be made with the same technique as the filler pass; however, a slightly wider weave pattern will be needed.

13. Visually inspect the weld and repeat this weld as needed with all three processes until you can consistently make welds free of defects. Turn off the welding machine and clean up your work area when you are finished welding.

INSTRUCTOR'S COMMENTS _____

_____

_____

■ PRACTICE 14-2

Name _____   Date _____

Class _____   Instructor _____   Grade _____

OBJECTIVE: After completing this practice, you are going to make a fillet weld in the 2F position using each of the three wire welding processes.

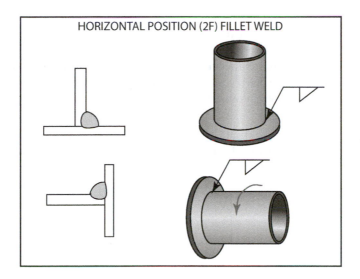

EQUIPMENT AND MATERIALS NEEDED FOR THIS PRACTICE

1. Semiautomatic welding power source approved for GMA and FCA welding.

2. Assorted hand tools, spare parts, and any other required materials.

3. Proper PPE.

4. E70T-1 and/or E70T-5 and ER70S-2 and/or ER70S-3 electrodes with 0.035-in. and/or 0.045-in. (0.9-mm and/or 1.2-mm) diameter.

5. One piece of schedule 40 mild steel pipe 3 in. to 6 in. (76 mm to 150 mm) in diameter.

6. One 4 in. or 7 in. (100 mm or 170 mm) 1/4 in. (6 mm) mild steel plate.

INSTRUCTIONS

1. Using a 15° to 20° backhand gun angle and a 35° to 40° work angle, start the weld on one of the tack welds, and make a stringer weld all the way to the center of the next tack weld. As you get to the tack weld, speed up your travel rate to taper the size of the weld down to make it easier to restart the next weld bead. Chip and wire brush the weld, and visually inspect it for uniformity and any undercut or overlap. The weld bead should have equal legs of approximately 3/16 in. (4.8 mm). The weld should be equal on both the plate and side of the pipe.

2. If there is undercut along the top toe of the weld or if the weld leg is not as large on the top side, you will need to decrease the work angle to direct more metal onto the pipe side. You may also want to make a slight J weave pattern to help correct this problem. Discuss with your instructor what you need to do to improve your root penetration before making the weld on the opposite side.

3. Make any corrections to your technique and complete the weld all the way around the pipe.

4. The second weld pass will be made using the spray metal transfer method. This weld bead will be placed around the lower side of the first weld so that approximately 2/3 to 3/4 of the root weld face is covered by this weld pass. Chip and wire brush the weld each time you stop to reposition yourself. Visually inspect the weld and discuss with your instructor what you need to do to improve your weld. Complete the weld all the way around the pipe.

5. The third weld pass will complete the cover pass, and it will also be made using the spray metal transfer method. This weld should be made approximately 1/2 to 2/3 of the way up on the second weld's face. Chip and wire brush the weld each time you stop to reposition yourself. Visually inspect the weld and discuss with your instructor what you need to do to improve your weld. Complete the weld all the way around the pipe.

6. Visually inspect the weld and repeat this weld as needed with all three processes until you can consistently make welds free of defects. Turn off the welding machine and clean up your work area when you are finished welding.

INSTRUCTOR'S COMMENTS _____

_____

_____

■ PRACTICE 14-3

Name _____   Date _____

Class _____   Instructor _____   Grade _____

OBJECTIVE: After completing this practice, you are going to make a fillet weld in the 5F fixed position using each of the three wire welding processes.

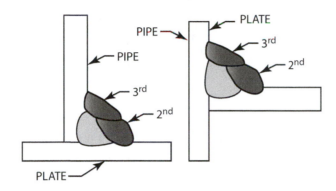

EQUIPMENT AND MATERIALS NEEDED FOR THIS PRACTICE

1. Semiautomatic welding power source approved for GMA and FCA welding.

2. Assorted hand tools, spare parts, and any other required materials.

3. Proper PPE.

4. E70T-1 and/or E70T-5 and ER70S-2 and/or ER70S-3 electrodes with 0.035-in. and/or 0.045-in. (0.9-mm and/or 1.2-mm) diameter.

5. One piece of schedule 40 mild steel pipe 3 in. to 6 in. (76 mm to 150 mm) in diameter.

6. One 4 in. or 7 in. (100 mm or 170 mm) 1/4 in. (6 mm) mild steel plate.

INSTRUCTIONS

1. The weld will be made in an uphill progression, starting on the tack weld at the 6:00 o'clock position. Most welders stop on top of the tack weld at the 9:00 o'clock position. They do this so they can get in a better position to complete the weld.

2. The gun angle relative to the pipe surface will be constantly changing as the weld progresses upward. Figure 14-18 shows how the gun angle starts at a slight forehand angle at 6:00 o'clock and becomes perpendicular to the pipe surface around 7:00 o'clock. It transitions to a steep forehand angle of around 30° between 7:00 and 9:00 o'clock. Between 9:00 and 11:00 o'clock, the angle transitions back to perpendicular and remains perpendicular through the 12:00 o'clock position where the weld ends.

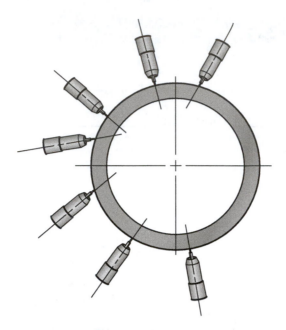

3. The weld bead should be made as an equal legged 3/16 in. (4.8 mm) fillet weld. Chip and wire brush the weld; and after you visually inspect it, discuss with your instructor what you need to do to improve your root penetration before making the weld on the opposite side.

4. Visually inspect the weld and repeat this weld as needed with all three processes until you can consistently make welds free of defects. Turn off the welding machine and clean up your work area when you are finished welding.

INSTRUCTOR'S COMMENTS _____

_____

_____

■ PRACTICE 14-4

Name _____   Date _____

Class _____   Instructor _____   Grade _____

OBJECTIVE: After completing this practice, you will make a pipe butt joint in the 1G horizontal rolled position using each of the three wire welding processes.

EQUIPMENT AND MATERIALS NEEDED FOR THIS PRACTICE

1. Semiautomatic welding power source approved for GMA and FCA welding.

2. Assorted hand tools, spare parts, and any other required materials.

3. Proper PPE.

4. E70T-1 and/or E70T-5 and ER70S-2 and/or ER70S-3 electrodes with 0.035-in. and/or 0.045-in. (0.9-mm and/or 1.2-mm) diameter.

5. Two or more piece of schedule 40 mild steel pipe 6 in. to 8 in. (150 mm to 200 mm) in diameter.

INSTRUCTIONS

1. Tack weld two pieces of pipe together with four 1-in. (25 mm) long welds at the 12:00, 3:00, 6:00, and 9:00 o'clock positions. Grind the ends of the tack welds before placing the pipe horizontally on the welding table or pipe stand with the top tack welds between the 10:30 and 1:30 o'clock positions.

2. Start the root pass weld at the 1:30 o'clock position and weld to the 10:30 o'clock position with a 14° to 20° forehand angle in the direction of travel. The root pass is made as a stringer bead with the GMA welding process unless a backing ring is used; then it can be made with the FCA welding process. As you pass over the top of the pipe and start downhill, change to a slight backhand angle that will increase to approximately a 45° angle. Watch the leading edge of the molten weld pool to see if you have a keyhole, which indicates you are getting 100% penetration.

3. Visually inspect the root surface to see if you had 100% penetration. Mark any areas that did not have complete penetration so you can make changes in the next section to be welded. Discuss with your instructor what you need to do to improve your root penetration.

4. Rotate the pipe so you can make the next root weld. Make the necessary changes in the machine settings before starting the next weld. Repeat this process until you have made the root pass all the way around the pipe.

5. Clean the root and grind any areas that need additional cleanup.

6. **NOTE:** The small weld coupon may become too hot to allow you to make a good weld. Be sure to give the coupon time to cool between weld passes. You may want to have two or three coupons being worked on at the same time so you can alternate between the three processes as a way of giving the coupon time to cool.

7. The hot pass is made with the same procedure of welding from 1:30 to 10:30 o'clock positions as the root pass. Watch the sides of the groove and the leading edge of the molten weld pool to make sure it is being fused.

8. Clean the weld face and visually inspect it for uniformity and sidewall fusion. Grind any trapped slag, and taper the end of the weld to make it easier to restart the weld bead. Discuss with your instructor what you need to do to improve your root penetration.

9. Rotate the pipe so you can make the next hot pass weld. Make the necessary changes in the machine settings before starting the next weld. Repeat this process until you have made the hot pass all the way around the pipe.

10. Grind any areas of the hot pass that have buildup greater than 1/8 in. (3 mm) higher than the surrounding weld face before starting the filler passes.

11. The first filler pass is made along one side so that its toe is about two-thirds of the way across the hot pass. If the toe of the filler pass is made so it meets the opposite sidewall, the V it forms will be likely to trap slag, and it is hard to get the next filler pass to penetrate the narrow gap it forms.

12. The end of the filler passes can be tapered down by slightly increasing the welding travel speed and reducing the weave pattern. This will make it easier to restart the next weld bead. When you have reached the 10:30 o'clock position, stop welding and clean off the flux, and visually inspect the weld. Discuss with your instructor what you need to do to improve your root penetration.

13. Rotate the pipe so you can make the next filler pass. Make the necessary changes in the machine settings before starting the next weld. Repeat this process until you have filled the weld groove to within approximately 1/16 to 1/8-in. (1.6 mm to 3 mm) below the pipe's outer surface. Clean the weld and grind any starts or stops that are more than 1/16 in. (1.6 mm) higher than the surrounding weld surface. Also grind any areas of the weld that are overfilled that would affect the uniformity of the cover weld pass.

14. Watch the edge of the groove as the first cover pass weld is made so that it does not overlap the edge more than 1/16 in. (1.6 mm). Bracing yourself on the table or bracing your hand on the pipe will help you make the cover pass more consistent. The transition for the pipe surface to the weld must be smooth and uniform, so do not spend too much time along the side of the groove because that could cause excessive reinforcement and an abrupt toe of the weld.

15. Clean each weld before the next weld is started. When the cover weld has been completed, clean the weld, do not grind it, and inspect it for discontinuities or defects.

16. Repeat this weld as needed with all three processes until you can consistently make welds free of defects. Turn off the welding machine and clean up your work area when you are finished welding.

INSTRUCTOR'S COMMENTS _____
_____
_____

■ PRACTICE 14-5

Name _____   Date _____

Class _____   Instructor _____   Grade _____

OBJECTIVE: After completing this practice, you are going to make a V-grooved weld in the 2G vertical fixed position using each of the three wire welding processes.

EQUIPMENT AND MATERIALS NEEDED FOR THIS PRACTICE

1. Semiautomatic welding power source approved for GMA and FCA welding.

2. Assorted hand tools, spare parts, and any other required materials.

3. Proper PPE.

4. E70T-1 and/or E70T-5 and ER70S-2 and/or ER70S-3 electrodes with 0.035-in. and/or 0.045-in. (0.9-mm and/or 1.2-mm) diameter.

5. Two or more piece of schedule 40 mild steel pipe 6 in. to 8 in. (150 mm to 200 mm) in diameter.

INSTRUCTIONS

1. Prepare the pipe and tack it together as before, then place it vertically on the table or on a pipe stand.

2. You will use a 14° to 20° forehand angle in the direction of travel for the root pass and the same forehand angle with a 5° to 10° upward work angle for all of the additional weld passes, Figure 14-2. This upward work angle will help to prevent the weld bead's bottom edge from sagging.

3. Keeping the weld beads small makes it easier to control the molten weld pool. Place the weld beads in the groove as illustrated in Figure 14-3.

4. Repeat this weld as needed with all three processes until you can consistently make welds free of defects. Turn off the welding machine and clean up your work area when you are finished welding.

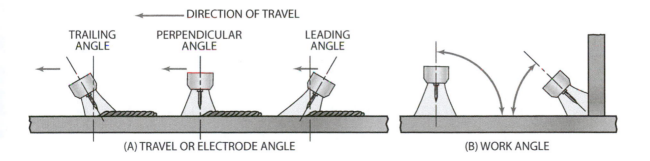

DIRECTION OF TRAVEL

TRAILING ANGLE   PERPENDICULAR ANGLE   LEADING ANGLE

(A) TRAVEL OR ELECTRODE ANGLE   (B) WORK ANGLE

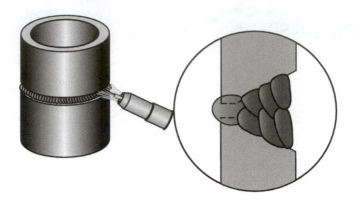

INSTRUCTOR'S COMMENTS _____

_____

_____

■ PRACTICE 14-6

Name _____   Date _____

Class _____   Instructor _____   Grade _____

OBJECTIVE: After completing this practice, you are going to make a V-grooved weld in the 5G fixed position using each of the three wire welding processes.

EQUIPMENT AND MATERIALS NEEDED FOR THIS PRACTICE

1. Semiautomatic welding power source approved for GMA and FCA welding.

2. Assorted hand tools, spare parts, and any other required materials.

3. Proper PPE.

4. E70T-1 and/or E70T-5 and ER70S-2 and/or ER70S-3 electrodes with 0.035-in. and/or 0.045-in. (0.9-mm and/or 1.2-mm) diameter.

5. Two or more piece of schedule 40 mild steel pipe 6 in. to 8 in. (150 mm to 200 mm) in diameter.

INSTRUCTIONS

1. Prepare the pipe and tack it together as before, and place it horizontally on a pipe stand. Any time the weld has to be stopped, the weld must be cleaned and the end of the weld bead feathered even if the weld stops on a tack weld.

2. These welds will be made in an uphill progression, starting near the 6:00 o'clock position. Most welders stop on top of the tack weld at the 9:00 o'clock position. They do this so they can get in a better position to complete the weld.

3. This weld transitions from the overhead position to the vertical position and then to the flat position. So the gun angle relative to the pipe surface must be constantly changing as the weld progresses upward. Refer back to Figure 14-18 in Practice 14-3 to see how the gun angle starts at a slight forehand angle at 6:00 o'clock and becomes perpendicular to the pipe surface around 7:00 o'clock. It transitions to a steep forehand angle of around 30° between 7:00 and 9:00 o'clock. Between 9:00 and 11:00 o'clock, the angle transitions back to perpendicular and remains perpendicular through the 12:00 o'clock position where the weld ends.

4. Clean the weld and feather the starting and stopping points. Chip and wire brush the root pass before visually inspecting it for any problem areas with lack of fusion, excessive buildup, burnthrough, etc. Discuss with your instructor what you need to do to improve your root penetration before making the weld on the opposite side.

5. Grind the root pass to remove any trapped slag and deep undercut before starting the hot pass. The beginning and ending points of all of the weld should be staggered so they are not all in the same area and overlap. Slightly increasing the voltage or decreasing the wire feed speed can help to make the weld pool a little more fluid and help increase the hot pass penetration. Use the same technique of transitioning the electrode angle as you used on the root pass to make the hot pass. Use a

stringer technique with little or no weave for most of the weld. A slight weave may be needed to fill any area that was ground out to remove root pass undercut.

6. Chip and wire brush the root pass before visually inspecting the hot pass for any problem areas with lack of fusion, excessive buildup, etc. Discuss with your instructor what you need to do to improve your root penetration before making the weld on the opposite side.

7. When the hot pass is completed, clean and wire brush it before showing it to your instructor for their advice.

8. Grind any areas of the hot pass that have a weld buildup greater than 1/8 in. (3 mm) before starting the filler passes.

9. Slightly reduce the voltage or increase the wire feed speed so that the weld has good fusion but not deep penetration since the filler passes do not need to have deep penetration.

10. The first filler pass can be made by rocking the welding gun so that the arc is directed from side to side or by using a slight side-to-side weave, Figure 14-4. This can make a concave weld that can make it easier to prevent slag entrapment and/or undercut along the toe of the weld. If this technique is used, it is important not to make the weave too large because that can cause excessive buildup and overlap.

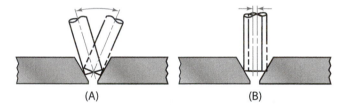

(A)   (B)

11. Whether you used a slight weave or stringer bead for the first filler pass, all of the following passes should be made as stringer beads. Remember to grind the ends of the welds before restarting the next weld. Also clean and grind any parts of a weld bead that has undercut, overlap, or slag trapped before making the next weld pass.

12. The groove must be left with 1/16 in. to 1/8 in. (1.6 mm to 3 mm) below the surface so that there will be room for the cover pass. If necessary, grind away any areas of the filler passes that are too high or uneven that would affect the appearance of the cover passes.

13. Again, follow the edge of the V-groove for the first cover pass. Keep the weld bead small and uniform. Overlap each weld pass about 1/3 of the way. Judge the width of the weld beads so that the last weld pass will not overlap the edge of the V-groove more than 1/8 in. (3 mm). You may have a problem if the last weld pass needs to be much wider than the other cover pass welds, because trying to make it larger can cause overlap or excessive buildup.

14. Visually inspect the weld and repeat this weld as needed with all three processes until you can consistently make welds free of defects. Turn off the welding machine and clean up your work area when you are finished welding.

INSTRUCTOR'S COMMENTS _____

_____

_____

■ PRACTICE 14-7

Name _____ Date _____

Class _____ Instructor _____ Grade _____

OBJECTIVE: After completing this practice, you are going to make a V-grooved weld in the 6G fixed position using each of the three wire welding processes.

EQUIPMENT AND MATERIALS NEEDED FOR THIS PRACTICE

1. Semiautomatic welding power source approved for GMA and FCA welding.

2. Assorted hand tools, spare parts, and any other required materials.

3. Proper PPE.

4. E70T-1 and/or E70T-5 and ER70S-2 and/or ER70S-3 electrodes with 0.035-in. and/or 0.045-in. (0.9-mm and/or 1.2-mm) diameter.

5. Two or more piece of schedule 40 mild steel pipe 6 in. to 8 in. (150 mm to 200 mm) in diameter.

INSTRUCTIONS

1. Prepare the pipe and tack it together as before, and place it at a 45° angle on a pipe stand. Any time the weld has to be stopped, the weld must be cleaned and the end of the weld bead feathered even if the weld stops on a tack weld.

2. The root pass is made without changing the work angle, because getting good root penetration is more important than trying to prevent the weld face from sagging to the downhill side of the groove. Any sagging can be removed by postweld grinding.

3. Once the root pass is made, the remaining welds in the 6G position become more challenging because both the electrode angle and the work angle must be constantly changed to maintain a consistent weld bead. That is because parts of the weld are much like the overhead portion of the 5G weld. Other parts are more like the 2G weld.

4. Slightly increase the voltage or decrease the wire feed speed to help increase the weld penetration for the hot pass. Start the hot pass weld on the opposite side of the bottom that the root pass was started. Use a slight forehand angle and around a 5° upward work angle when starting to weld. Transition to a perpendicular work angel at the 6:00 o'clock position with the same 5° upward work angle. As the weld progresses toward the 9:00 o'clock position, increase the forward gun angle to around 30° and keep the same 5° work angle. Between the 9:00 and 11:00 o'clock positions, transition to a perpendicular gun angle while keeping enough work angle to force the weld face to be uniform and not sagging toward the downhill side of the joint.

5. Clean and visually inspect the hot pass. Discuss with your instructor what you need to do to improve your hot pass before making the weld on the opposite side.

6. The filler passes will be made with a similar technique as the hot pass but with a little less penetration, so set the voltage and/or amperage for a less penetrating weld. The first filler welds will be made on the downhill side of the groove and cover about two-thirds of the hot pass weld surface. Each of the next filler welds will be made on the upper side of the previous filler welds, Figure 14-5. Remember to stagger the starts and stops as you make the filler welds.

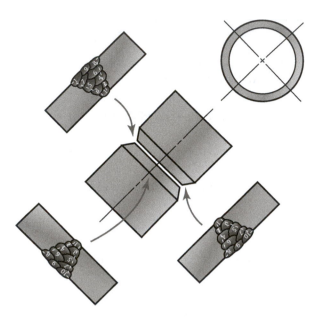

7. Grind any areas of the filler pass that are not uniform in buildup or are excessively built up before starting the cover passes. The cover passes are made with the same technique as the filler passes but with possibly a little greater work angle of between 5° and 10°. Keep the cover passes small so they can be controlled and kept uniform.

8. Visually inspect the weld, and repeat this weld as needed with all three processes until you can consistently make welds free of defects. Turn off the welding machine, and clean up your work area when you are finished welding.

INSTRUCTOR'S COMMENTS _____

_____

_____

CHAPTER 14: QUIZ 1

Name _____   Date _____

Class _____   Instructor _____   Grade _____

INSTRUCTIONS

Carefully read Chapter 14 in the textbook and answer each question.

MATCHING

In the space provided to the left of Column A, write the letter from Column B that best answers or completes the statement in Column A.

Column A	Column B
_____ 1. GMA and FCA pipe welding processes are often used together in the field to make _____ welds.	a. hot pass
_____ 2. The ends of pipe are _____ to between 30° and 35° to form a 60° to 70° V-groove when they are put together.	b. featheredge
_____ 3. Once the root face has been prepared, the inside and outside pipe surfaces must be ground clean back _____ from the joint.	c. diameter
_____ 4. Pipe is not always round; in fact, large _____ pipe is almost always not round.	d. tack
_____ 5. There are welding gages, called _____ gages, that are specifically designed to check for the alignment of pipe.	e. hammer
_____ 6. According to the API code, for pipes of the same nominal diameter, the _____ must not be more than 1/8 in. (3 mm).	f. more
_____ 7. One way to adjust the pipe so that it can be aligned properly is to place the coupon on an anvil and strike it with a _____.	g. flanges
_____ 8. It is important that tack welds are made with the same welding procedure as the finish welds so that they do not become weld _____ in the finished weld.	h. pipe
_____ 9. The number and size of tack welds will vary depending on the pipe's diameter and wall _____.	i. cap
_____ 10. Larger, thicker pipes will require _____ tack welds to keep the root opening from being changed due to welding stresses.	j. defects
_____ 11. The ends of the tack welds must be ground to a _____ so that the root weld can achieve 100% penetration at their ends.	k. beveled
_____ 12. The root pass welds must start and end on the _____ welds, because it is difficult to get 100% weld penetration the instant that both GMA and FCA welds start.	l. fusion

—— 13. Starting the weld on top of the tack will eliminate the potential lack of _____ at the start of the weld.

m. offset

—— 14. The _____ is used to reshape the face of the root weld, burn out any small pieces of trapped slag, and in some cases push the root penetration a little deeper if the root pass did not have 100% penetration all the way around the pipe.

n. high-low

—— 15. The bulk of the weld metal added to a groove weld is made up of the _____.

o. plate

—— 16. The cover pass or passes are sometimes called the _____.

p. thickness

—— 17. Undercut shall not exceed the lesser of 10% of the base metal thickness or _____, and the weld must be free of overlap.

q. 1 in. (25 mm)

—— 18. Pipe sockets are joined to pipe with _____ welds.

r. 1/32 in. (0.8 mm)

—— 19. _____ make it easier to change out items such as check valves, gate valves, tanks, etc.

s. fillet

—— 20. Pipe may also be joined to _____ to make it easier to attach the pipe to the wall or floor for a handrail, equipment base, or it may be used to close off the end of a pipe.

t. filler passes

CHAPTER 14: QUIZ 2

Name _____   Date _____

Class _____   Instructor _____   Grade _____

INSTRUCTIONS

Carefully read Chapter 14 in the textbook and answer each question.

IDENTIFICATION

Identify the numbered items on the drawing by writing the letter next to the identifying term in the space provided.

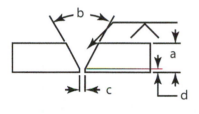
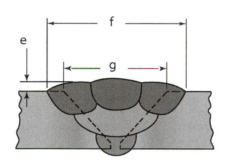
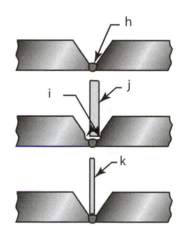

1. _____   GROOVE FACE GROUND AWAY

2. _____   R = 5/32", ±1/16"

3. _____   COVER PASSES MAX. WIDTH = $\frac{1"}{8}$ (3 mm) WIDER THAN THE GROOVE

4. _____   MAX. COVER PASS REINFORCEMENT $\frac{1"}{8}$ (3 mm)

5. _____   STANDARD WIDTH GRINDING DISK

6. _____   GROOVE WIDTH

7. _____   a = 60°, +10° -5°

8. _____   THIN GRINDING DISK

9. _____   f = 1/16 minimum

10. _____   THICKNESS

11. _____   TACK WELD

CHAPTER 14: QUIZ 3

Name _____ Date _____

Class _____ Instructor _____ Grade _____

INSTRUCTIONS
Carefully read Chapter 14 in the textbook and answer each question.

IDENTIFICATION
Identify the numbered items on the drawing by writing the letter next to the identifying term in the space provided.

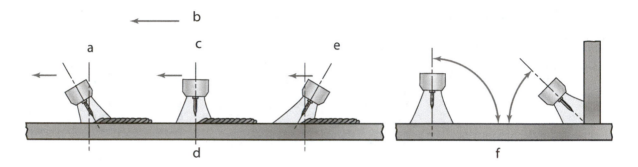

1. _____ PERPENDICULAR ANGLE

2. _____ CLEAT OR DOG

3. _____ WEDGE

4. _____ LEADING ANGLE

5. _____ WORK ANGLE

6. _____ TRAVEL OR ELECTRODE ANGLE

7. _____ HAMMER

8. _____ BACKING STRIP

9. _____ TACK WELD

10. _____ TRAILING ANGLE

11. _____ ANVIL

12. _____ DIRECTION OF TRAVEL

13. _____ TACK WELD ONE SIDE

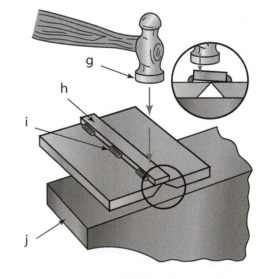

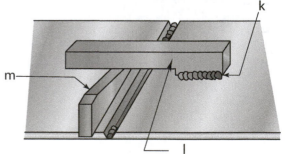

Gas Metal Arc and Flux Cored Arc Welding
AWS SENSE Certification

Electrode			Welding Power			Shielding Gas		Base Metal	
Type	Size	Amps	Wire-Feed Speed IPM (cm/min)	Volts	Polarity	Type	Flow	Type	Thickness
E70S-6	0.035 in. (0.9 mm)	80 to 112	126 to 179 (320 to 455)	16 to 19	DCEP	75% Ar 25% CO_2	20 to 40	Low carbon steel	1/4 in. to 1/2 in. (6 mm to 13 mm)
E70S-6	0.045 in. (1.2 mm)	100 to 140	100 to 140 (254 to 356)	18 to 20	DCEP	75% Ar 25% CO_2	20 to 40	Low carbon steel	1/4 in. to 1/2 in. (6 mm to 13 mm)

Electrode			Welding Power			Shielding Gas		Base Metal	
Type	Size	Amps	Wire-Feed Speed IPM (cm/min)	Volts	Polarity	Type	Flow	Type	Thickness
E70S-3	0.035 in. (0.9 mm)	180 to 280	288 to 448 (732 to 1138)	24 to 28	DCEP	Ar plus 2% O_2	20 to 40	Low carbon steel	1/4 in. to 1/2 in. (6 mm to 13 mm)
E70S-3	0.045 in. (1.2 mm)	210 to 350	210 to 350 (533 to 889)	24 to 29	DCEP	Ar plus 2% O_2	20 to 40	Low carbon steel	1/4 in. to 1/2 in. (6 mm to 13 mm)

Electrode			Welding Power			Shielding Gas		Base Metal	
Type	Size	Amps	Wire-Feed Speed IPM (cm/min)	Volts	Polarity	Type	Flow	Type	Thickness
E71T-1	0.045 in. (1.2 mm)	150 to 210	240 to 440 (610 to 1118)	24 to 28	DCEP	CO_2	20 to 40	Low carbon steel	1/4 in. to 1/2 in. (6 mm to 13 mm)
E71T-1	0.052 in. (1.3 mm)	190 to 300	250 to 500 (635 to 1270)	24 to 29	DCEP	CO_2	20 to 40	Low carbon steel	1/4 in. to 1/2 in. (6 mm to 13 mm)

Electrode Type	Diameter	Volts	Amps	Wire-Feed Speed IPM (cm/min)	Electrode Stickout inch (mm)
Self-Shielded E71T-11	0.035 in. (0.9 mm)	14-15	30	1.3 (50)	3/8 (10)
		15-16	60	1.8 (70)	3/8 (10)
		16-17	115	2.8 (110)	3/8 (10)
		18-19	155	5.1 (200)	3/8 (10)
	0.045 in. (1.2 mm)	15-16	120	1.8 (70)	3/8 (10)
		16-17	140	2.3 (90)	3/8 (10)
		17-18	160	2.8 (110)	3/8 (10)
		18-19	170	3.3 (130)	3/8 (10)
	0.068 in. (1.7 mm)	15-16	125	1.0 (40)	1/2 (13)
		18-19	190	1.9 (75)	1/2 (13)
		20-21	270	3.3 (130)	1/2 (13)
		23-24	300	4.4 (175)	1/2 (13)

■ PRACTICE 15-1

Name _____   Date _____

Class _____   Instructor _____   Grade _____

OBJECTIVE: After completing this practice, you will make welded butt joints, tee joints, and lap joints in all four positions on 10 gauge to 14 gauge thickness carbon sheet steel.

EQUIPMENT AND MATERIALS NEEDED FOR THIS PRACTICE

1. Semiautomatic welding power system approved for GMA.

2. Assorted hand tools, spare parts, and any other required materials.

3. Proper PPE.

4. ER70S-2 and/or ER70S-3 electrodes with 0.035-in. and/or 0.045-in. (0.9-mm and/or 1.2-mm) diameter.

5. 75% Ar + 25% CO_2 or Ar + 2% to 5% O_2 shielding gas.

6. Two or more pieces of 6 in. long 1-1/2 in. wide (152 mm by 38 mm) 10 to 14 gauge carbon sheet steel.

INSTRUCTIONS

1. One at a time, tack weld the sheets together to form the joints and place them in the desired position on a welding stand.

2. Starting at one end, run a 1/8 in. (3 mm) bead along the joint. Watch the molten weld pool and bead for signs it is increasing or decreasing in size. Make any necessary changes in your technique to keep the weld consistent.

3. When the welds have cooled, use a hack saw or thermal cutting torch to cut out 1-in. (25-mm) wide strips as shown in Figure 15-2.

4. Using a vice and a hammer, bend the strip of welded joint as shown in Figure 15-3.

5. Check for complete root fusion on the lap and tee joints and 100% penetration on the butt joint.

6. Repeat each type of joint in each position as needed until consistently good beads are obtained. Turn off the welding machine and shielding gas and clean up your work area when you are finished welding.

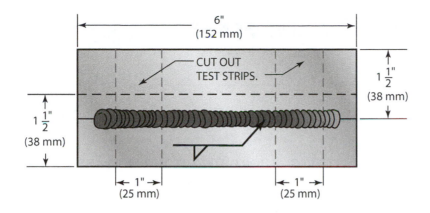

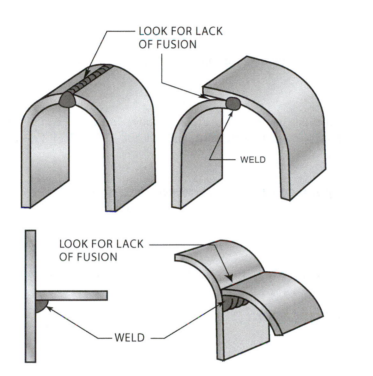

INSTRUCTOR'S COMMENTS _____

_____

_____

■ PRACTICE 15-2 AWS SENSE

Name _____ Date _____

Class _____ Instructor _____ Grade _____

GAS METAL ARC WELDING—SHORT-CIRCUIT METAL TRANSFER (GMAW-S)

WORKMANSHIP SAMPLE

Welding Procedure Specification (WPS) No.: Practice 15-2.

TITLE:

Welding GMAW-S of carbon steel plate to carbon steel plate.

SCOPE:

This procedure is applicable for square 2G, 3G, and 4G groove and 2F, 3F, and 4F fillet welds within the range of 10 gauge (3.4 mm) through 14 gauge (1.9 mm) for the AWS SENSE Level I short-circuit metal transfer GMA welding.

WELDING MAY BE PERFORMED IN THE FOLLOWING POSITIONS:

All.

BASE METAL:

The base metal shall conform to carbon steel M-1, P-1, and S-1 Group 1 or 2.

BACKING MATERIAL SPECIFICATION:

None.

FILLER METAL:

The filler metal shall conform to AWS specification no. E70S-6 from AWS specification A5.18. This filler metal falls into F-number F-6 and A-number A-1.

SHIELDING GAS:

The shielding gas, or gases, shall conform to the following compositions and purity: 25% CO_2 + 75% Ar at 20 to 40 cfh.

JOINT DESIGN AND TOLERANCES:

Refer to the drawing in Figure 15-4 for the joint layout specifications.

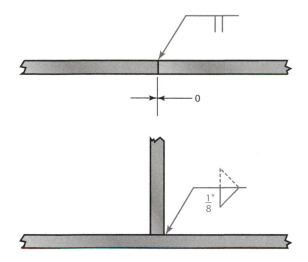

PREPARATION OF BASE METAL:

All parts may be mechanically cut or machine PAC unless specified as manual PAC.

All hydrocarbons and other contaminants, such as cutting fluids, grease, oil, and primers, must be cleaned off all parts and filler metals before welding. This cleaning can be done with any suitable solvents or detergents. The groove face and inside and outside plate surfaces within 1 in. (25 mm) of the joint must be mechanically cleaned of slag, rust, and mill scale. Cleaning must be done with a wire brush or grinder down to bright metal.

ELECTRICAL CHARACTERISTICS:

Set the voltage, amperage, wire-feed speed, and shielding gas flow according to Table 15-1.

PREHEAT:

The parts must be heated to a temperature higher than 50°F (10°C) before any welding is started.

BACKING GAS:

N/A.

SAFETY:

Proper protective clothing and equipment must be used. The area must be free of all hazards that may affect the welder or others in the area. The welding machine, welding leads, work clamp, electrode holder, and other equipment must be in safe working order.

WELDING TECHNIQUE:

Using a 1/2-in. (13-mm) or larger gas nozzle for all welding, first tack weld the plates together according to the drawing, Figure 15-4. Use the E70S-6 filler metal to fuse the plates together. Clean any silicon slag, being sure to remove any trapped silicon slag along the sides of the weld.

All welds are to be made with one pass. All welds must be placed in the orientation shown in the drawing.

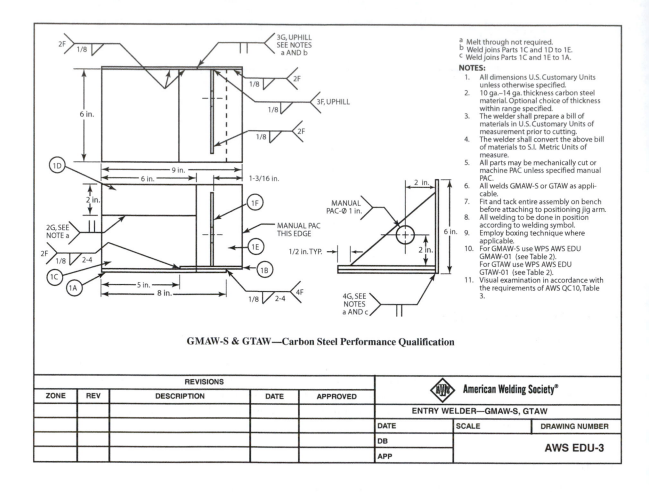

GMAW-S & GTAW—Carbon Steel Performance Qualification

a Melt through not required.
b Weld joins Parts 1C and 1D to 1E.
c Weld joins Parts 1C and 1E to 1A.

NOTES:

1. All dimensions U.S. Customary Units unless otherwise specified.
2. 10 ga.–14 ga. thickness carbon steel material. Optional choice of thickness within range specified.
3. The welder shall prepare a bill of materials in U.S. Customary Units of measurement prior to cutting.
4. The welder shall convert the above bill of materials to S.I. Metric Units of measure.
5. All parts may be mechanically cut or machine PAC unless specified manual PAC.
6. All welds GMAW-S or GTAW as applicable.
7. Fit and tack entire assembly on bench before attaching to positioning jig arm.
8. All welding to be done in position according to welding symbol.
9. Employ boxing technique where applicable.
10. For GMAW-S use WPS AWS EDU GMAW-01 (see Table 2). For GTAW use WPS AWS EDU GTAW-01 (see Table 2).
11. Visual examination in accordance with the requirements of AWS QC10, Table 3.

REVISIONS					American Welding Society®		
ZONE	REV	DESCRIPTION	DATE	APPROVED	ENTRY WELDER—GMAW-S, GTAW		
					DATE	SCALE	DRAWING NUMBER
					DB		AWS EDU-3
					APP		

INTERPASS TEMPERATURE:

The plate should not be heated to a temperature higher than 350°F (175°C) during the welding process. After each weld pass is completed, allow it to cool but never to a temperature below 50°F (10°C). The weldment must not be quenched in water.

CLEANING:

The weld beads may be cleaned by a hand wire brush, a chipping hammer, or a punch and hammer. All weld cleaning must be performed with the test plate in the welding position.

VISUAL INSPECTION:

Visually inspect the weld for uniformity and discontinuities.

- There shall be no cracks, no incomplete fusion.
- There shall be no incomplete joint penetration in groove welds except as permitted for partial joint penetration welds.
- The Test Supervisor shall examine the weld for acceptable appearance and shall be satisfied that the welder is skilled in using the process and procedure specified for the test.
- Undercut shall not exceed the lesser of 10% of the base metal thickness or 1/32 in. (0.8 mm).
- Where visual examination is the only criterion for acceptance, all weld passes are subject to visual examination at the discretion of the Test Supervisor.
- The frequency of porosity shall not exceed one in each 4 in. (100 mm) of weld length, and the maximum diameter shall not exceed 3/32 in. (2.4 mm).
- Welds shall be free from overlap.

SKETCHES:

GMAW Short-Circuit Metal Transfer Workmanship Sample drawing.

INSTRUCTOR'S COMMENTS _____

_____

_____

■ PRACTICE 15-3

Name _____ Date _____

Class _____ Instructor _____ Grade _____

OBJECTIVE: After completing this practice, you will make a V-groove welds in the flat position using the spray metal transfer technique.

EQUIPMENT AND MATERIALS NEEDED FOR THIS PRACTICE

1. Semiautomatic welding power system approved for GMA.

2. Assorted hand tools, spare parts, and any other required materials.

3. Proper PPE.

4. ER70S-2 and/or ER70S-3 electrodes with 0.035-in. and/or 0.045-in. (0.9-mm and/or 1.2-mm) diameter.

5. Ar + 2% to 5% O_2 shielding gas.

6. Two or more pieces of 6 in. long 3 in. wide and 3/8 thick (152 mm × 76 mm × 9.5 mm) steel plate.

INSTRUCTIONS

1. The 30° bevel angles are to be flame or plasma cut on the edges of the plate before the parts are assembled.

2. Grind a 1/16-in. (1.6 mm) to 1/8-in. (3 mm) root face on the beveled edge.

3. The groove face and inside and outside plate surface within 1 in. (25 mm) of the joint must be mechanically cleaned of slag, rust, and mill scale. Cleaning must be done with a wire brush or grinder down to bright metal.

4. Set the voltage, amperage, wire-feed speed, and shielding gas flow according to Table 15-2.

5. Tack weld the plates with a root gap between the plates that measures approximately 1/16 in. to 1/8 in. (1.6 mm to 3 mm).

6. Place the plates in position at a comfortable height and location. Be sure you have complete and free movement along the full length of the weld joint. It is often a good idea to make a practice pass along the joint with the welding gun without power to make sure nothing will interfere with your making the weld. Be sure the welding cable is free and will not get caught on anything during the weld.

7. Grind the ends of the tack welds with a thin grinding disk.

8. Using 20° to 30° backhand technique, start the weld at the very end of the groove. Allow the weld pool to expand to approximately 1/4 in. to 3/8 in. (6 mm to 9.5 mm) wide before starting to travel along the groove. Watch the leading edge of the weld pool for a keyhole that indicates that you are getting 100% penetration.

9. When the weld reaches the opposed end, pause long enough for the weld crater to fill up, so the end of the weld is all the way at the edge of the plates.

10. Clean off all of the slag with a wire brush, chipping hammer, or punch.

11. It should take three or more weld passes to fill the groove. Make sure not to make the weld any wider than 1/8 in. (3 mm) than the groove and no more than 1/8 in. (3 mm) of reinforcement.

INSTRUCTOR'S COMMENTS _____

_____

_____

■ PRACTICE 15-4

Name _____   Date _____

Class _____   Instructor _____   Grade _____

OBJECTIVE: After completing this practice, you will make a tee joint with fillet welds in the horizontal position using the spray metal transfer technique.

EQUIPMENT AND MATERIALS NEEDED FOR THIS PRACTICE

1. Semiautomatic welding power system approved for GMA.

2. Assorted hand tools, spare parts, and any other required materials.

3. Proper PPE.

4. ER70S-3 electrodes with 0.035-in. and/or 0.045-in. (0.9-mm and/or 1.2-mm) diameter.

5. Ar + 2% to 5% O_2 shielding gas.

6. Two or more pieces of 6 in. long 3 in. wide and 3/8 thick (152 mm × 76 mm × 9.5 mm) steel plate.

INSTRUCTIONS

1. The plate surfaces within 1 in. (25 mm) of the joint must be mechanically cleaned of slag, rust, and mill scale. Cleaning must be done with a wire brush or grinder down to bright metal.

2. Set the voltage, amperage, wire-feed speed, and shielding gas flow according to Table 15-2.

3. Tack weld the vertical piece in the center of the base plate. Make sure it is square and in the center of the base plate. Make tack welds on both sides of the tee joint.

4. Place the plates in position at a comfortable height and location. Be sure you have complete and free movement along the full length of the weld joint.

5. Using 20° to 30° backhand technique start the weld at the very end of the joint. Allow the weld pool to expand to ¼-in. (6 mm) in size before starting to travel along the joint. Watch the leading edge of the weld pool to make sure you are getting complete root fusion.

6. When the weld reaches the opposite end, pause long enough for the weld crater to fill up, so the end of the weld is all the way at the edge of the plates.

7. Clean off all of the slag with a wire brush, chipping hammer, or punch.

8. Make any adjustments in your machine setting and techniques that your instructor might suggest before making the second weld on the opposite side.

INSTRUCTOR'S COMMENTS _____

_____

_____

■ PRACTICE 15-5 AWS SENSE

Name _____ Date _____

Class _____ Instructor _____ Grade _____

GAS METAL ARC WELDING (GMAW) SPRAY TRANSFER WORKMANSHIP SAMPLE

Welding Procedure Specification (WPS) No.: Practice 15-5.

TITLE:

Welding GMAW of carbon steel plate to carbon steel plate.

SCOPE:

This procedure is applicable for V-groove and fillet welds on 3/8 in. (9.5 mm) thick plate for the AWS SENSE Level I flat groove and horizontal fillet welds using spray transfer GMA welding.

WELDING MAY BE PERFORMED IN THE FOLLOWING POSITIONS:

1G and 2F.

BASE METAL:

The base metal shall conform to carbon steel M-1, P-1, and S-1, Group 1 or 2.

BACKING MATERIAL SPECIFICATION:

None.

FILLER METAL:

The filler metal shall conform to ER70S-3, 0.035-in. (0.90-mm) to 0.045-in. (1.2-mm) diameter E70S-3 from AWS specification A5.18. This filler metal falls into F-number F-6 and A-number A-1.

SHIELDING GAS:

The shielding gas, or gases, shall conform to the following compositions and purity: Ar + 2% to 5% O_2 at 20 to 40 cfh.

JOINT DESIGN AND TOLERANCES:

Refer to the drawing in **Figure 15-13** for the joint layout specifications.

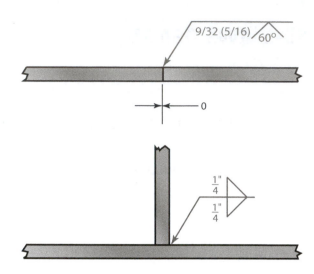

PREPARATION OF BASE METAL:

The 30° bevel angles are to be flame or plasma cut on the edges of the plate before the parts are assembled. The beveled surface must be smooth and free of notches. Any roughness or notches deeper than 1/64 in. (0.4 mm) must be ground smooth.

All hydrocarbons and other contaminants, such as cutting fluids, grease, oil, and primers, must be cleaned off all parts and filler metals before welding. This cleaning can be done with any suitable solvents or detergents. The groove face and inside and outside plate surface within 1 in. (25 mm) of the joint must be mechanically cleaned of slag, rust, and mill scale. Cleaning must be done with a wire brush or grinder down to bright metal.

ELECTRICAL CHARACTERISTICS:

Set the voltage, amperage, wire-feed speed, and shielding gas flow according to Table 15-2.

PREHEAT:

The parts must be heated to a temperature higher than 50°F (10°C) before any welding is started.

BACKING GAS:

N/A.

SAFETY:

Proper protective clothing and equipment must be used. The area must be free of all hazards that may affect the welder or others in the area. The welding machine, welding leads, work clamp, electrode holder, and other equipment must be in safe working order.

WELDING TECHNIQUE:

Using a 3/4-in. (19-mm) or larger gas nozzle for all welding, first tack weld the plates together, Figure 15-14. Use the E70S-X arc welding electrodes to make the welds.

Using the E70S-X arc welding electrodes, make a series of stringer filler welds, no thicker than 1/4 in. (6.4 mm) in the groove until the joint is filled.

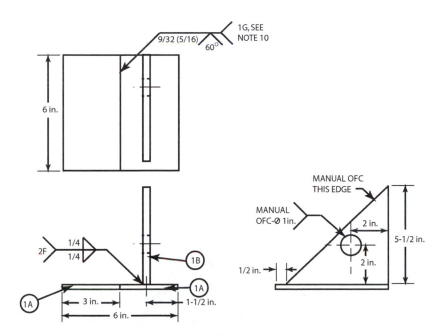

INTERPASS TEMPERATURE:

The plate should not be heated to a temperature higher than 350°F (175°C) during the welding process. After each weld pass is completed, allow it to cool but never to a temperature below 50°F (10°C). The weldment must not be quenched in water.

CLEANING:

Any slag must be cleaned off between passes. The weld beads may be cleaned by a hand wire brush, a chipping hammer, a punch, and hammer, or a needle scaler. All weld cleaning must be performed with the test plate in the welding position.

VISUAL INSPECTION:

Visually inspect the weld for uniformity and discontinuities.

1. There shall be no cracks no incomplete fusion.

2. There shall be no incomplete joint penetration in groove welds except as permitted for partial joint penetration welds.

3. The Test Supervisor shall examine the weld for acceptable appearance and shall be satisfied that the welder is skilled in using the process and procedure specified for the test.

4. Undercut shall not exceed the lesser of 10% of the base metal thickness or 1/32 in. (0.8 mm).

5. Where visual examination is the only criterion for acceptance, all weld passes are subject to visual examination at the discretion of the Test Supervisor.

6. The frequency of porosity shall not exceed one in each 4 in. (100 mm) of weld length and the maximum diameter shall not exceed 3/32 in. (2.4 mm).

7. Welds shall be free from overlap.

SKETCHES:

Gas Metal Arc Welding Spray Transfer (GMAW)

Workmanship Sample drawing, Figure 15-15.

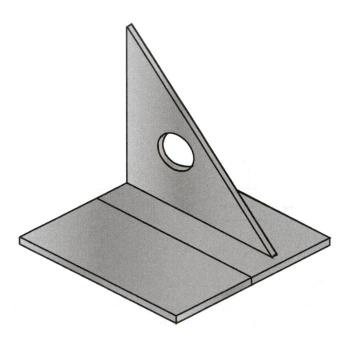

INSTRUCTOR'S COMMENTS _____

_____

_____

■ PRACTICE 15-6 AWS SENSE

Name _____ Date _____

Class _____ Instructor _____ Grade _____

GAS METAL ARC WELDING (GMAW) SPRAY TRANSFER WORKMANSHIP SAMPLE

Welding Procedure Specification (WPS) No.: Practice 15-6.

TITLE:

Welding GMAW of carbon steel plate to carbon steel plate.

SCOPE:

This procedure is applicable for V-groove weld on 1 in. (25 mm) thick plate with backing for the AWS SENSE Level II flat groove weld using spray transfer GMA welding.

WELDING MAY BE PERFORMED IN THE FOLLOWING POSITIONS:

1G and 2F.

BASE METAL:

The base metal shall conform to carbon steel M-1, P-1, and S-1, Group 1 or 2.

BACKING MATERIAL SPECIFICATION:

The backing metal strip shall conform to carbon steel M-1, P-1, and S-1, Group 1 or 2.

FILLER METAL:

The filler metal shall conform to ER70S-3, 0.035-in. (0.90-mm) to 0.045-in. (1.2-mm) diameter E70S-3 from AWS specification A5.18. This filler metal falls into F-number F-6 and A-number A-1.

SHIELDING GAS:

The shielding gas, or gases, shall conform to the following compositions and purity: Ar + 2% to 5% O_2 at 20 to 40 cfh.

JOINT DESIGN AND TOLERANCES:

Refer to the drawing in Figure 15-16 for the joint layout specifications.

Joint Details:

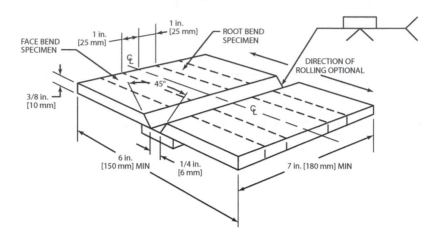

PREPARATION OF BASE METAL:

The 22 1/2° to 30° bevel angles are to be flame or plasma cut on the edges of the plate before the parts are assembled. The beveled surface must be smooth and free of notches. Any roughness or notches deeper than 1/64 in. (0.4 mm) must be ground smooth.

All hydrocarbons and other contaminants, such as cutting fluids, grease, oil, and primers, must be cleaned off all parts and filler metals before welding. This cleaning can be done with any suitable solvents or detergents. The groove face and inside and outside plate surface within 1 in. (25 mm) of the joint must be mechanically cleaned of slag, rust, and mill scale. Cleaning must be done with a wire brush or grinder down to bright metal.

ELECTRICAL CHARACTERISTICS:

Set the voltage, amperage, wire-feed speed, and shielding gas flow according to Table 15-2.

PREHEAT:

The parts must be heated to a temperature higher than 50°F (10°C) before any welding is started.

BACKING GAS:

N/A.

SAFETY:

Proper protective clothing and equipment must be used. The area must be free of all hazards that may affect the welder or others in the area. The welding machine, welding leads, work clamp, electrode holder, and other equipment must be in safe working order.

WELDING TECHNIQUE:

Using a 3/4-in. (19-mm) or larger gas nozzle for all welding, first tack weld the plates together. Use the E70S-X arc welding electrodes to make the welds.

Using the E70S-3 arc welding electrodes, make a series of stringer filler welds, no thicker than 1/4 in. (6.4 mm) in the groove until the joint is filled.

INTERPASS TEMPERATURE:

The plate should not be heated to a temperature higher than 350°F (175°C) during the welding process. After each weld pass is completed, allow it to cool but never to a temperature below 50°F (10°C). The weldment must not be quenched in water.

CLEANING:

Any slag must be cleaned off between passes. The weld beads may be cleaned by a hand wire brush, a chipping hammer, a punch, and hammer, or a needle scaler. All weld cleaning must be performed with the test plate in the welding position.

VISUAL INSPECTION:

There shall be no cracks and no incomplete fusion. There shall be no incomplete joint penetration in groove welds except as permitted for partial joint penetration welds.

- The Test Supervisor shall examine the weld for acceptable appearance and shall be satisfied that the welder is skilled in using the process and procedure specified for the test.
- Undercut shall not exceed the lesser of 10% of the base metal thickness or 1/32 in. (0.8 mm).
- Where visual examination is the only criterion for acceptance, all weld passes are subject to visual examination at the discretion of the Test Supervisor.
- The frequency of porosity shall not exceed one in each 4 in. (100 mm) of weld length, and the maximum diameter shall not exceed 3/32 in. (2.4 mm).
- Welds shall be free from overlap.

BEND TEST:

The weld is to be mechanically tested only after it has passed the visual inspection. Be sure that the test specimens are properly marked to identify the welder, the position, and the process.

SPECIMEN PREPARATION:

For 1/2-in. (13-mm) test plates, two side bend specimens are to be located in accordance with the requirements in Figure 15-15.

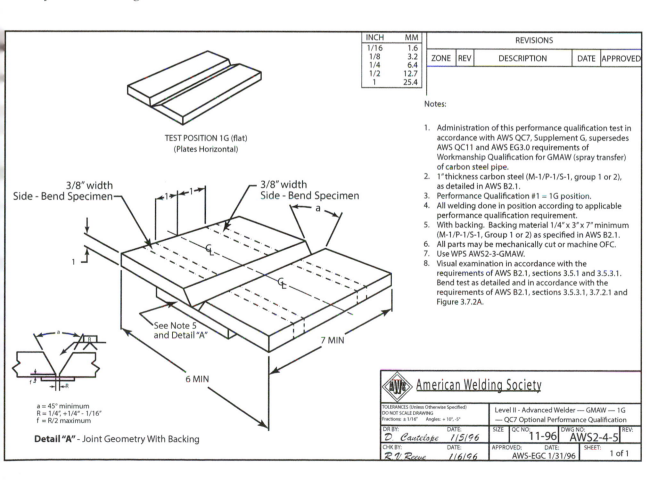

INCH	MM
1/16	1.6
1/8	3.2
1/4	6.4
1/2	12.7
1	25.4

REVISIONS

ZONE	REV	DESCRIPTION	DATE	APPROVED

TEST POSITION 1G (flat)
(Plates Horizontal)

3/8" width
Side - Bend Specimen

3/8" width
Side - Bend Specimen

See Note 5
and Detail "A"

7 MIN

6 MIN

Notes:

1. Administration of this performance qualification test in accordance with AWS QC7, Supplement G, supersedes AWS QC11 and AWS EG3.0 requirements of Workmanship Qualification for GMAW (spray transfer) of carbon steel pipe.
2. 1" thickness carbon steel (M-1/P-1/S-1, group 1 or 2), as detailed in AWS B2.1.
3. Performance Qualification #1 = 1G position.
4. All welding done in position according to applicable performance qualification requirement.
5. With backing. Backing material 1/4" x 3" x 7" minimum (M-1/P-1/S-1, Group 1 or 2) as specified in AWS B2.1.
6. All parts may be mechanically cut or machine OFC.
7. Use WPS AWS2-3-GMAW.
8. Visual examination in accordance with the requirements of AWS B2.1, sections 3.5.1 and 3.5.3.1. Bend test as detailed in and in accordance with the requirements of AWS B2.1, sections 3.5.3.1, 3.7.2.1 and Figure 3.7.2A.

a = 45° minimum
R = 1/4", +1/4" - 1/16"
f = R/2 maximum

Detail "A" - Joint Geometry With Backing

American Welding Society

TOLERANCES (Unless Otherwise Specified)
DO NOT SCALE DRAWING
Fractions: ±1/16" Angles: + 10°, -5°

Level II - Advanced Welder — GMAW — 1G
— QC7 Optional Performance Qualification

DR BY:	DATE:	SIZE	QC NO:	DWG NO:	REV:
D. Cantelope	1/5/96		11-96	AWS2-4-5	
CHK BY:	DATE:	APPROVED:	DATE:	SHEET:	
R V Reeve	1/6/96		AWS-EGC 1/31/96	1 of 1	

- *Transverse side bend.* The weld is perpendicular to the longitudinal axis of the specimen and is bent so that the side of the weld face becomes the tension surface of the specimen.

ACCEPTANCE CRITERIA FOR BEND TEST:

For acceptance, the convex surface of the side of the weld bend specimen shall meet both of the following requirements:

- No single indication shall exceed 1/8 in. (3.2 mm) measured in any direction on the surface.
- The sum of the greatest dimensions of all indications on the surface that exceed 1/32 in. (0.8 mm) but are less than or equal to 1/8 in. (3.2 mm) shall not exceed 3/8 in. (9.5 mm).

SKETCHES:

Gas Metal Arc Welding Spray Transfer (GMAW) Workmanship Sample drawing.

INSTRUCTOR'S COMMENTS _____

_____

_____

■ PRACTICE 15-7

Name _____ Date _____

Class _____ Instructor _____ Grade _____

OBJECTIVE: After completing this practice, you will make a V-groove welds in the flat and vertical position using the FCAW-S and FCAW-G techniques.

EQUIPMENT AND MATERIALS NEEDED FOR THIS PRACTICE

1. Semiautomatic welding power system approved for GMA.

2. Assorted hand tools, spare parts, and any other required materials.

3. Proper PPE.

4. E70T-1 and/or E70T-11 electrodes with 0.035-in. and/or 0.045-in. (0.9-mm and/or 1.2-mm) diameter.

5. Ar + 2% to 5% O_2 shielding gas.

6. Two or more pieces of 6 in. long 3 in. wide and 3/8 thick (152 mm × 76 mm × 9.5 mm) steel plate.

INSTRUCTIONS

1. The 30° bevel angles are to be flame or plasma cut on the edges of the plate before the parts are assembled.

2. Grind a 1/16-in. (1.6-mm) to 1/8-in. (3-mm) root face on the beveled edge.

3. The groove face and inside and outside plate surface within 1 in. (25 mm) of the joint must be mechanically cleaned of slag, rust, and mill scale. Cleaning must be done with a wire brush or grinder down to bright metal.

4. Set the voltage, amperage, wire-feed speed, and shielding gas flow according to Table 15-3.

5. Tack weld the plates with a root gap between the plates that measures approximately 1/16 in. (1.6 mm).

6. Place the plates in position at a comfortable height and location. Be sure you have complete and free movement along the full length of the weld joint.

7. Using 20° to 30° backhand technique start the weld at the very end of the groove. Allow the weld pool to expand to fill the full width of the groove wide before starting to travel along the groove.

8. A slight weave pattern will help you control the molten weld pool in the 3G position.

9. When the weld reaches the opposed end, pause long enough for the weld crater to fill up, so the end of the weld is all the way at the edge of the plates.

10. Clean off all of the slag with a wire brush, chipping hammer, or punch.

11. Make sure not to make the weld any wider than 1/8 in. (3 mm) than the groove and no more than 1/8 in. (3 mm) of reinforcement.

INSTRUCTOR'S COMMENTS _____

_____

_____

■ PRACTICE 15-8

Name _____ Date _____

Class _____ Instructor _____ Grade _____

OBJECTIVE: After completing this practice, you will make a beveled grooved butt joint in the horizontal position using both the FCAW-G and FCAW-S processes.

EQUIPMENT AND MATERIALS NEEDED FOR THIS PRACTICE

1. Semiautomatic welding power system approved for GMA.

2. Assorted hand tools, spare parts, and any other required materials.

3. Proper PPE.

4. E70T-1 and/or E70T-11 electrodes with 0.035-in. and/or 0.045-in. (0.9-mm and/or 1.2-mm) diameter.

5. Ar + 2% to 5% O_2 shielding gas.

6. Two or more pieces of 6 in. long 3 in. wide and 3/8 thick (152 mm × 76 mm × 9.5 mm) steel plate.

INSTRUCTIONS

1. The plate surfaces within 1 in. (25 mm) of the joint must be mechanically cleaned of slag, rust, and mill scale. Cleaning must be done with a wire brush or grinder down to bright metal.

2. Set the voltage, amperage, wire-feed speed, and shielding gas flow according to Table 15-3 for FCAW-G and Table 15-4 for FCAW-S.

3. Tack weld the plates with a root gap between the plates that measures approximately 1/16 in. (1.6 mm).

4. Place the plates in position at a comfortable height and location. Be sure you have complete and free movement along the full length of the weld joint.

5. Using 20° to 30° backhand stringer bead technique, start the weld at the very end of the groove. Allow the root weld pool to expand to approximately ¼ in. (6 mm) in width before starting to travel along the groove.

6. When the weld reaches the opposite end, pause long enough for the weld crater to fill up, so the end of the weld is all the way at the edge of the plates.

7. Clean off all of the slag with a wire brush, chipping hammer, or punch.

8. Use the same backhand technique but with a J weave pattern for the second weld. Place this weld bead on the bottom plate along its outside edge extending 2/3 of the way up the beveled surface.

9. Clean the weld.

10. The cover pass is made with the same backhand technique with a J weave pattern and extends from the top edge of the bevel to almost the bottom edge.

INSTRUCTOR'S COMMENTS _____

_____

_____

■ PRACTICE 15-9

Name _____   Date _____

Class _____   Instructor _____   Grade _____

OBJECTIVE: After completing this practice, you will make a beveled grooved butt joint in the overhead position using both the FCAW-G and FCAW-S processes.

EQUIPMENT AND MATERIALS NEEDED FOR THIS PRACTICE

1. Semiautomatic welding power system approved for GMA.

2. Assorted hand tools, spare parts, and any other required materials.

3. Proper PPE.

4. E70T-1 and/or E70T-11 electrodes with 0.035-in. and/or 0.045-in. (0.9-mm and/or 1.2-mm) diameter.

5. Ar + 2% to 5% O_2 shielding gas.

6. Two or more pieces of 6 in. long 3 in. wide and 3/8 thick (152 mm × 76 mm × 9.5 mm) steel plate.

INSTRUCTIONS

1. The plate surfaces within 1 in. (25 mm) of the joint must be mechanically cleaned of slag, rust, and mill scale. Cleaning must be done with a wire brush or grinder down to bright metal.

2. Set the voltage, amperage, wire-feed speed, and shielding gas flow according to Table 15-3 for FCAW-G and Table 15-4 for FCAW-S.

3. Tack weld the plates with a root gap between the plates that measures approximately 1/16 in. (1.6 mm).

4. Place the plates in position at a comfortable height and location. Be sure you have complete and free movement along the full length of the weld joint.

5. Using 10° to 15° backhand stringer bead technique, start the weld at the very end of the groove. Allow the root weld pool to expand to approximately ¼ in. (6 mm) in width before starting to travel along the groove.

6. When the weld reaches the opposite end, pause long enough for the weld crater to fill up, so the end of the weld is all the way at the edge of the plates.

7. Clean off all of the slag with a wire brush, chipping hammer, or punch.

8. Using 15° to 25° backhand technique with a zigzag or semi-circle for the second weld, placing this weld bead on the top of the root pass.

9. Clean the weld.

10. The cover pass is made with a 15° to 25° backhand technique with a zigzag or semi-circle for the second weld, placing this weld bead on the top of the filler pass. This weld can be made in two passes if you have trouble controlling the larger molten weld pool in the overhead position.

INSTRUCTOR'S COMMENTS _____

_____

_____

■ PRACTICE 15-10

Name _____   Date _____

Class _____   Instructor _____   Grade _____

OBJECTIVE: After completing this practice, you will make fillet welded lap joints and tee joints in all four positions using two or more pieces of mild steel plate welds with both the FCAW-G and FCAW-S processes.

EQUIPMENT AND MATERIALS NEEDED FOR THIS PRACTICE

1. Semiautomatic welding power system approved for GMA.

2. Assorted hand tools, spare parts, and any other required materials.

3. Proper PPE.

4. E70T-1 and/or E70T-11 electrodes with 0.035-in. and/or 0.045-in. (0.9-mm and/or 1.2-mm) diameter.

5. Ar + 2% to 5% O_2 shielding gas.

6. Two or more pieces of 6 in. long 3 in. wide and 3/8 thick (152 mm × 76 mm × 9.5 mm) steel plate.

INSTRUCTIONS

1. One at a time, tack weld the sheets together to form the joints and place them in the desired position on a welding stand.

2. Starting at one end, run a 1/4 in. (6 mm) equal leg fillet weld along the joint. Watch the molten weld pool and bead for signs it is increasing or decreasing in size. Make any necessary changes in your technique to keep the weld consistent.

3. Check for complete root fusion on the lap and tee joints and 100% penetration on the butt joint.

INSTRUCTOR'S COMMENTS _____

_____

_____

■ PRACTICE 15-11

Name _____   Date _____

Class _____   Instructor _____   Grade _____

AWS SENSE ENTRY-LEVEL WELDER WORKMANSHIP SAMPLE FOR FLUX CORED ARC WELDING, GAS-SHIELDED (FCAW-G)

Welding Procedure Specification (WPS) No.: Practice 15-11.

TITLE:

Welding FCAW of plate to plate.

SCOPE:

This procedure is applicable for V-groove, bevel, and fillet welds within the range of 1/8 in. (3.2 mm) through 1-1/2 in. (38 mm) for the AWS SENSE Level I all-position groove and fillet welds using the FCAW-G welding process.

WELDING MAY BE PERFORMED IN THE FOLLOWING POSITIONS:

All.

BASE METAL:

The base metal shall conform to carbon steel M-1, P-1, and S-1, Group 1 or 2.

BACKING MATERIAL SPECIFICATION:

None.

FILLER METAL:

The filler metal shall conform to AWS specification no. E71T-1 from AWS specification A5.20. This filler metal falls into F-number F-6 and A-number A-1.

SHIELDING GAS:

The shielding gas, or gases, shall conform to the following compositions and purity: CO_2 at 20 to 40 cfh.

JOINT DESIGN AND TOLERANCES:

Refer to the drawing and specifications in Figure 15-23 for the workmanship sample layout.

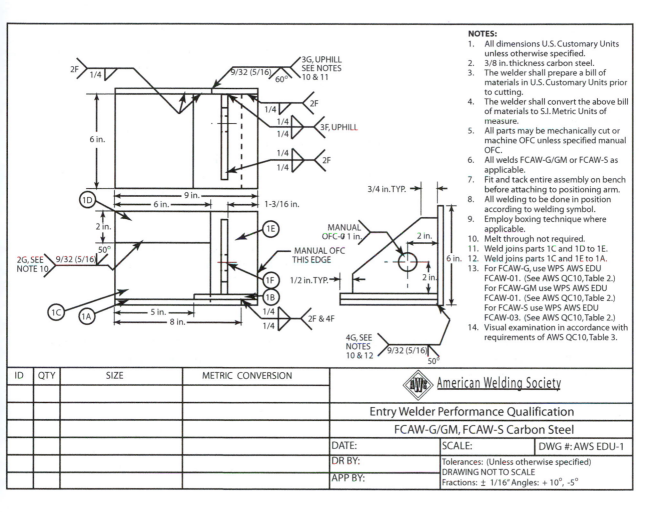

NOTES:
1. All dimensions U.S. Customary Units unless otherwise specified.
2. 3/8 in. thickness carbon steel.
3. The welder shall prepare a bill of materials in U.S. Customary Units prior to cutting.
4. The welder shall convert the above bill of materials to S.I. Metric Units of measure.
5. All parts may be mechanically cut or machine OFC unless specified manual OFC.
6. All welds FCAW-G/GM or FCAW-S as applicable.
7. Fit and tack entire assembly on bench before attaching to positioning arm.
8. All welding to be done in position according to welding symbol.
9. Employ boxing technique where applicable.
10. Melt through not required.
11. Weld joins parts 1C and 1D to 1E.
12. Weld joins parts 1C and 1E to 1A.
13. For FCAW-G, use WPS AWS EDU FCAW-01. (See AWS QC10, Table 2.) For FCAW-GM use WPS AWS EDU FCAW-01. (See AWS QC10, Table 2.) For FCAW-S use WPS AWS EDU FCAW-03. (See AWS QC10, Table 2.)
14. Visual examination in accordance with requirements of AWS QC10, Table 3.

ID	QTY	SIZE	METRIC CONVERSION	American Welding Society		
				Entry Welder Performance Qualification		
				FCAW-G/GM, FCAW-S Carbon Steel		
				DATE:	SCALE:	DWG #: AWS EDU-1
				DR BY:	Tolerances: (Unless otherwise specified) DRAWING NOT TO SCALE	
				APP BY:	Fractions: ± 1/16" Angles: + 10°, -5°	

PREPARATION OF BASE METAL:

The bevels are to be flame or plasma cut on the edges of the plate before the parts are assembled. The beveled surface must be smooth and free of notches. Any roughness or notches deeper than 1/64 in. (0.4 mm) must be ground smooth.

All hydrocarbons and other contaminants, such as cutting fluids, grease, oil, and primers, must be cleaned off all parts and filler metals before welding. This cleaning can be done with any suitable solvents or detergents. The groove face and inside and outside plate surface within 1 in. (25 mm) of the joint must be mechanically cleaned of slag, rust, and mill scale. Cleaning must be done with a wire brush or grinder down to bright metal.

ELECTRICAL CHARACTERISTICS:

Set the voltage, amperage, wire-feed speed, and shielding gas flow according to Table 15-3.

PREHEAT:

The parts must be heated to a temperature higher than 50°F (10°C) before any welding is started.

BACKING GAS:

N/A.

SAFETY:

Proper protective clothing and equipment must be used. The area must be free of all hazards that may affect the welder or others in the area. The welding machine, welding leads, work clamp, electrode holder, and other equipment must be in safe working order.

WELDING TECHNIQUE:

Using a 1/2-in. (13-mm) or larger gas nozzle and a distance from contact tube to work of approximately 3/4 in. (19 mm) for all welding, first tack weld the plates together according to Figure 15-22. There should be a root gap of about 1/8 in. (3.2 mm) between the plates with V-grooved or beveled edges. Use an E71T-1 arc welding electrode to make a weld. If multiple pass welds are going to be made, a root pass weld should be made to fuse the plates together. Clean the slag from the root pass, being sure to remove any trapped slag along the sides of the weld.

Using an E71T-1 arc welding electrode, make a series of stringer or weave filler welds, no thicker than 1/4 in. (6.4 mm) in the groove until the joint is filled. The 1/4-in. (6.4-mm) fillet welds are to be made with one pass.

INTERPASS TEMPERATURE:

The plate should not be heated to a temperature higher than 350°F (175°C) during the welding process. After each weld pass is completed, allow it to cool but never to a temperature below 50°F (10°C). The weldment must not be quenched in water.

CLEANING:

The slag must be cleaned off between passes. The weld beads may be cleaned by a hand wire brush, a chipping hammer, a punch, and hammer, or a needle scaler. All weld cleaning must be performed with the test plate in the welding position. A grinder may not be used to remove weld control problems such as undercut, overlap, or trapped slag.

INSPECTION:

Visually inspect the weld for uniformity and discontinuities. There shall be no cracks, no incomplete fusion, and no overlap. Undercut shall not exceed the lesser of 10% of the base metal thickness or 1/32 in. (0.8 mm). The frequency of porosity shall not exceed one in each 4 in. (100 mm) of weld length, and the maximum diameter shall not exceed 3/32 in. (2.4 mm).

SKETCHES:

Flux Cored Arc Welding (FCAW) Gas Shielded Workmanship Sample drawing.

INSTRUCTOR'S COMMENTS _____

_____

_____

■ PRACTICE 15-12

Name _____   Date _____

Class _____   Instructor _____   Grade _____

AWS SENSE ENTRY-LEVEL WELDER WORKMANSHIP SAMPLE FOR FLUX CORED ARC WELDING SELF-SHIELDED (FCAW-S)

Welding Procedure Specification (WPS) No.: Practice 15-12.

TITLE:

Welding FCAW of plate to plate.

SCOPE:

This procedure is applicable for V-groove, bevel, and fillet welds within the range of 1/8 in. (3.2 mm) through 1-1/2 in. (38 mm) for the AWS SENSE Level I all position groove and fillet welds using the FCAW-S welding process.

WELDING MAY BE PERFORMED IN THE FOLLOWING POSITIONS:

All.

BASE METAL:

The base metal shall conform to carbon steel M-1, P-1, and S-1, Group 1 or 2.

BACKING MATERIAL SPECIFICATION:

None.

FILLER METAL:

The filler metal shall conform to ER70S-3, 0.035 in. (0.90 mm) to 0.045 in. (1.2 mm) diameter E71T-11 from AWS specification A5.20. This filler metal falls into F-number F-6 and A-number A-1.

SHIELDING GAS:

None.

JOINT DESIGN AND TOLERANCES:

Refer to the drawing and specifications in Figure 15-23 for the workmanship sample layout.

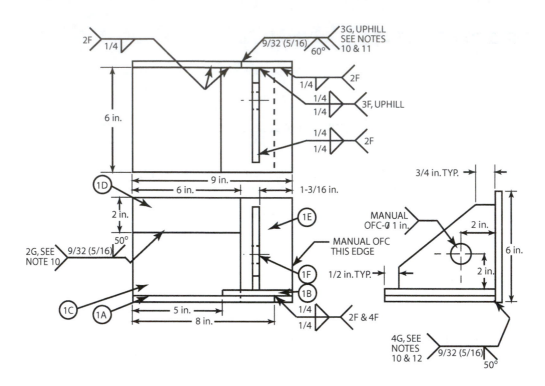

PREPARATION OF BASE METAL:

The bevels are to be flame or plasma cut on the edges of the plate before the parts are assembled. The beveled surface must be smooth and free of notches. Any roughness or notches deeper than 1/64 in. (0.4 mm) must be ground smooth.

All hydrocarbons and other contaminants, such as cutting fluids, grease, oil, and primers, must be cleaned off all parts and filler metals before welding. This cleaning can be done with any suitable solvents or detergents. The groove face and inside and outside plate surface within 1 in. (25 mm) of the joint must be mechanically cleaned of slag, rust, and mill scale. Cleaning must be done with a wire brush or grinder down to bright metal.

ELECTRICAL CHARACTERISTICS:

Set the voltage, amperage, and wire-feed speed flow according to Table 15-4.

PREHEAT:

The parts must be heated to a temperature higher than 50°F (10°C) before any welding is started.

BACKING GAS:

N/A.

SAFETY:

Proper protective clothing and equipment must be used. The area must be free of all hazards that may affect the welder or others in the area. The welding machine, welding leads, work clamp, electrode holder, and other equipment must be in safe working order.

WELDING TECHNIQUE:

Using a 1/2-in. (13-mm) or larger gas nozzle and a distance from contact tube to work of approximately 3/4 in. (19 mm) for all welding, first tack weld the plates together according to Figure 15-23. There should be about a root gap of about 1/8 in. (3.2 mm) between the plates with V-grooved or beveled edges. Use an E71T-11 arc welding electrode to make a weld. If multiple pass welds are going to be made, a root pass weld should be made to fuse the plates together. Clean the slag from the root pass, being sure to remove any trapped slag along the sides of the weld.

Using an E71T-11 arc welding electrode, make a series of stringer or weave filler welds, no thicker than 1/4 in. (6.4 mm) in the groove until the joint is filled. The 1/4-in. (6.4-mm) fillet welds are to be made with one pass.

INTERPASS TEMPERATURE:

The plate should not be heated to a temperature higher than 350°F (175°C) during the welding process. After each weld pass is completed, allow it to cool but never to a temperature below 50°F (10°C). The weldment must not be quenched in water.

CLEANING:

The slag must be cleaned off between passes. The weld beads may be cleaned by a hand wire brush, a chipping hammer, a punch, and hammer, or a needle scaler. All weld cleaning must be performed with the test plate in the welding position. A grinder may not be used to remove weld control problems such as undercut, overlap, or trapped slag.

VISUAL INSPECTION CRITERIA FOR ENTRY WELDERS:

There shall be no cracks and no incomplete fusion. There shall be no incomplete joint penetration in groove welds except as permitted for partial joint penetration welds.

- The Test Supervisor shall examine the weld for acceptable appearance and shall be satisfied that the welder is skilled in using the process and procedure specified for the test.
- Undercut shall not exceed the lesser of 10% of the base metal thickness or 1/32 in. (0.8 mm).
- Where visual examination is the only criterion for acceptance, all weld passes are subject to visual examination at the discretion of the Test Supervisor.
- The frequency of porosity shall not exceed one in each 4 in. (100 mm) of weld length, and the maximum diameter shall not exceed 3/32 in. (2.4 mm).
- Welds shall be free from overlap.

SKETCHES:

Flux Cored Arc Welding (FCAW) Self-Shielded Workmanship Sample drawing.

INSTRUCTOR'S COMMENTS _____

_____

_____

■ PRACTICE 15-13

Name _____ Date _____

Class _____ Instructor _____ Grade _____

GAS METAL ARC WELDING—SHORT-CIRCUIT METAL TRANSFER (GMAW-S) WORKMANSHIP SAMPLE

Welding Procedure Specification (WPS) No.: Practice 15-13.

TITLE:

Welding GMAW-S of carbon steel plate to carbon steel plate.

SCOPE:

This procedure is applicable for square 2G, 5G, and 6G groove on 4 in. to 8 in. (100 mm to 150 mm) schedule 40 carbon steel pipe with and without backing for the AWS SENSE Level II short-circuit metal transfer GMA pipe welding.

WELDING MAY BE PERFORMED IN THE FOLLOWING POSITIONS:

2G, 5G, and 6G

BASE METAL:

The base metal shall conform to carbon steel M-1, P-1, and S-1 Group 1 or 2.

BACKING MATERIAL SPECIFICATION:

Backing metal shall conform to carbon steel M-1, P-1, and S-1 Group 1 or 2.

FILLER METAL:

The filler metal shall conform to AWS specification no. E70S-X from AWS specification A5.18. This filler metal falls into F-number F-6 and A-number A-1.

SHIELDING GAS:

The shielding gas, or gases, shall conform to the following compositions and purity: 25% CO_2 + 75% Ar at 20 to 40 cfh or Ar + 2% to 5% O_2 at 20 to 40 cfh.

JOINT DESIGN AND TOLERANCES:

Refer to the drawing in Figure 15-25 for the joint layout specifications.

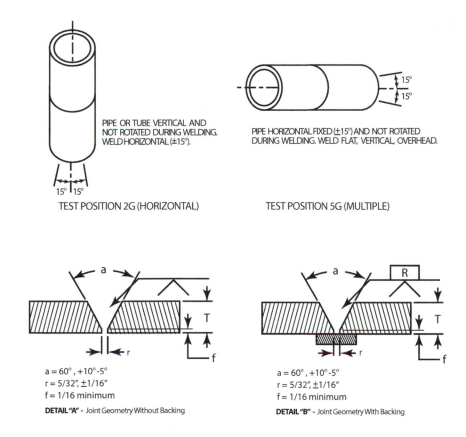

PIPE OR TUBE VERTICAL AND
NOT ROTATED DURING WELDING.
WELD HORIZONTAL (±15°).

PIPE HORIZONTAL FIXED (±15°) AND NOT ROTATED
DURING WELDING. WELD FLAT, VERTICAL, OVERHEAD.

TEST POSITION 2G (HORIZONTAL)

TEST POSITION 5G (MULTIPLE)

a = 60°, +10°-5°
r = 5/32", ±1/16"
f = 1/16 minimum
DETAIL "A" - Joint Geometry Without Backing

a = 60°, +10°-5°
r = 5/32", ±1/16"
f = 1/16 minimum
DETAIL "B" - Joint Geometry With Backing

PREPARATION OF BASE METAL:

All parts may be mechanically cut or machine PAC unless specified as manual PAC.

All hydrocarbons and other contaminants, such as cutting fluids, grease, oil, and primers, must be cleaned off all parts and filler metals before welding. This cleaning can be done with any suitable solvents or detergents. The groove face and inside and outside plate surface within 1 in. (25 mm) of the joint must be mechanically cleaned of slag, rust, and mill scale. Cleaning must be done with a wire brush or grinder down to bright metal.

ELECTRICAL CHARACTERISTICS:

Set the voltage, amperage, wire-feed speed, and shielding gas flow according to Table 15-1.

PREHEAT:

The parts must be heated to a temperature higher than 50°F (10°C) before any welding is started.

BACKING GAS:

N/A.

SAFETY:

Proper protective clothing and equipment must be used. The area must be free of all hazards that may affect the welder or others in the area. The welding machine, welding leads, work clamp, electrode holder, and other equipment must be in safe working order.

WELDING TECHNIQUE:

Using a 1/2-in. (13-mm) or larger gas nozzle for all welding, first tack weld the plates together according to the drawing, Figure 15-25. Make four tack welds of approximately 1 in. (25 mm) long. Grind the ends of the tack welds to a featheredge with a narrow grinding disk.

Use the E70S-X filler metal to fuse the plates together. Clean any silicon slag, being sure to remove any trapped silicon slag along the sides of the weld.

All welds are to be made with one pass. All welds must be placed in the orientation shown in the drawing.

INTERPASS TEMPERATURE:

The plate should not be heated to a temperature higher than 350°F (175°C) during the welding process. After each weld pass is completed, allow it to cool but never to a temperature below 50°F (10°C). The weldment must not be quenched in water.

CLEANING:

The weld beads may be cleaned by a hand wire brush, a chipping hammer, a punch, hammer, and/or a narrow grinding disk. All weld cleaning must be performed with the test plate in the welding position.

VISUAL INSPECTION:

There shall be no cracks and no incomplete fusion. There shall be no incomplete joint penetration in groove welds except as permitted for partial joint penetration welds.

- The Test Supervisor shall examine the weld for acceptable appearance and shall be satisfied that the welder is skilled in using the process and procedure specified for the test.

- Undercut shall not exceed the lesser of 10% of the base metal thickness or 1/32 in. (0.8 mm).

- Where visual examination is the only criterion for acceptance, all weld passes are subject to visual examination at the discretion of the Test Supervisor.

- The frequency of porosity shall not exceed one in each 4 in. (100 mm) of weld length, and the maximum diameter shall not exceed 3/32 in. (2.4 mm).

- Welds shall be free from overlap.

BEND TEST:

The weld is to be mechanically tested only after it has passed the visual inspection. Be sure that the test specimens are properly marked to identify the welder, the position, and the process.

SPECIMEN PREPARATION:

Four specimens are to be located in accordance with the requirements in Figure 15-26. Two are to be prepared for a transverse face bend, and the other two are to be prepared for a transverse root bend, Figure 15-27.

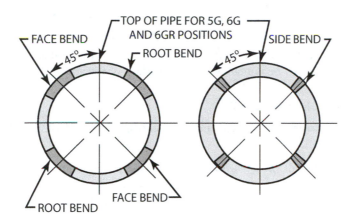

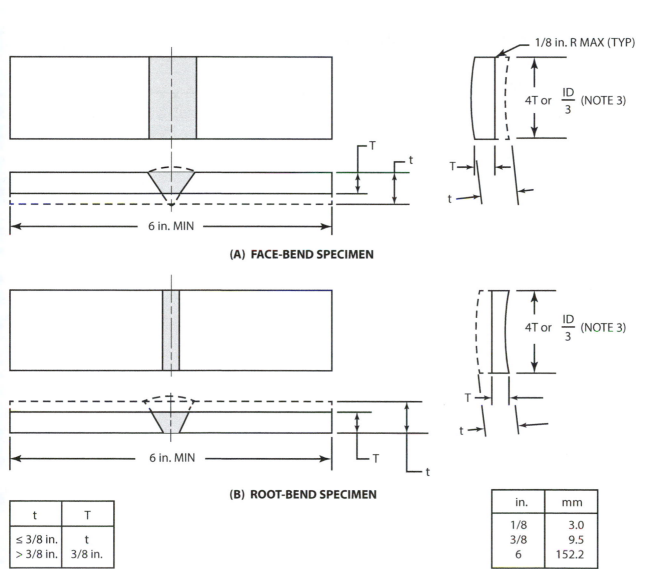

(A) FACE-BEND SPECIMEN

(B) ROOT-BEND SPECIMEN

t	T
≤ 3/8 in.	t
> 3/8 in.	3/8 in.

in.	mm
1/8	3.0
3/8	9.5
6	152.2

- Transverse face bend. The weld is perpendicular to the longitudinal axis of the specimen and is bent so that the weld face becomes the tension surface of the specimen.
- Transverse root bend. The weld is perpendicular to the longitudinal axis of the specimen and is bent so that the weld root becomes the tension surface of the specimen.

ACCEPTANCE CRITERIA FOR BEND TEST:

For acceptance, the convex surface of the face and root bend specimens shall meet both of the following requirements:

- No single indication shall exceed 1/8 in. (3.2 mm) measured in any direction on the surface.
- The sum of the greatest dimensions of all indications on the surface that exceed 1/32 in. (0.8 mm) but are less than or equal to 1/8 in. (3.2 mm) shall not exceed 3/8 in. (9.5 mm).

SKETCHES:

GMAW Short-Circuit Metal Transfer Workmanship Sample drawing.

INSTRUCTOR'S COMMENTS _____

_____

_____

■ PRACTICE 15-14

Name _____   Date _____

Class _____   Instructor _____   Grade _____

AWS SENSE ENTRY-LEVEL WELDER WORKMANSHIP SAMPLE FOR FLUX CORED ARC WELDING, GAS-SHIELDED (FCAW-G)

Welding Procedure Specification (WPS) No.: Practice 15-14.

TITLE:

Welding FCAW of plate to plate.

SCOPE:

This procedure is applicable for square 2G, 5G, and 6G groove on 4 in. to 8 in. (100 mm to 200 mm) schedule 40 carbon steel pipe with and without backing for the AWS SENSE Level II short-circuit metal transfer FCAW-G pipe welding.

WELDING MAY BE PERFORMED IN THE FOLLOWING POSITIONS:

All.

BASE METAL:

The base metal shall conform to carbon steel M-1, P-1, and S-1, Group 1 or 2.

BACKING MATERIAL SPECIFICATION:

None.

FILLER METAL:

The filler metal shall conform to AWS specification no. E71T-1 from AWS specification A5.20. This filler metal falls into F-number F-6 and A-number A-1.

SHIELDING GAS:

The shielding gas, or gases, shall conform to the following compositions and purity: 25% CO_2 + 75% Ar at 20 to 40 cfh.

JOINT DESIGN AND TOLERANCES:

Refer to the drawing and specifications in Figure 15-29 for the workmanship sample layout.

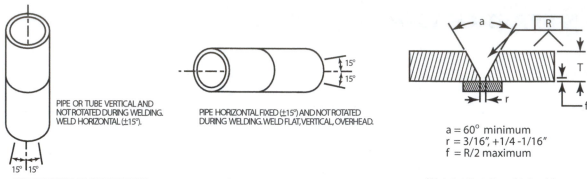

PIPE OR TUBE VERTICAL AND
NOT ROTATED DURING WELDING.
WELD HORIZONTAL (±15°).

15° | 15°

TEST POSITION 2G (HORIZONTAL)

PIPE HORIZONTAL FIXED (±15°) AND NOT ROTATED
DURING WELDING. WELD FLAT, VERTICAL, OVERHEAD.

15°
15°

TEST POSITION 5G (MULTIPLE)

a = 60° minimum
r = 3/16″, +1/4 -1/16″
f = R/2 maximum

(A) Joint Detail - with backing

PREPARATION OF BASE METAL:

The bevels are to be flame or plasma cut on the edges of the plate before the parts are assembled. The beveled surface must be smooth and free of notches. Any roughness or notches deeper than 1/64 in. (0.4 mm) must be ground smooth.

All hydrocarbons and other contaminants, such as cutting fluids, grease, oil, and primers, must be cleaned off all parts and filler metals before welding. This cleaning can be done with any suitable solvents or detergents. The groove face and inside and outside plate surface within 1 in. (25 mm) of the joint must be mechanically cleaned of slag, rust, and mill scale. Cleaning must be done with a wire brush or grinder down to bright metal.

ELECTRICAL CHARACTERISTICS:

Set the voltage, amperage, wire-feed speed, and shielding gas flow according to Table 15-3.

PREHEAT:

The parts must be heated to a temperature higher than 50°F (10°C) before any welding is started.

BACKING GAS:

N/A.

SAFETY:

Proper protective clothing and equipment must be used. The area must be free of all hazards that may affect the welder or others in the area. The welding machine, welding leads, work clamp, electrode holder, and other equipment must be in safe working order.

WELDING TECHNIQUE:

Using a 1/2-in. (13-mm) or larger gas nozzle and a distance from contact tube to work of approximately 3/4 in. (19 mm) for all welding, first tack weld the plates together according to Figure 15-29. There should be a root gap of about 1/8 in. (3.2 mm) between the plates with V-grooved or beveled edges. Use an E71T-1 arc welding electrode to make a weld. If multiple pass welds are going to be made, a root pass weld should be made to fuse the plates together. Clean the slag from the root pass, being sure to remove any trapped slag along the sides of the weld.

Using an E71T-1 arc welding electrode, make a series of stringer or weave filler welds, no thicker than 1/4 in. (6.4 mm) in the groove until the joint is filled. The 1/4-in. (6.4-mm) fillet welds are to be made with one pass.

INTERPASS TEMPERATURE:

The plate should not be heated to a temperature higher than 350°F (175°C) during the welding process. After each weld pass is completed, allow it to cool but never to a temperature below 50°F (10°C). The weldment must not be quenched in water.

CLEANING:

The slag must be cleaned off between passes. The weld beads may be cleaned by a hand wire brush, a chipping hammer, a punch, and hammer, or a needle scaler. All weld cleaning must be performed with the test plate in the welding position. A grinder may not be used to remove weld control problems such as undercut, overlap, or trapped slag.

VISUAL INSPECTION:

There shall be no cracks and no incomplete fusion. There shall be no incomplete joint penetration in groove welds except as permitted for partial joint penetration welds.

- The Test Supervisor shall examine the weld for acceptable appearance and shall be satisfied that the welder is skilled in using the process and procedure specified for the test.
- Undercut shall not exceed the lesser of 10% of the base metal thickness or 1/32 in. (0.8 mm).
- Where visual examination is the only criterion for acceptance, all weld passes are subject to visual examination at the discretion of the Test Supervisor.
- The frequency of porosity shall not exceed one in each 4 in. (100 mm) of weld length, and the maximum diameter shall not exceed 3/32 in. (2.4 mm).
- Welds shall be free from overlap.

BEND TEST:

The weld is to be mechanically tested only after it has passed the visual inspection. Be sure that the test specimens are properly marked to identify the welder, the position, and the process.

SPECIMEN PREPARATION:

Four specimens are to be located in accordance with the requirements in Figure 15-26. Two are to be prepared for a transverse face bend, and the other two are to be prepared for a transverse root bend, Figure 15-27.

- Transverse face bend. The weld is perpendicular to the longitudinal axis of the specimen and is bent so that the weld face becomes the tension surface of the specimen.
- Transverse root bend. The weld is perpendicular to the longitudinal axis of the specimen and is bent so that the weld root becomes the tension surface of the specimen.

ACCEPTANCE CRITERIA FOR BEND TEST:

For acceptance, the convex surface of the face and root bend specimens shall meet both of the following requirements:

- No single indication shall exceed 1/8 in. (3.2 mm) measured in any direction on the surface.
- The sum of the greatest dimensions of all indications on the surface that exceed 1/32 in. (0.8 mm) but are less than or equal to 1/8 in. (3.2 mm) shall not exceed 3/8 in. (9.5 mm).

SKETCHES:

Flux Cored Arc Welding (FCAW) Gas Shielded Workmanship Sample drawing.

INSTRUCTOR'S COMMENTS _____

_____

_____

CHAPTER 15: QUIZ 1

Name _____   Date _____

Class _____   Instructor _____   Grade _____

INSTRUCTIONS

Carefully read Chapter 15 in the textbook and answer each question.

MATCHING

In the space provided to the left of Column A, write the letter from Column B that best answers or completes the statement in Column A.

	Column A	Column B
_____	1. Welders are often required to take a _____ test.	a. cleaned
_____	2. All _____ parts must be machine torch cut with OFC or PAC or mechanically cut.	b. 25% CO_2 + 75% Ar
_____	3. _____ FCA welds will be made with 0.035 or 0.045 diameter E71T-11.	c. workmanship
_____	4. Cut a 1-in. (25-mm) wide strip, then using a vice and a hammer, bend the strip of welded joint to check for 100% _____ on the butt joint.	d. quenched
_____	5. What shielding gas is to be used with Short-Circuit Metal Transfer Workmanship Samples? _____	e. during the welding process
_____	6. All _____ and other contaminants, such as cutting fluids, grease, oil, and primers, must be cleaned off all parts and filler metals before welding.	f. assembled
_____	7. The groove face and inside and outside plate surfaces within 1 in. (25 mm) of the joint must be mechanically _____ of slag, rust, and mill scale.	g. qualification
_____	8. The parts must be heated to a _____ higher than 50°F (10°C) before any welding is started.	h. narrow grinding disk
_____	9. The plate should not be heated to a temperature higher than 350°F (175°C) _____.	i. slag
_____	10. All weld cleaning must be performed with the test plate in the _____.	j. hydrocarbons
_____	11. Visually inspect the weld for uniformity and discontinuities; there shall be no _____ fusion.	k. single indication
_____	12. The 30° bevel angles are to be flame or plasma cut on the edges of the plate before the parts are _____.	l. Ar +2% to 5% O_2

—— 13. Watch the leading edge of the V-groove weld pool for a _____ that indicates that you are getting 100% penetration.

m. Self-shielded

—— 14. What shielding gas is to be used with Gas Metal Arc Spray Transfer Workmanship Samples? _____

n. bright metal

—— 15. The weldment must not be _____ in water.

o. welding position

—— 16. When visually inspecting a weld for uniformity and discontinuities, you shall not find _____ exceeding the lesser of 10% of the base metal thickness or 1/32 in. (0.8 mm).

p. undercut

—— 17. Cleaning must be done with a wire brush or grinder down to _____.

q. penetration

—— 18. Any _____ must be cleaned off between passes.

r. keyhole

—— 19. In a bend test, no _____ shall exceed 1/8 in. (3.2 mm) measured in any direction on the surface.

s. cracks or incomplete

—— 20. The weld beads may be cleaned by a hand wire brush, a chipping hammer, a punch, hammer, and/or a _____.

t. temperature

CHAPTER 15: QUIZ 2

Name _____ Date _____

Class _____ Instructor _____ Grade _____

INSTRUCTIONS

Carefully read Chapter 15 in the textbook and answer each question.

IDENTIFICATION

Identify the numbered items on the drawing by writing the letter next to the identifying term in the space provided.

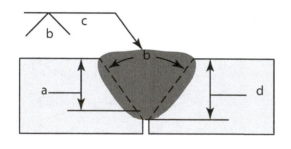

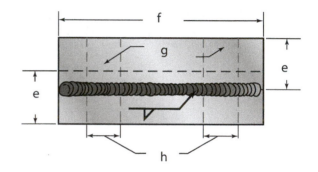

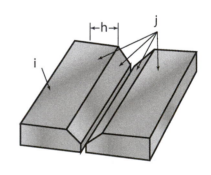

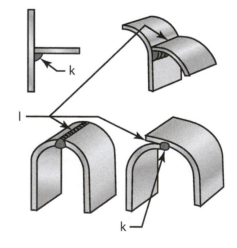

1. _____ WELD

2. _____ 9/32 (5/16)

3. _____ 60°

4. _____ DEPTH OF CHAMFER 9/32"

5. _____ EFFECTIVE THROAT 5/16"

6. _____ CLEAN TO BRIGHT METAL

7. _____ 1" (25 mm)

8. _____ 6" (152 mm)

9. _____ MILL SCALE

10. _____ $1\frac{1}{2}"$ (38 mm)

11. _____ CUT OUT TEST STRIPS

12. _____ LOOK FOR LACK OF FUSION

Gas Tungsten Arc Welding Equipment, Setup, Operation, and Filler Metals

CURRENT SELECTION FOR GAS TUNGSTEN ARC WELDING

Material	Alternating Current — Stabilized			Direct Current — Straight Polarity			Reverse Polarity			Tungsten Selection
	Ex	Go	NR	Ex	Go	NR	Ex	Go	NR	
Magnesium up to 1/8 in. thick	X					X	X			Pure
Magnesium above 3/16 in. thick	X					X		X		Zirconium
Magnesium castings	X					X	X			Thoriated
Aluminum up to 3/32 in. thick	X					X	X			Pure
Aluminum over 3/32 in. thick	X					X		X		Zirconium
Aluminum castings	X				X			X		Thoriated
Stainless steel		X		X					X	Thoriated
Brass alloys		X		X					X	Thoriated
Silicon copper			X						X	Pure Zirconium
Silver		X		X					X	Thoriated
High-chromium, nickel-base, high-temperature alloys		X		X					X	Thoriated
Silver cladding	X					X			X	Thoriated
Hardfacing	X			X					X	
Cast iron		X		X					X	
Low-carbon steel, 0.015 to 0.030 in.		X		X					X	
Low-carbon steel, 0.030 to 0.125 in.			X	X					X	Thoriated Pure Zirconium
High-carbon steel, 0.030 in.		X		X					X	
High-carbon steel, 0.030 in. and up		X		X					X	
Deoxidized copper			X	X					X	

GTAW TROUBLESHOOTING GUIDE

Problem	Remedy
Arc Blow	1. Make sure the ground is properly connected. 2. Changing the angle of the electrode relative to the work may stabilize the arc. 3. Make sure the base metal is clean. 4. Make sure tungsten electrode is clean. 5. Use a smaller tungsten. 6. Bring nozzle closer to work.
Brittle Welds	1. Check the type of filler metal being used. 2. Use preheat and postheat while welding. 3. Check the welding procedure. A. Multilayer welds will tend to anneal hard heat-affected zone.

Cracked Welds	1. Check for inferior welds. Make sure all welds are sound and the fusion is good.
	2. Check the type of filler metal being used.
	3. Check the welding procedure.
	   A. Fill all craters at the end of the weld passes.
	   B. Use preheat and postheat.
	4. Do not use too small a weld between heavy plates.
	5. Check for proper preparation of joints.
	6. Check for excessive rigidity of joints.

Inferior Weld Appearance	1. Change welding technique. 2. Use proper current. 3. Check for proper shielding gas. 4. Check for drafts blowing shielding gas away. 5. Check for proper filler metal. 6. Make sure base metal is clean.
Distortion	1. Make sure parts are properly tack welded together. 2. Clamp parts to resist shrinkage. 3. Removing of rolling or forming strains by stress relieving before welding is sometimes helpful. 4. Distribute welding to prevent excessive local heating.
Porosity	1. Check shielding gas. 2. Check for drafts blowing shielding gas away. 3. Do not weld on wet metal. 4. Make sure base metal is clean. 5. Check shielding gas hoses; plastic hoses are best. 6. Check welding procedure and current settings.

Rapid Tungsten Electrode Consumption	1. Check polarity.
	2. Decrease current.
	3. Use a larger tungsten electrode.
	4. Check shielding gas.
	   A. Increase gas flow.
	5. Check for good collet contact.
	   A. Change collet.
	   B. Use ground-finished tungsten.

Problem	Remedy
Tungsten Inclusions	1. Use a larger tungsten electrode or reduce current. 2. Do not allow the tungsten to come into contact with the molten weld pool while welding. 3. Use high-frequency starting device. 4. Do not weld with tungsten electrodes that have become cracked or split on the ends.
Lack of Penetration	1. Increase current. 2. Decrease welding speed. 3. Decrease the size of the filler metal. 4. Adjust root opening.
Undercutting	1. Decrease current. 2. Make smaller weld beads. 3. Use a closer tungsten-to-work distance. 4. Change welding technique.

CHAPTER 16: QUIZ 1

Name _____ Date _____

Class _____ Instructor _____ Grade _____

INSTRUCTIONS

Carefully read Chapter 16 in the textbook and answer each question.

MATCHING

In the space provided to the left of Column A, write the letter from Column B that best answers or completes the statement in Column A.

Column A	Column B
_____ 1. The gas tungsten arc welding (GTAW) process is sometimes referred to as _____.	a. pointed
_____ 2. To use this process, an arc is established between a _____ tungsten electrode and the base metal, which is called the work.	b. balanced wave
_____ 3. The high melting temperature and good _____ make tungsten the best choice for a nonconsumable electrode.	c. TIG, or heliarc
_____ 4. In general, tungsten is a good conductor of _____.	d. rounded
_____ 5. Because of the intense heat of the arc, some _____ of the electrode will occur.	e. slight ball
_____ 6. With DCEN, a _____ tip concentrates the arc as much as possible and improves arc starting with either a high-voltage electrical discharge or a touch start.	f. electrical conductivity
_____ 7. With conventional alternating current (AC), the tip is subjected to more heat than with DCEN, and the tip is _____.	g. erosion
_____ 8. DCEP has the highest heat concentration on the electrode tip; for this reason, a _____ of molten tungsten is suspended at the end of a tapered electrode tip.	h. diamond coated grinding
_____ 9. Pure tungsten has a number of properties that make it an _____ for the GTA welding process.	i. current-carrying
_____ 10. Pure tungsten easily forms a balled tip for good arc stability, especially with _____ welding power.	j. lanthanum oxide (La_2O_3)
_____ 11. Thorium oxide (ThO_2), when added in percentages of up to 0.6% to tungsten, improves its _____ capacity.	k. end shape
_____ 12. Zirconiated tungstens also have the advantage over _____ tungsten in that they are not radioactive.	l. nonconsumable

———— 13. Cerium oxide electrodes were developed as replacements for thoriated tungstens because they are not made of a _____ material.

m. thoriated

———— 14. EWLa electrodes are available with three different levels of _____ dispersed in the tungsten.

n. palm of your hand

———— 15. The EWG classifications are a mixture of rare earth oxides, such as cerium, lanthanum, yttrium, and others, combined with approximately 97% of _____ to form these electrodes.

o. excellent nonconsumable electrode

———— 16. The desired _____ of tungsten can be obtained by grinding, breaking, or remelting the end, or using chemical compounds.

p. light pressure

———— 17. Both bench grinders and hand grinders use _____ disks.

q. coarse

———— 18. A _____ grinding stone will result in more tungsten breakage and a poorer finish.

r. radioactive

———— 19. When hand grinding, to prevent overheating of the tungsten, only _____ should be applied against the grinding wheel.

s. tungsten

———— 20. When holding one end of the tungsten against the grinding wheel while hand grinding, the other end of the tungsten must not be directed toward the _____.

t. heat

CHAPTER 16: QUIZ 2

Name _____ Date _____

Class _____ Instructor _____ Grade _____

INSTRUCTIONS

Carefully read Chapter 16 in the textbook and answer each question.

MATCHING

In the space provided to the left of Column A, write the letter from Column B that best answers or completes the statement in Column A.

Column A

Column B

——— 1. You can use a _____ to sharpen tungsten on any clean grinding disk or wheel.

a. grinding

——— 2. Start both the grinder and the drill before beginning to grind so that the starting torque of the motors does not cause the tungsten to _____ into the flap disk.

b. water

——— 3. Most tungsten, except for lanthanated tungsten, can be broken easily because it is _____, resulting in low impact strength.

c. rounded

——— 4. Once the tungsten has been broken squarely, the end must be melted back so that it becomes somewhat _____.

d. maximum

——— 5. The tapered tungsten with a _____ end, a shape sometimes used for DCEP welding, is made by first grinding or chemically pointing the electrode.

e. lower

——— 6. The tapered tungsten with a balled end, a shape sometimes used for DCEP welding, is made by first _____ or chemically pointing the electrode.

f. Air

——— 7. The water-cooled torch, as compared to the air-cooled torch, operates at a _____ temperature, resulting in a lower tungsten temperature and less erosion.

g. protective covering

——— 8. The air-cooled torch is more portable because it has fewer _____, and it may be easier to manipulate than the water-cooled torch.

h. nozzle

——— 9. The amperage listed on a torch is the _____ rating and cannot be exceeded without possible damage to the torch.

i. hard but brittle

——— 10. A _____-cooled torch has three hoses connecting it to the welding machine.

j. work

——— 11. _____-cooled torches may have one hose for shielding gas attached to the power cable.

k. balled

——— 12. Water hose fittings have left-hand threads, and gas hose fittings have _____ threads.

——— 13. A _____ can be used to prevent the hoses from becoming damaged by hot metal.

——— 14. The _____ or cup is used to direct the shielding gas directly on the welding zone.

——— 15. _____ nozzle diameters allow the welder to better see the molten weld pool and can be operated with lower gas flow rates.

——— 16. The useful life of a ceramic nozzle is affected by the current level and proximity to the _____.

——— 17. The added _____ with glass nozzles in tight, hard-to-reach places is often worth the added expense.

——— 18. The gas lens reduces the _____ in the shielding gas as it leaves the nozzle.

——— 19. The _____ may be merely a flow regulator used on a manifold system, or it may be a combination flow and pressure regulator used on an individual cylinder.

——— 20. Some flowmeters use a ball that floats on top of the flowing gas stream in a clear tube to indicate the shielding gas flow rate; other flowmeters use a _____.

l. hoses

m. gauge

n. Small

o. turbulence

p. gouge

q. flowmeter

r. drill

s. visibility

t. right-hand

CHAPTER 16: QUIZ 3

Name _____   Date _____

Class _____   Instructor _____   Grade _____

INSTRUCTIONS

Carefully read Chapter 16 in the textbook and answer each question.

MATCHING

In the space provided to the left of Column A, write the letter from Column B that best answers or completes the statement in Column A.

Column A

Column B

_____   1. In addition to the two standard types of welding currents—alternating current (AC) and direct current (DC)—a _____ version of alternating current is available for GTAW.

a. work

_____   2. Direct-current electrode negative (DCEN) concentrates approximately two-thirds of its welding heat on the work and the remaining one-third on the _____.

b. electrode

_____   3. Direct-current electrode positive (DCEP) concentrates only one-third of the arc heat on the plate and two-thirds of the heat on the _____.

c. hybrid

_____   4. Alternating current (AC) concentrates approximately half of its heat on the _____ and the other half on the tungsten.

d. low

_____   5. Conventional GTA welding machines use high frequency, high voltage, and very _____ current to maintain the arc during the DCEP phase of the AC cycle to stabilize the arc.

e. 50/50 balance

_____   6. EP and EN Time—the EP (electrode positive) portion of alternating current provides _____ and the EN (electrode negative) portion of the current provides heat on the weld metal.

f. denser

_____   7. Conventional GTA welding machines have a _____ between the EP and EN portions of the welding current.

g. current

_____   8. The traditional AC sine wave provides good _____, which allows the molten weld pool to flow easily into the joint or onto the metal surface.

h. tungsten

_____   9. A lower frequency control setting produces a much less _____ arc and a higher setting provides a much more concentrated, stiffer, and more focused arc.

i. spare parts

_____   10. The purpose of the shielding gas is to protect the molten weld pool and the tungsten electrode from the harmful effects of _____.

j. preflow

——— 11. Because argon is _____ than air, it effectively shields welds in deep grooves in the flat position.

——— 12. Helium offers the advantage of deeper _____.

——— 13. Hydrogen additions added to argon shielding gas are restricted to use on _____ welds because hydrogen is the primary cause of porosity in aluminum welds.

——— 14. The _____ start allows a controlled surge of welding current as the arc is started to establish a molten weld pool quickly.

——— 15. The _____ is the time during which gas flows to clear out any air in the nozzle or surrounding the weld zone.

——— 16. The _____ is the time during which the gas continues flowing after the welding current has stopped.

——— 17. The shielding gas flow rate should be as low as possible and still give adequate _____.

——— 18. A remote control can be used to start the weld, increase or decrease the _____, and stop the weld.

——— 19. When changing tungsten sizes or if parts are damaged, it is a good idea to carry _____ for your GTA welding torch.

——— 20. In order to get a good GTA weld on thick sections of metal, especially aluminum, you will need to _____ it.

k. preheat

l. coverage

m. surface cleaning

n. surface wetting

o. penetration

p. postflow

q. focused

r. hot

s. stainless steel

t. air

CHAPTER 16: QUIZ 4

Name _____   Date _____

Class _____   Instructor _____   Grade _____

INSTRUCTIONS
Carefully read Chapter 16 in the textbook and answer each question.

IDENTIFICATION
Identify the numbered items on the drawing by writing the letter next to the identifying term in the space provided.

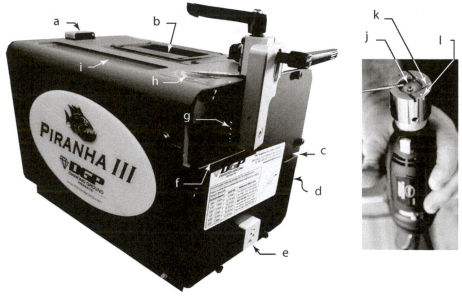

1. _____   CARRYING HANDLE

2. _____   POWER SWITCH

3. _____   DIAMOND GRINDING DISC

4. _____   GRINDING ANGLE

5. _____   TRAY FOR TUNGSTEN

6. _____   NOTCH FOR BREAKING TUNGSTENS

7. _____   SLOTS FOR SHARPENING DIFFERENTLY SIZED TUNGSTENS

8. _____   VACUUM PORT (NOT SHOWN)

9. _____   DUST COLLECTING TRAY

10. _____   SLOT FOR SCORING AND BREAKING TUNGSTEN

11. _____   PORTS TO GUIDE THE TUNGSTEN FOR SHARPENING

12. _____   HOLES FOR FLATTENING ENDS

CHAPTER 16: QUIZ 5

Name _____   Date _____

Class _____   Instructor _____   Grade _____

INSTRUCTIONS

Carefully read Chapter 16 in the textbook and answer each question.

IDENTIFICATION

Identify the numbered items on the drawing by writing the letter next to the identifying term in the space provided.

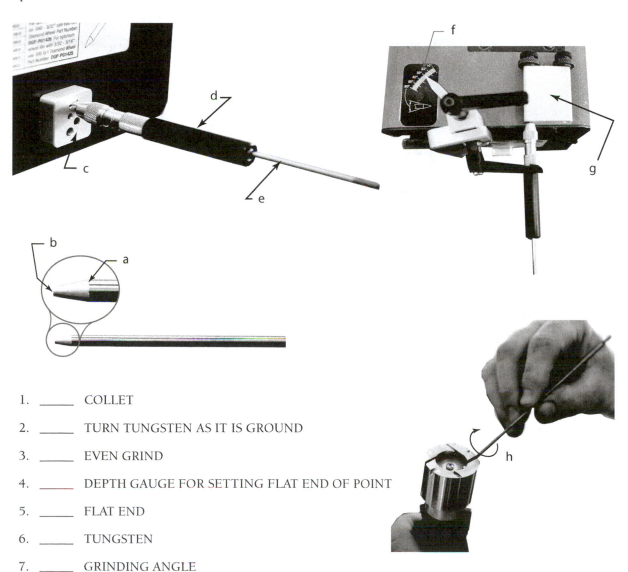

1. _____   COLLET

2. _____   TURN TUNGSTEN AS IT IS GROUND

3. _____   EVEN GRIND

4. _____   DEPTH GAUGE FOR SETTING FLAT END OF POINT

5. _____   FLAT END

6. _____   TUNGSTEN

7. _____   GRINDING ANGLE

8. _____   SIX DIFFERENTLY SIZED OPENINGS TO SNUGLY FIT TUNGSTENS FOR FLATTENING

CHAPTER 16: QUIZ 6

Name _____   Date _____

Class _____   Instructor _____   Grade _____

INSTRUCTIONS

Carefully read Chapter 16 in the textbook and answer each question.

IDENTIFICATION

Identify the numbered items on the drawing by writing the letter next to the identifying term in the space provided.

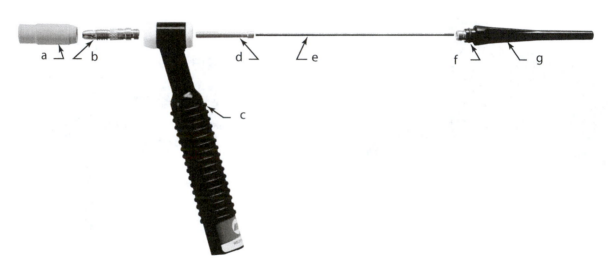

1. _____   TORCH BODY

2. _____   CAP

3. _____   COLLET

4. _____   TUNGSTEN ELECTRODE

5. _____   COLLET BODY

6. _____   NOZZLE

7. _____   O-RING

CHAPTER 16: QUIZ 7

Name _____   Date _____

Class _____   Instructor _____   Grade _____

INSTRUCTIONS

Carefully read Chapter 16 in the textbook and answer each question.

IDENTIFICATION

Identify the numbered items on the drawing by writing the letter next to the identifying term in the space provided.

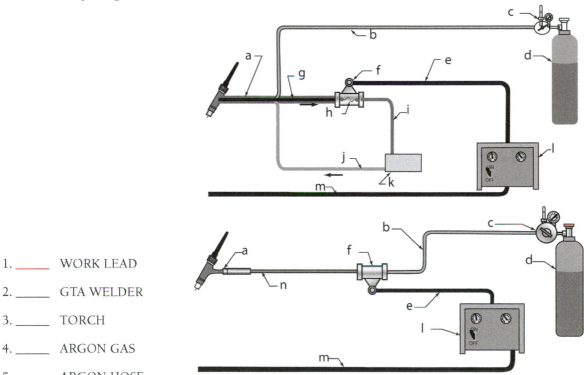

1. _____ WORK LEAD

2. _____ GTA WELDER

3. _____ TORCH

4. _____ ARGON GAS

5. _____ ARGON HOSE

6. _____ SAFETY FUSE

7. _____ CABLE ADAPTER

8. _____ FLOWMETER REGULATOR

9. _____ WELDING POWER CABLE

10. _____ POWER CABLE AND GAS HOSE

11. _____ COOL WATER SUPPLY HOSE

12. _____ POWER CABLE INSIDE OF THE WATER RETURN HOSE

13. _____ RECIRCULATING WATER PUMP

14. _____ WARM WATER RETURN HOSE

Gas Tungsten Arc Welding of Plate

OPERATING INSTRUCTIONS FOR THE GTAW PROCESS

1. With the power off, switch the machine to the GTA welding mode.

2. Select the desired type of current and amperage range:

Electrode Diameter		DCSP	DCRP	AC
in.	(mm)			
.04	(1)	15–60	Not recommended	10–50
1/16	(2)	70–100	10–20	50–90
3/32	(2.4)	90–200	15–30	80–130
1/8	(3)	150–350	25–40	100–200
5/32	(4)	300–450	40–55	160–300

3. Set the fine current adjustment to the proper range, depending on the size of the tungsten being used.

4. Place the high-frequency switch in the appropriate position, auto or start for DC or continuous for AC.

5. The remote control can be plugged in and the selector switch set if you are using the remote.

6. The collet and collet body should be installed and match the tungsten size being used.

7. Install the tungsten into the torch and tighten the back cap to hold the tungsten in place.

8. Select and install the desired nozzle size. Adjust the tungsten length so that it does not stick out more than the inside diameter of the nozzle.

9. Check the manufacturer's operating manual for the machine to ensure that all connections and settings are correct.

10. Turn on the power, depress the remote control, and check for water or gas leaks.

11. While postpurge is still engaged, set the gas flow by adjusting the valve on the flowmeter.

12. Position yourself so that you are comfortable and can see the torch, tungsten, and plate while the tungsten tip is held about 1/4 in. (6 mm) above the metal. Try to hold the torch at a vertical angle ranging from 0° to 15°. Too steep an angle will not give adequate gas coverage.

CAUTION **Avoid touching the metal table with any unprotected skin or jewelry. The high frequency can cause an uncomfortable shock.**

13. Lower your arc welding helmet and depress the remote control. A high-pitched, erratic arc should be immediately jumping across the gap between the tungsten and the plate. If the high-frequency arc is not established, lower the torch until it appears.

14. Slowly increase the current until the main welding arc appears.

15. Observe the color change of the tungsten as the arc appears.

16. Move the tungsten around in a small circle until a weld pool appears on the metal.

17. Slowly decrease the current and observe the change in the weld pool.

18. Reduce the current until the arc is extinguished.

19. Hold the torch in place over the weld until the postpurge stops.

20. Raise your hood and inspect the weld.

21. Turn off the welding machine and shielding gas and clean up your work area when you are finished welding.

■ PRACTICE 17-1

Name _____   Date _____

Base Metal _____   Thickness _____   Filler Dia. _____   Joint _____

Class _____   Instructor _____   Grade _____

Welding Principles and Applications	
MATERIAL: 1/8" X 6" MILD STEEL	
PROCESS: GTAW STRINGER BEAD FLAT POSITION	
NUMBER: PRACTICE 17-1	DRAWN BY: WENDY JEFFUS

OBJECTIVE: After completing this practice, you should be able to make stringer beads in the flat position on 16-gauge and 1/8-in. (3-mm) thick carbon steel without filler metal using the GTAW process.

EQUIPMENT AND MATERIALS NEEDED FOR THIS PRACTICE

1. A properly set-up and adjusted GTA welding machine.

2. Proper safety protection (welding hood, safety glasses, pliers, gloves, long-sleeved shirt, long pants, and leather boots or shoes). Refer to Chapter 2 in the text for more specific safety information.

3. One or more pieces of mild steel, 3-in. (76-mm) wide by 6-in. (152-mm) long, 16-gauge, and 1/8-in. (3-mm) thick.

INSTRUCTIONS

1. Starting at one end of the piece of metal that is 1/8-in. (3-mm) thick, hold the torch as close as possible to a 90° angle.

2. Lower your hood, strike an arc, and establish a weld pool.

3. Move the torch in a circular oscillation pattern down the plate toward the other end.

4. If the size of the weld pool changes, speed up or slow down the travel rate to keep the weld pool the same width for the entire length of the plate.

5. Repeat the process using both thicknesses of metal until you can consistently make the weld pool visually defect free.

6. Turn off the welding machine and shielding gas and clean up your work area when you are finished welding.

INSTRUCTOR'S COMMENTS _____

_____

_____

■ PRACTICE 17-2

Name _____   Date _____

Base Metal _____   Thickness _____   Filler Dia. _____   Joint _____

Class _____   Instructor _____   Grade _____

OBJECTIVE: After completing this practice, you should be able to make stringer beads in the flat position on 16-gauge and ¼-in. (6-mm)-thick stainless steel without filler metal using the GTAW process.

EQUIPMENT AND MATERIALS NEEDED FOR THIS PRACTICE

1. A properly set-up and adjusted GTA welding machine.

2. Proper safety protection (welding hood, safety glasses, pliers, gloves, long-sleeved shirt, long pants, and leather boots or shoes). Refer to Chapter 2 in the text for more specific safety information.

3. One or more pieces of mild steel, 3-in. (76-mm) wide by 6-in. (152-mm) long, 16-gauge, and 1/8-in. (3-mm) thick.

INSTRUCTIONS

1. Starting at one end of the piece of metal that is 1/8-in. (3-mm) thick, hold the torch as close as possible to a 90° angle.

2. Lower your hood, strike an arc, and establish a weld pool.

3. Move the torch in a circular oscillation pattern down the plate toward the other end.

4. If the size of the weld pool changes, speed up or slow down the travel rate to keep the weld pool the same width for the entire length of the plate.

5. To keep the formation of oxides on the bead to a minimum, a chill plate (a thick piece of metal used to absorb heat) may be required. Another method is to make the bead using as low of a heat input as possible. When the weld is finished, the weld bead should be no darker than dark blue.

6. Repeat the process using both thicknesses of metal until you can consistently make the weld visually defect free.

7. Turn off the welding machine and shielding gas and clean up your work area when you are finished welding.

INSTRUCTOR'S COMMENTS _____

_____

_____

■ PRACTICE 17-3

Name _____   Date _____

Base Metal _____   Thickness _____   Filler Dia. _____   Joint _____

Class _____   Instructor _____   Grade _____

OBJECTIVE: After completing this practice, you should be able to make stringer beads in the flat position on 16-gauge and 1/8-in. (3-mm) thick and ¼-in. (6-mm)-thick aluminum without the filler metal using the GTAW process.

EQUIPMENT AND MATERIALS NEEDED FOR THIS PRACTICE

1. A properly set-up and adjusted GTA welding machine.

2. Proper safety protection (welding hood, safety glasses, pliers, gloves, long-sleeved shirt, long pants, and leather boots or shoes). Refer to Chapter 2 in the text for more specific safety information.

3. One or more pieces of aluminum, 3-in. (76-mm) wide by 6-in. (152-mm) long, 16-gauge, and 1/8-in. (3-mm) thick.

INSTRUCTIONS

1. Starting at one end of the piece of metal that is 1/8-in. (3-mm) thick, hold the torch as close as possible to a 90° angle.

2. Lower your hood, strike an arc, and establish a weld pool.

3. Move the torch in a circular oscillation pattern down the plate toward the other end.

4. If the size of the weld pool changes, speed up or slow down the travel rate to keep the weld pool the same width for the entire length of the plate.

5. A high current setting will allow faster travel speeds. The faster speed helps to control excessive penetration. Hot cracking may occur on some types of aluminum after a surfacing weld. This is not normally a problem when filler metal is added. If hot cracking should occur during this practice, do not be concerned.

6. Repeat the process using both thicknesses of metal until you can consistently make the weld visually defect free.

7. Turn off the welding machine and shielding gas and clean up your work area when you are finished welding.

INSTRUCTOR'S COMMENTS _____

_____

_____

■ PRACTICE 17-4

Name _____ Date _____

Base Metal _____ Thickness _____ Filler Dia. _____ Joint _____

Class _____ Instructor _____ Grade _____

Welding Principles and Applications

MATERIAL: 1/16" x 6" MILD STEEL & STAINLESS STEEL
1/8" x 6" MILD STEEL & STAINLESS STEEL
1/4" x 6" ALUMINUM

PROCESS:
GTAW STRINGER BEAD FLAT POSITION

NUMBER: PRACTICE 17-4 | DRAWN BY: WENDY JEFFUS

OBJECTIVE: After completing this practice, you should be able to weld stringer beads on carbon steel, stainless steel, and aluminum in the flat position using the correct filler metal using the GTAW process.

EQUIPMENT AND MATERIALS NEEDED FOR THIS PRACTICE

1. A properly set-up and adjusted GTA welding machine.

2. Proper safety protection.

3. Filler rods, 36-in. (0.9-m) long by 1/16-in. (1.6-mm), 3/32-in. (2.4-mm), and 1/8-in. (3-mm) diameter.

4. One or more pieces of mild steel, stainless steel, and aluminum, 3-in. (76-mm) wide by 6-in. (152-mm) long by 1/16-in. (2-mm) and 1/8-in. (3-mm) thick, and aluminum plate 1/4-in. (6-mm) thick.

INSTRUCTIONS

1. Starting with the metal that is 1/8-in. (3-mm) thick and the filler rod having a 3/32 in. (2.4 mm) diameter, strike an arc and establish a weld pool.

2. Move the torch in a circle as in the first three practices. When the torch is on one side, add filler rod to the other side of the molten weld pool. The end of the filler rod should be dipped into the front of the weld pool but should not be allowed to melt and drip into the weld pool.

3. Change to another size filler rod and determine its effect on the weld pool.

4. Maintain a smooth and uniform rhythm as filler metal is added. This will help to keep the bead uniform.

5. Vary the rhythms to determine which one is easiest for you. If the filler rod sticks, move the torch toward the rod until it melts free.

6. When the full 6-in. (152-mm) long weld bead is completed, cool and inspect it for uniformity and defects.

7. Repeat the process using all thicknesses and types of metal until you can consistently make the weld visually defect free.

8. Turn off the welding machine and shielding gas and clean up your work area when you are finished welding.

INSTRUCTOR'S COMMENTS _____

_____

_____

■ PRACTICE 17-5

Name _____ Date _____

Base Metal _____ Thickness _____ Filler Dia. _____ Joint _____

Class _____ Instructor _____ Grade _____

6"
(152 mm)

$1\frac{1}{2}"$
(38 mm)

$1\frac{1}{2}"$
(38 mm)

Welding Principles and Applications

MATERIAL: 1/16" x 6" MILD STEEL & STAINLESS STEEL
1/8" x 6" MILD STEEL & STAINLESS STEEL
1/4" x 6" ALUMINUM

PROCESS:
GTAW OUTSIDE CORNER JOINT 1G

NUMBER:
PRACTICE 17-5

DRAWN BY:
WENDY JEFFUS

OBJECTIVE: After completing this practice, you should be able to GTAW weld an outside corner joint on carbon steel, stainless steel, and aluminum in the flat position.

EQUIPMENT AND MATERIALS NEEDED FOR THIS PRACTICE

1. A properly set-up and adjusted GTA welding machine.

2. Proper safety protection.

3. Filler rods, 36-in. (0.9-m) long by 1/16-in. (1.6-mm), 3/32-in. (2.4-mm), and 1/8-in. (3-mm) diameter.

4. One or more pieces of mild steel, stainless steel, and aluminum, 1-1/2-in. (38-mm) wide by 6-in. (152-mm) long by 1/16-in. (2-mm) and 1/8-in. (3-mm) thick, and aluminum plate 1/4-in. (6-mm) thick.

INSTRUCTIONS

1. Place one of the pieces of metal flat on the table and hold or brace the other piece of metal vertically on it forming a corner joint.

2. Tack weld both ends of the plates together.

3. Set the plates up and add two or three more tack welds on the joint as required.

4. Clean the weld area before making the weld. Then, starting at one end, make a weld bead of uniform height and width, adding filler metal as needed to the outside corner.

5. Repeat each weld as needed with all types of base metal and filler rod sizes until all welds can be done with consistently high quality.

6. Turn off the welding machine and shielding gas and clean up your work area when you are finished welding.

INSTRUCTOR'S COMMENTS _____

_____

_____

■ PRACTICE 17-6

Name _____ Date _____

Base Metal _____ Thickness _____ Filler Dia. _____ Joint _____

Class _____ Instructor _____ Grade _____

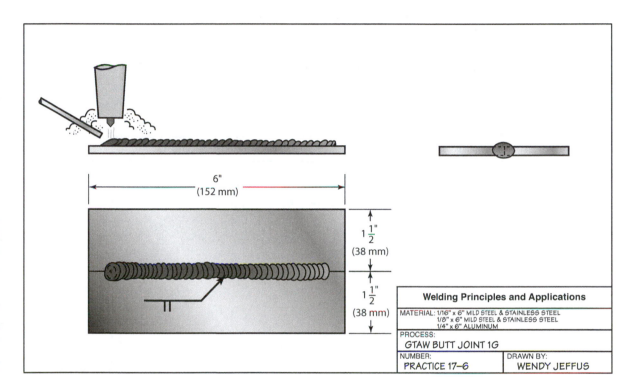

Welding Principles and Applications

MATERIAL: 1/16" x 6" MILD STEEL & STAINLESS STEEL
1/8" x 6" MILD STEEL & STAINLESS STEEL
1/4" x 6" ALUMINUM

PROCESS:
GTAW BUTT JOINT 1G

NUMBER: PRACTICE 17–6

DRAWN BY: WENDY JEFFUS

OBJECTIVE: After completing this practice, you should be able to weld a butt joint on carbon steel, stainless steel, and aluminum in the 1G position using the GTAW process.

EQUIPMENT AND MATERIALS NEEDED FOR THIS PRACTICE

1. A properly set-up and adjusted GTA welding machine.

2. Proper safety protection.

3. Filler rods, 36-in. (0.9-m) long by 1/16-in. (1.6-mm), 3/32-in. (2.4-mm), and 1/8-in. (3-mm) diameter.

4. Two or more pieces of mild steel, stainless steel, and aluminum, 1-1/2-in. (38-mm) wide by 6-in. (152-mm) long by 1/16-in. (2-mm) and 1/8-in. (3-mm) thick, and aluminum plate 1/4-in. (6-mm) thick.

INSTRUCTIONS

1. Place the metal flat on the table and tack weld both ends together.

2. Two or three additional tack welds can be made along the joint as needed.

3. Starting at one end, make a uniform width weld along the joint. Add filler metal as required to make a uniform height weld.

4. Repeat the process using all thicknesses and types of metal until you can consistently make the weld visually defect free and of high quality.

5. Turn off the welding machine and shielding gas and clean up your work area when you are finished welding.

INSTRUCTOR'S COMMENTS _____

_____

_____

■ PRACTICE 17-7

Name _____ Date _____

Base Metal _____ Thickness _____ Filler Dia. _____ Joint _____

Class _____ Instructor _____ Grade _____

OBJECTIVE: After completing this practice, you should be able to make a butt joint while controlling both distortion and penetration in the flat position using the GTAW process.

EQUIPMENT AND MATERIALS NEEDED FOR THIS PRACTICE

1. A properly set-up and adjusted GTA welding machine.

2. Proper safety protection.

3. Filler rods, 36-in. (0.9-m) long by 1/16-in. (1.6-mm), 3/32-in. (2.4-mm), and 1/8-in. (3-mm) diameter.

4. Two or more pieces of mild steel, stainless steel, and aluminum, 1-1/2-in. (38-mm) wide by 6-in. (152-mm) long by 1/16-in. (2-mm) and 1/8-in. (3-mm) thick, and aluminum plate 1/4-in. (6-mm) thick.

INSTRUCTIONS

1. Place the metal flat on the table and tack weld both ends together in the butt joint configuration.

2. Two or three additional tacks can be made along the joint as needed.

3. Using a back-stepping weld sequence, make a series of welds approximately 1-in. (25-mm) long along the joint.

4. Be sure to fill each weld crater adequately to reduce crater cracking.

5. Repeat the process using all thicknesses and types of metal until you can consistently make distortion-free and defect-free welds.

6. Turn off the welding machine and shielding gas and clean up your work area when finished with this practice.

INSTRUCTOR'S COMMENTS _____

_____

_____

■ PRACTICE 17-8

Name _____ Date _____

Base Metal _____ Thickness _____ Filler Dia. _____ Joint _____

Class _____ Instructor _____ Grade _____

TURN PLATE OVER TO WELD THIS SIDE.

$\frac{1}{2}$"*
(13 mm)

45

6"
(152 mm)

$1\frac{1}{2}$"**
(38 mm)

$1\frac{1}{2}$"**
(38 mm)

Welding Principles and Applications

MATERIAL:	1/16" x 6" MILD STEEL & STAINLESS STEEL 1/8" x 6" MILD STEEL & STAINLESS STEEL 1/4" x 6" ALUMINUM
PROCESS: GTAW LAP JOINT 1F	
NUMBER: PRACTICE 17–8	DRAWN BY: WENDY JEFFUS

**This dimension will decrease as the old weld is cut out so the metal can be reused.

*$\frac{1}{2}$" (13 mm) is the maximum to conserve metal.

OBJECTIVE: After completing this practice, you should be able to weld lap joints on carbon steel, stainless steel, and aluminum in the 1F position using the GTAW process.

EQUIPMENT AND MATERIALS NEEDED FOR THIS PRACTICE

1. A properly set-up and adjusted GTA welding machine.

2. Proper safety protection.

3. Filler rods, 36-in. (0.9-m) long by 1/16-in. (1.6-mm), 3/32-in. (2.4-mm), and 1/8-in. (3-mm) diameter.

4. Two or more pieces of mild steel, stainless steel, and aluminum, 1-1/2-in. (38-mm) wide by 6-in. (152-mm) long by 1/16-in. (2-mm) and 1/8-in. (3-mm) thick, and aluminum plate 1/4-in. (6-mm) thick.

INSTRUCTIONS

1. Place the two pieces of metal flat on the table with an overlap of 1/4 in. (6 mm) to 3/8 in. (10 mm).

2. Hold the pieces of metal tightly together and tack weld them.

3. Starting at one end, make a uniform fillet weld along the joint.

4. Both sides of the joint should be welded.

5. Repeat the process using all thicknesses of metal until you can consistently make the weld visually defect free and of high quality.

6. Turn off the welding machine and shielding gas and clean up your work area when you are finished welding.

INSTRUCTOR'S COMMENTS _____

_____

_____

■ PRACTICE 17-9

Name _____ Date _____

Base Metal _____ Thickness _____ Filler Dia. _____ Joint _____

Class _____ Instructor _____ Grade _____

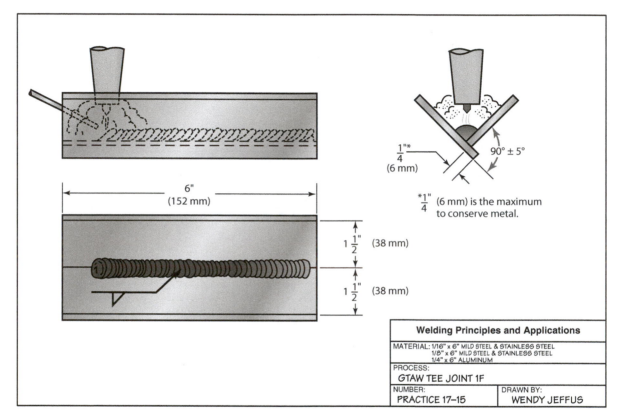

Welding Principles and Applications

MATERIAL: 1/16" x 6" MILD STEEL & STAINLESS STEEL
1/8" x 6" MILD STEEL & STAINLESS STEEL
1/4" x 6" ALUMINUM

PROCESS:
GTAW TEE JOINT 1F

NUMBER: **PRACTICE 17–15**   DRAWN BY: **WENDY JEFFUS**

OBJECTIVE: After completing this practice, you should be able to weld tee joints on carbon steel, stainless steel, and aluminum in the 1F position using the GTAW process.

EQUIPMENT AND MATERIALS NEEDED FOR THIS PRACTICE

1. A properly set-up and adjusted GTA welding machine.

2. Proper safety protection.

3. Filler rods, 36-in. (0.9-m) long by 1/16-in. (1.6-mm), 3/32-in. (2.4-mm), and 1/8-in. (3-mm) diameter.

4. Two or more pieces of mild steel, stainless steel, and aluminum, 1-1/2-in. (38-mm) wide by 6-in. (152-mm) long by 1/16 in. (2 mm) and 1/8-in. (3-mm) thick, and aluminum plate 1/4-in. (6-mm) thick.

INSTRUCTIONS

1. Place one of the pieces of metal flat on the table and hold or brace the other piece of metal vertically on it in the tee joint configuration.

2. Tack weld both ends of the plates together.

3. Set up the plates in the flat position and add two or three more tack welds to the joint as required.

4. On the metal that is 1/16 in. (1.5 mm) thick, it may not be possible to weld both sides, but on the thicker material a fillet weld can usually be made on both sides. The exception to this is if carbide precipitation occurs on the stainless steel during welding.

5. Starting at one end, make a uniform height and width weld, adding filler metal as needed.

6. Repeat the process using all thicknesses of metal until you can consistently make the weld visually defect free.

7. Turn off the welding machine and shielding gas and clean up your work area when you are finished welding.

INSTRUCTOR'S COMMENTS _____

_____

_____

■ PRACTICE 17-10

Name _____   Date _____

Base Metal _____   Thickness _____   Filler Dia. _____   Joint _____

Class _____   Instructor _____   Grade _____

OBJECTIVE: After completing this practice, you should be able to make a stringer bead at a 45° vertical up angle using mild steel, stainless steel, and aluminum with the GTAW process.

EQUIPMENT AND MATERIALS NEEDED FOR THIS PRACTICE

1. A properly set-up and adjusted GTA welding machine.

2. Proper safety protection.

3. Filler rods, 36-in. (0.9-m) long by 1/16-in. (1.6-mm), 3/32-in. (2.4-mm), and 1/8-in. (3-mm) diameter.

4. One or more pieces of mild steel, stainless steel, and aluminum, 1-1/2-in. (38-mm) wide by 6-in. (152-mm) long by 1/16-in. (2-mm) and 1/8-in. (3-mm) thick, and aluminum plate 1/4-in. (6-mm) thick.

INSTRUCTIONS

1. Position the workpiece at a 45° angle to the table.

2. Starting at the bottom and welding in an upward direction, add filler metal to the top edge of the weld pool, and move the torch in a circle or "C" pattern.

3. If the weld pool size starts to increase, the "C" pattern can be increased in length or the power can be decreased.

4. Watch the weld pool and establish a rhythm of torch movement and filler metal addition, which will keep the bead size and penetration uniform.

5. Repeat the process using all thicknesses and types of metal until you can consistently make the weld visually defect free.

6. Turn off the welding machine and shielding gas and clean up your work area when finished with this practice.

INSTRUCTOR'S COMMENTS _____

_____

_____

■ PRACTICE 17-11

Name _____ Date _____

Base Metal _____ Thickness _____ Filler Dia. _____ Joint _____

Class _____ Instructor _____ Grade _____

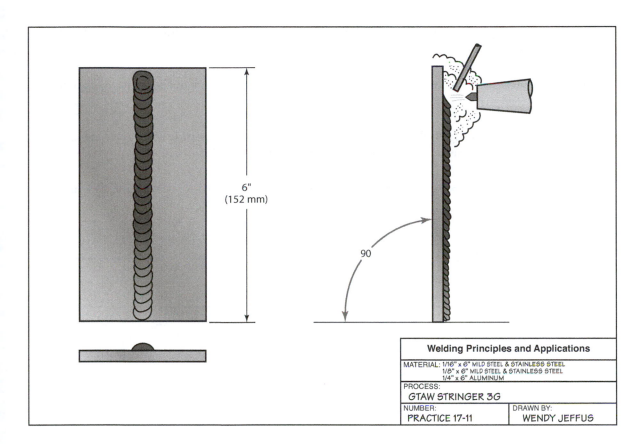

6"
(152 mm)

90

Welding Principles and Applications

MATERIAL: 1/16" x 6" MILD STEEL & STAINLESS STEEL
1/8" x 6" MILD STEEL & STAINLESS STEEL
1/4" x 6" ALUMINUM

PROCESS:
GTAW STRINGER 3G

NUMBER:
PRACTICE 17-11

DRAWN BY:
WENDY JEFFUS

OBJECTIVE: After completing this practice, you should be able to make stringer beads on carbon steel, stainless steel, and aluminum in the 3G position using the GTAW process.

EQUIPMENT AND MATERIALS NEEDED FOR THIS PRACTICE

1. A properly set-up and adjusted GTA welding machine.

2. Proper safety protection.

3. Filler rods, 36-in. (0.9-m) long by 1/16-in. (1.6 mm), 3/32-in. (2.4-mm), and 1/8-in. (3-mm) diameter.

4. One or more pieces of mild steel, stainless steel, and aluminum, 1-1/2-in. (38-mm) wide by 6-in. (152-mm) long by 1/16-in. (2-mm) and 1/8-in. (3-mm) thick, and aluminum plate 1/4-in. (6-mm) thick.

INSTRUCTIONS

1. Start with the plate at a 45° angle to the table.

2. Starting at the bottom and welding in an upward direction, add the filler metal at the top edge of the weld pool and move the torch in a circle or "C" pattern.

3. If the weld pool size starts to increase, the "C" pattern can be increased in length or the amperage can be decreased.

4. Watch the weld pool and establish a rhythm of torch movement and addition of filler rod to keep the weld uniform.

5. Gradually increase the angle as you develop skill until the weld is being made in the vertical up position.

6. Repeat the process using all thicknesses of metal until you can consistently make the weld visually defect free and the same height and width throughout the length of the bead.

7. Turn off the welding machine and shielding gas and clean up your work area when you are finished welding.

INSTRUCTOR'S COMMENTS _____

_____

_____

■ PRACTICE 17-12

Name _____ Date _____

Base Metal _____ Thickness _____ Filler Dia. _____ Joint _____

Class _____ Instructor _____ Grade _____

OBJECTIVE: After completing this practice, you should be able to make butt joints on carbon steel, stainless steel, and aluminum in a 45° vertical angle position using the GTAW process.

EQUIPMENT AND MATERIALS NEEDED FOR THIS PRACTICE

1. A properly set-up and adjusted GTA welding machine.

2. Proper safety protection.

3. Filler rods, 36-in. (0.9-m) long by 1/16-in. (1.6-mm), 3/32-in. (2.4-mm), and 1/8-in. (3-mm) diameter.

4. Two or more pieces of mild steel, stainless steel, and aluminum, 1-1/2-in. (25-mm) wide by 6-in. (152-mm) long by 1/16-in. (2-mm) and 1/8-in. (3-mm) thick, and aluminum plate 1/4-in. (6-mm) thick.

INSTRUCTIONS

1. Tack weld the plates together and then place the plates on the welding table at a 45° vertical angle. Start the weld at the bottom and weld in an upward direction.

2. If the weld pool size starts to increase, the "C" pattern can be increased in length or the amperage can be decreased.

3. Watch the weld pool and establish a rhythm of torch movement and addition of filler rod to keep the weld uniform.

4. Repeat the process using all thicknesses and types of metal until you can consistently make the weld visually defect free and the same height and width throughout the length of the bead.

5. Turn off the welding machine and shielding gas and clean up your work area when you are finished welding.

INSTRUCTOR'S COMMENTS _____

_____

_____

■ PRACTICE 17-13

Name _____   Date _____

Base Metal _____   Thickness _____   Filler Dia. _____   Joint _____

Class _____   Instructor _____   Grade _____

OBJECTIVE: After completing this practice, you should be able to make butt joints on carbon steel, stainless steel, and aluminum in the 3G position using the GTAW process.

EQUIPMENT AND MATERIALS NEEDED FOR THIS PRACTICE

1. A properly set-up and adjusted GTA welding machine.

2. Proper safety protection.

3. Filler rods, 36-in. (0.9-m) long by 1/16-in. (1.6-mm), 3/32-in. (2.4-mm), and 1/8-in. (3-mm) diameter.

4. Two or more pieces of mild steel, stainless steel, and aluminum, 1-1/2-in. (25-mm) wide by 6-in. (152-mm) long by 1/16-in. (2-mm) and 1/8-in. (3-mm) thick, and aluminum plate 1/4-in. (6-mm) thick.

INSTRUCTIONS

1. Tack weld the plates together and then place the plates on the welding table at a 45° vertical angle. Start the weld at the bottom and weld in an upward direction.

2. If the weld pool size starts to increase, the "C" pattern can be increased in length or the amperage can be decreased.

3. Watch the weld pool and establish a rhythm of torch movement and addition of filler rod to keep the weld uniform.

4. As you develop skill, continue increasing the angle until the weld is being made in the vertical up position.

5. Repeat the process using all thicknesses and types of metal until you can consistently make the weld visually defect free and the same height and width throughout the length of the bead.

6. Turn off the welding machine and shielding gas and clean up your work area when you are finished welding.

INSTRUCTOR'S COMMENTS _____

_____

_____

■ PRACTICE 17-14

Name _____ Date _____

Base Metal _____ Thickness _____ Filler Dia. _____ Joint _____

Class _____ Instructor _____ Grade _____

OBJECTIVE: After completing this practice, you should be able to make fillet weld lap joints on carbon steel, stainless steel, and aluminum in the 45° vertical up position using the GTAW process.

EQUIPMENT AND MATERIALS NEEDED FOR THIS PRACTICE

1. A properly set-up and adjusted GTA welding machine.

2. Proper safety protection.

3. Filler rods, 36-in. (0.9-m) long by 1/16-in. (1.6-mm), 3/32 in. (2.4-mm), and 1/8-in. (3-mm) diameter.

4. Two or more pieces of mild steel, stainless steel, and aluminum, 1-1/2-in. (38-mm) wide by 6-in. (152-mm) long by 1/16-in. (2-mm) and 1/8-in. (3-mm) thick, and aluminum plate 1/4-in. (6-mm) thick.

INSTRUCTIONS

1. Tack weld the plates together in the lap joint configuration with an overlap of 1/4 in. (6 mm) to 3/8 in. (10 mm). Then place them on the welding table at a 45° vertical angle. Start the weld at the bottom and weld in an upward direction.

2. Maintain a uniform rhythm so that a nice-looking weld bead is formed.

3. It may be necessary to move the torch in and around the base of the weld pool to ensure adequate root fusion. The filler metal should be added along the top edge of the weld pool to ensure adequate fusion.

4. Repeat the process using all thicknesses and types of metal until you can consistently make the weld visually defect free.

5. As you develop skill, increase the angle until the weld is being made in the vertical up position.

6. Turn off the welding machine and shielding gas and clean up your work area when you are finished welding.

INSTRUCTOR'S COMMENTS _____

_____

_____

■ PRACTICE 17-15

Name _____   Date _____

Base Metal _____   Thickness _____   Filler Dia. _____   Joint _____

Class _____   Instructor _____   Grade _____

OBJECTIVE: After completing this practice, you should be able to make fillet weld lap joints on carbon steel, stainless steel, and aluminum in the 3F position using the GTAW process.

EQUIPMENT AND MATERIALS NEEDED FOR THIS PRACTICE

1. A properly set-up and adjusted GTA welding machine.

2. Proper safety protection.

3. Filler rods, 36-in. (0.9-m) long by 1/16-in. (1.6-mm), 3/32-in. (2.4-mm), and 1/8-in. (3-mm) diameter.

4. Two or more pieces of mild steel, stainless steel, and aluminum, 1-1/2-in. (38-mm) wide by 6-in. (152-mm) long by 1/16-in. (2-mm) and 1/8-in. (3-mm) thick, and aluminum plate 1/4-in. (6-mm) thick.

INSTRUCTIONS

1. Tack weld the plates together in the lap joint configuration with an overlap of 1/4 in. (6 mm) to 3/8 in. (10 mm). Then place them on the welding table at a 45° vertical angle. Start the weld at the bottom and weld in an upward direction.

2. Maintain a uniform rhythm so that a nice-looking weld bead is formed.

3. It may be necessary to move the torch in and around the base of the weld pool to ensure adequate root fusion. The filler metal should be added along the top edge of the weld pool to ensure adequate fusion.

4. As you develop skill, increase the angle until the weld is being made in the vertical up position.

5. Repeat the process using all thicknesses and types of metal until you can consistently make the weld visually defect free.

6. Turn off the welding machine and shielding gas and clean up your work area when you are finished welding.

INSTRUCTOR'S COMMENTS _____

_____

_____

■ PRACTICE 17-16

Name _____ Date _____

Base Metal _____ Thickness _____ Filler Dia. _____ Joint _____

Class _____ Instructor _____ Grade _____

OBJECTIVE: After completing this practice, you should be able to make a tee joint on carbon steel, stainless steel, and aluminum at a 45° vertical angle using the GTAW process.

EQUIPMENT AND MATERIALS NEEDED FOR THIS PRACTICE

1. A properly set-up and adjusted GTA welding machine.

2. Proper safety protection.

3. Filler rods, 36-in. (0.9-m) long by 1/16-in. (1.6-mm), 3/32-in. (2.4-mm), and 1/8-in. (3-mm) diameter.

4. Two or more pieces of mild steel, stainless steel, and aluminum, 1-1/2-in. (38-mm) wide by 6-in. (152-mm) long by 1/16-in. (2-mm) and 1/8-in. (3-mm) thick, and aluminum plate 1/4-in. (6-mm) thick.

INSTRUCTIONS

1. Tack weld the plates together in the tee joint configuration then place them on the welding table at a 45° vertical angle. Start the weld at the bottom and weld in an upward direction.

2. The edge of the side plate will heat up more quickly than the back plate. This may lead to undercutting along the edge of the weld.

3. To control undercutting, keep the arc on the back plate and add the filler metal to the weld pool near the side plate.

4. Repeat the process using all thicknesses and types of metal until you can consistently make the weld with 100% penetration.

5. Turn off the welding machine and shielding gas and clean up your work area when you are finished welding.

INSTRUCTOR'S COMMENTS _____

_____

_____

■ PRACTICE 17-17

Name _____   Date _____

Base Metal _____   Thickness _____   Filler Dia. _____   Joint _____

Class _____   Instructor _____   Grade _____

OBJECTIVE: After completing this practice, you should be able to make a tee joint on carbon steel, stainless steel, and aluminum in the 3F position using the GTAW process.

EQUIPMENT AND MATERIALS NEEDED FOR THIS PRACTICE

1. A properly set-up and adjusted GTA welding machine.

2. Proper safety protection.

3. Filler rods, 36-in. (0.9-m) long by 1/16-in. (1.6-mm), 3/32-in. (2.4-mm), and 1/8-in. (3-mm) diameter.

4. Two or more pieces of mild steel, stainless steel, and aluminum, 1-1/2-in. (38-mm) wide by 6-in. (152-mm) long by 1/16-in. (2-mm) and 1/8-in. (3-mm) thick, and aluminum plate 1/4-in. (6-mm) thick.

INSTRUCTIONS

1. Tack weld the plates together in the tee joint configuration then place them on the welding table at a 45° vertical angle. Start the weld at the bottom and weld in an upward direction.

2. The edge of the side plate will heat up more quickly than the back plate. This may lead to undercutting along the edge of the weld.

3. To control undercutting, keep the arc on the back plate and add the filler metal to the weld pool near the side plate.

4. Gradually increase the plate angle as your skill develops until the weld is being made in the vertical up position.

5. Repeat the process using all thicknesses and types of metal until you can consistently make the weld visually defect free.

6. Turn off the welding machine and shielding gas and clean up your work area when you are finished welding.

INSTRUCTOR'S COMMENTS _____

_____

_____

■ PRACTICE 17-18

Name _____ Date _____

Base Metal _____ Thickness _____ Filler Dia. _____ Joint _____

Class _____ Instructor _____ Grade _____

OBJECTIVE: After completing this practice, you should be able to make a stringer bead on carbon steel, stainless steel, and aluminum at a 45° reclining angle using the GTAW process.

EQUIPMENT AND MATERIALS NEEDED FOR THIS PRACTICE

1. A properly set-up and adjusted GTA welding machine.

2. Proper safety protection.

3. Filler rods, 36-in. (0.9-m) long by 1/16-in. (1.6-mm), 3/32-in. (2.4-mm), and 1/8-in. (3-mm) diameter.

4. One or more pieces of mild steel, stainless steel, and aluminum, 1-1/2-in. (38-mm) wide by 6-in. (152-mm) long by 1/16-in. (2-mm) and 1/8-in. (3-mm) thick, and aluminum plate 1/4-in. (6-mm) thick.

INSTRUCTIONS

1. Place the plate on the welding table at a 45° reclining angle. Add the filler metal along the top leading edge of the weld pool.

2. Do not let the weld pool become too large so that surface tension will hold it in place.

3. Use a rhythmic movement of the torch and filler rod so that the bead will be uniform in width and contour.

4. Repeat the process using all thicknesses and types of metal until you can consistently make the weld visually defect free.

5. Turn off the welding machine and shielding gas and clean up your work area when you are finished welding.

INSTRUCTOR'S COMMENTS _____

_____

_____

■ PRACTICE 17-19

Name _____ Date _____

Base Metal _____ Thickness _____ Filler Dia. _____ Joint _____

Class _____ Instructor _____ Grade _____

OBJECTIVE: After completing this practice, you should be able to make a stringer bead on carbon steel, stainless steel, and aluminum in the horizontal position using the GTAW process.

EQUIPMENT AND MATERIALS NEEDED FOR THIS PRACTICE

1. A properly set-up and adjusted GTA welding machine.

2. Proper safety protection.

3. Filler rods, 36-in. (0.9-m) long by 1/16-in. (1.6-mm), 3/32-in. (2.4-mm), and 1/8-in. (3-mm) diameter.

4. One or more pieces of mild steel, stainless steel, and aluminum, 1-1/2-in. (38-mm) wide by 6-in. (152-mm) long by 1/16-in. (2-mm) and 1/8-in. (3-mm) thick, and aluminum plate 1/4-in. (6-mm) thick.

INSTRUCTIONS

1. Place the plate on the welding table at a 45° reclining angle. Add the filler metal along the top leading edge of the weld pool.

2. Do not let the weld pool become too large so that surface tension will hold it in place.

3. Use a rhythmic movement of the torch and filler rod so that the bead will be uniform in width and contour.

4. Gradually increase the plate angle as your skill develops until the weld is being made in the horizontal position on a vertical plate.

5. Repeat the process using all thicknesses and types of metal until you can consistently make the weld visually defect free.

6. Turn off the welding machine and shielding gas and clean up your work area when you are finished welding.

INSTRUCTOR'S COMMENTS _____

_____

_____

■ PRACTICE 17-20

Name _____   Date _____

Base Metal _____   Thickness _____   Filler Dia. _____   Joint _____

Class _____   Instructor _____   Grade _____

OBJECTIVE: After completing this practice, you should be able to make a butt joint on carbon steel, stainless steel, and aluminum in the horizontal 2G position using the GTAW process.

EQUIPMENT AND MATERIALS NEEDED FOR THIS PRACTICE

1. A properly set-up and adjusted GTA welding machine.

2. Proper safety protection.

3. Filler rods, 36-in. (0.9-m) long by 1/16-in. (1.6-mm), 3/32-in. (2.4-mm), and 1/8-in. (3-mm) diameter.

4. Two or more pieces of mild steel, stainless steel, and aluminum, 1-1/2-in. (38-mm) wide by 6-in. (152-mm) long by 1/16-in. (2-mm) and 1/8-in. (3-mm) thick, and aluminum plate 1/4-in. (6-mm) thick.

INSTRUCTIONS

1. Tack the plates together in the butt joint configuration and place them on the weld table at a 45° reclining angle.

2. Add the filler metal to the top plate and keep the bead small in size.

3. Use a rhythmic movement of the torch and filler rod so that the bead will be uniform in width and contour.

4. Gradually increase the plate angle as your skill develops until the weld is being made in the horizontal position on a vertical plate.

5. Repeat the process using all thicknesses and types of metal until you can consistently make the weld visually defect free.

6. Turn off the welding machine and shielding gas and clean up your work area when you are finished welding.

INSTRUCTOR'S COMMENTS _____

_____

_____

■ PRACTICE 17-21

Name _____ Date _____

Base Metal _____ Thickness _____ Filler Dia. _____ Joint _____

Class _____ Instructor _____ Grade _____

OBJECTIVE: After completing this practice, you should be able to make a lap joint on carbon steel, stainless steel, and aluminum in the horizontal 2F position using the GTAW process.

EQUIPMENT AND MATERIALS NEEDED FOR THIS PRACTICE

1. A properly set-up and adjusted GTA welding machine.

2. Proper safety protection.

3. Filler rods, 36-in. (0.9-m) long by 1/16-in. (1.6-mm), 3/32 in. (2.4-mm), and 1/8-in. (3-mm) diameter.

4. Two or more pieces of mild steel, stainless steel, and aluminum, 1-1/2 in. (38-mm) wide by 6 in. (152-mm) long by 1/16-in. (2-mm) and 1/8-in. (3-mm) thick, and aluminum plate 1/4-in. (6-mm) thick.

INSTRUCTIONS

1. Tack the plates together in the lap joint configuration with an overlap of 1/4 in. (6 mm) to 3/8 in. (10 mm). Place them on the weld table in the horizontal position.

2. Add the filler metal along the top edge of the weld pool and keep the bead small in size. Avoid heating the top plate too much which will result in undercutting.

3. The bottom plate will act as a shelf to support the molten weld pool.

4. Repeat the process using all thicknesses and types of metal until you can consistently make the weld visually defect free.

5. Turn off the welding machine and shielding gas and clean up your work area when you are finished welding.

INSTRUCTOR'S COMMENTS _____

_____

_____

■ PRACTICE 17-22

Name _____   Date _____

Base Metal _____   Thickness _____   Filler Dia. _____   Joint _____

Class _____   Instructor _____   Grade _____

OBJECTIVE: After completing this practice, you should be able to make a tee joint on carbon steel, stainless steel, and aluminum in the horizontal 2F position using the GTAW process.

EQUIPMENT AND MATERIALS NEEDED FOR THIS PRACTICE

1. A properly set-up and adjusted GTA welding machine.

2. Proper safety protection.

3. Filler rods, 36-in. (0.9-m) long by 1/16-in. (1.6-mm), 3/32-in. (2.4 mm), and 1/8-in. (3-mm) diameter.

4. Two or more pieces of mild steel, stainless steel, and aluminum, 1-1/2 in. (38 mm) wide by 6 in. (152 mm) long by 1/16-in. (2-mm) and 1/8-in. (3-mm) thick, and aluminum plate 1/4-in. (6-mm) thick.

INSTRUCTIONS

1. Tack the plates together in the tee joint configuration and place them on the weld table in the horizontal position.

2. Add the filler metal along the top leading edge of the weld pool. Beware of undercutting.

3. The bottom plate will act as a shelf to support the molten weld pool.

4. Repeat the process using all thicknesses and types of metal until you can consistently make the weld visually defect free.

5. Turn off the welding machine and shielding gas and clean up your work area when you are finished welding.

INSTRUCTOR'S COMMENTS _____

_____

_____

■ PRACTICE 17-23

Name _____ Date _____

Base Metal _____ Thickness _____ Filler Dia. _____ Joint _____

Class _____ Instructor _____ Grade _____

OBJECTIVE: After completing this practice, you should be able to make a stringer bead on carbon steel, stainless steel, and aluminum in the overhead position using the GTAW process.

EQUIPMENT AND MATERIALS NEEDED FOR THIS PRACTICE

1. A properly set-up and adjusted GTA welding machine.

2. Proper safety protection.

3. Filler rods, 36-in. (0.9-m) long by 1/16-in. (1.6-mm), 3/32-in. (2.4-mm), and 1/8-in. (3-mm) diameter.

4. One or more pieces of mild steel, stainless steel, and aluminum, 1-1/2-in. (38-mm) wide by 6 in. (152 mm) long by 1/16-in. (2-mm) and 1/8-in. (3-mm) thick, and aluminum plate 1/4-in. (6-mm) thick.

INSTRUCTIONS

1. Place the plate in the overhead position.

2. Keep the bead small and add the filler metal to the leading edge of the weld pool. Surface tension will hold the weld pool in place unless it gets too large.

3. A wide bead with little buildup will be easier to control and less likely to undercut along the edges.

4. Repeat the process using all thicknesses and types of metal until you can consistently make the weld visually defect free.

5. Turn off the welding machine and shielding gas and clean up your work area when you are finished welding.

INSTRUCTOR'S COMMENTS _____

_____

_____

■ PRACTICE 17-24

Name _____   Date _____

Base Metal _____   Thickness _____   Filler Dia. _____   Joint _____

Class _____   Instructor _____   Grade _____

OBJECTIVE: After completing this practice, you should be able to make a butt joint on carbon steel, stainless steel, and aluminum in the overhead 4G position using the GTAW process.

EQUIPMENT AND MATERIALS NEEDED FOR THIS PRACTICE

1. A properly set-up and adjusted GTA welding machine.

2. Proper safety protection.

3. Filler rods, 36-in. (0.9-m) long by 1/16-in. (1.6-mm), 3/32-in. (2.4-mm), and 1/8-in. (3-mm) diameter.

4. One or more pieces of mild steel, stainless steel, and aluminum, 1-1/2-in. (38-mm) wide by 6-in. (152-mm) long by 1/16-in. (2-mm) and 1/8-in. (3-mm) thick, and aluminum plate 1/4-in. (6-mm) thick.

INSTRUCTIONS

1. Tack weld the pieces together in the butt joint configuration and place them in the overhead position.

2. Keep the bead small and add the filler metal to the leading edge of the weld pool. Surface tension will hold the weld pool in place unless it gets too large.

3. The complete bead should have a uniform width.

4. Repeat the process using all thicknesses and types of metal until you can consistently make the weld visually defect free.

5. Turn off the welding machine and shielding gas and clean up your work area when you are finished welding.

INSTRUCTOR'S COMMENTS _____

_____

_____

■ PRACTICE 17-25

Name _____ Date _____

Base Metal _____ Thickness _____ Filler Dia. _____ Joint _____

Class _____ Instructor _____ Grade _____

OBJECTIVE: After completing this practice, you should be able to make a fillet weld on a lap joint on carbon steel, stainless steel, and aluminum in the overhead 4F position using the GTAW process.

EQUIPMENT AND MATERIALS NEEDED FOR THIS PRACTICE

1. A properly set-up and adjusted GTA welding machine.

2. Proper safety protection.

3. Filler rods, 36-in. (0.9-m) long by 1/16-in. (1.6-mm), 3/32-in. (2.4-mm), and 1/8-in. (3-mm) diameter.

4. Two or more pieces of mild steel, stainless steel, and aluminum, 1-1/2-in. (38-mm) wide by 6-in. (152-mm) long by 1/16-in. (2-mm) and 1/8-in. (3-mm) thick, and aluminum plate 1/4-in. (6-mm) thick.

INSTRUCTIONS

1. Tack weld the plates together in the lap joint configuration with an overlap of 1/4 in. (6 mm) to 3/8 in. (10 mm). Place them in the overhead position.

2. Concentrate the heat and filler metal on the top plate.

3. Gravity and an occasional sweep of the torch along the bottom plate will pull the weld pool down.

4. Control undercutting along the top edge of the weld by putting most of the filler metal along the top edge.

5. Repeat the process using all thicknesses and types of metal until you can consistently make the weld visually defect free.

6. Turn off the welding machine and shielding gas and clean up your work area when you are finished welding.

INSTRUCTOR'S COMMENTS _____

_____

_____

■ PRACTICE 17-26

Name _____ Date _____

Base Metal _____ Thickness _____ Filler Dia. _____ Joint _____

Class _____ Instructor _____ Grade _____

OBJECTIVE: After completing this practice, you should be able to make a fillet weld on a tee joint on carbon steel, stainless steel, and aluminum in the overhead 4F position using the GTAW process.

EQUIPMENT AND MATERIALS NEEDED FOR THIS PRACTICE

1. A properly set-up and adjusted GTA welding machine.

2. Proper safety protection.

3. Filler rods, 36-in. (0.9-m) long by 1/16-in. (1.6-mm), 3/32-in. (2.4-mm), and 1/8-in. (3-mm) diameter.

4. Two or more pieces of mild steel, stainless steel, and aluminum, 1-1/2-in. (38-mm) wide by 6-in. (152-mm) long by 1/16-in. (2-mm) and 1/8-in. (3-mm) thick, and aluminum plate 1/4-in. (6-mm) thick.

INSTRUCTIONS

1. Tack weld the plates together in the tee joint configuration and place them in the overhead position.

2. Concentrate the heat and filler metal on the top plate. Beware of undercutting.

3. A "J" weave pattern will help pull down any needed metal to the side plate.

4. Repeat the process using all thicknesses and types of metal until you can consistently make the weld visually defect free.

5. Turn off the welding machine and shielding gas and clean up your work area when you are finished welding.

INSTRUCTOR'S COMMENTS _____

_____

_____

CHAPTER 17: QUIZ 1

Name _____ Date _____

Class _____ Instructor _____ Grade _____

INSTRUCTIONS

Carefully read Chapter 17 in the textbook and answer each question.

MATCHING

In the space provided to the left of Column A, write the letter from Column B that best answers or completes the statement in Column A.

Column A

Column B

_____ 1. With gas tungsten arc welding, welders can have a clear unobstructed view of the molten weld pool because GTA welding is fluxless, slagless, and _____.

a. quality

_____ 2. The fine _____ of the weld that is possible with GTA welding makes it an ideal process for very close-tolerance, high-quality welds.

b. shielded

_____ 3. The proper setup of GTA equipment can often affect the _____ of the weld performed.

c. settings

_____ 4. The torch should be held as close to _____ as possible in relation to the plate surface.

d. shielding gas

_____ 5. The closer the torch is held to perpendicular, the better the weld is _____.

e. shortest

_____ 6. As the velocity of the shielding gas increases, a low-pressure area develops behind the _____.

f. base metal

_____ 7. When the low-pressure area becomes strong enough, _____ is pulled into the shielding gas.

g. mild steel

_____ 8. The filler rod end must be kept inside the protective zone of the _____.

h. smokeless

_____ 9. If the end of the rod becomes oxidized, then it should be _____ before restarting.

i. carbon

_____ 10. The rod should enter the shielding gas as close to the _____ as possible.

j. filler metal

_____ 11. The tungsten becomes contaminated when it touches the molten weld pool or when it is touched by the _____.

k. air

_____ 12. Failure to remove the contamination properly will result in the _____ of the weld.

l. improper

—— 13. The amperage set on a machine and the actual welding current are often _____.

m. problem

—— 14. In addition to the difference between indicated and actual welding amperage, there is a more significant difference between amperage and welding _____.

n. not the same

—— 15. A chart and a series of tests can be used to set the lower and upper limits for the amperage _____.

o. control

—— 16. Even experienced welders make changes in the setup, current, or welding technique as they try to resolve a _____.

p. power

—— 17. The weld quality can be adversely affected by _____ gas flow settings.

q. cup

—— 18. The lowest possible gas flow rates and the _____ preflow or postflow time can help reduce the cost of welding by saving the expensive shielding gas.

r. failure

—— 19. Low carbon and _____ are two basic steel classifications.

s. cut off

—— 20. _____ is the primary alloy in these classifications of steel, and it ranges from 0.15% or less for low carbon and 0.15% to 0.30% for mild steel.

t. perpendicular

CHAPTER 17: QUIZ 2

Name _____   Date _____

Class _____   Instructor _____   Grade _____

INSTRUCTIONS

Carefully read Chapter 17 in the textbook and answer each question.

MATCHING

In the space provided to the left of Column A, write the letter from Column B that best answers or completes the statement in Column A.

Column A	Column B
_____ 1. The setup and manipulation techniques required for stainless steel are nearly the same as those for low carbon and _____, and skills transfer is easy.	a. rounded
_____ 2. To make a weld on stainless steel, you must do a better job of precleaning the base metal and filler metal; make sure you have adequate shielding gas coverage and do not _____ the weld.	b. carbide precipitation
_____ 3. The most common sign that there is a problem with a stainless steel weld is the bead _____ after the weld.	c. flux
_____ 4. The greater the contamination, the _____ the color.	d. cleaned
_____ 5. Using a low arc current setting with faster travel speeds is important when welding stainless steel because some stainless steels are subject to _____.	e. surface tension
_____ 6. Black crusty spots on weld beads are often caused by improper cleaning of the filler rod or failure to keep the end of the rod inside the _____.	f. mechanical
_____ 7. Aluminum is GTA welded using a _____ tip tungsten, Figure 17-12, with the welding machine set for ACHF welding current.	g. overheat
_____ 8. The molten aluminum weld pool has high _____, which allows large weld beads to be controlled easily.	h. chill
_____ 9. The high thermal conductivity of aluminum may make starting a weld on thick sections difficult without first _____ the base metal.	i. time
_____ 10. The processes of cleaning and keeping the metal clean take a lot of _____.	j. cut out
_____ 11. Keep your hands and gloves clean and oil-free so the base metal or filler rods do not become _____.	k. color
_____ 12. Both the base metal and the filler metal used in the GTAW process must be thoroughly _____ before welding.	l. recontaminated
_____ 13. Contamination left on the metal will be deposited in the weld because there is no _____ to remove it.	m. manufacturer's
_____ 14. Oxides, oils, and dirt are the most common types of _____.	n. darker

_____ 15. _____ metal cleaning may be done by grinding, wire brushing, scraping, machining, sand blasting, or filing.

_____ 16. _____ cleaning may be done by using acids, alkalies, solvents, or detergents.

_____ 17. Failure to follow the _____ recommendations for using their products may result in chemical burns, fires, fumes, or other safety hazards that could lead to serious injury.

_____ 18. If anyone comes in contact with any chemicals, immediately refer to the _____ for the proper corrective action.

_____ 19. To keep the formation of oxides on the bead to a minimum, a _____ plate (a thick piece of metal used to absorb heat) may be required.

_____ 20. Mild steel weld coupons can have the welds removed using an OFC torch, but welds on stainless and aluminum coupons will have to be _____ with a PAC torch.

o. mild steels

p. shielding gas

q. contaminants

r. preheating

s. chemical

t. material safety data sheet (MSDS)

CHAPTER 17: QUIZ 3

Name _____ Date _____

Class _____ Instructor _____ Grade _____

INSTRUCTIONS
Carefully read Chapter 17 in the textbook and answer each question.

IDENTIFICATION
Identify the numbered items on the drawing by writing the letter next to the identifying term in the space provided.

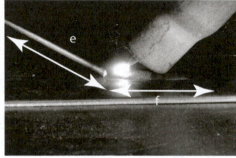

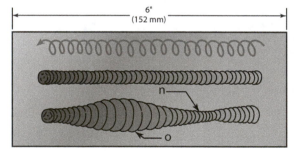
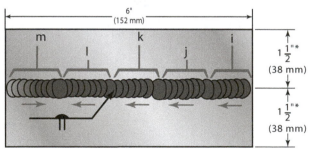

1. _____ ROD ANGLE

2. _____ OXIDE ON ROD

3. _____ WATCH BUILDUP HERE

4. _____ TOO SLOW AND/OR HOT

5. _____ WATCH PENETRATION HERE

6. _____ MOVE THE TORCH BACK AND FORTH

7. _____ OXIDE ON WELD

8. _____ 1ST WELD

9. _____ 2ND WELD

10. _____ 3RD WELD

11. _____ 4TH WELD

12. _____ 5TH WELD

13. _____ TOO FAST AND/OR COLD

14. _____ MOVE THE FILLER METAL BACK AND FORTH

15. _____ AIR

Gas Tungsten Arc Welding of Pipe

■ PRACTICE 18-1

Name _____ Date _____

Pipe Diameter _____ Filler Metal Type _____ Diameter _____

Class _____ Instructor _____ Grade _____

OBJECTIVE: After completing this practice, you should be able to properly tack weld 3-in. (76-mm) to 10-in. (254-mm) diameter schedule 40 carbon steel pipe using the GTAW process.

EQUIPMENT AND MATERIALS NEEDED FOR THIS PRACTICE

1. A properly set-up and adjusted GTA welding machine.

2. Proper safety protection (welding hood, safety glasses, wire brush, leather gloves, pliers, long-sleeved shirt, long pants, and leather boots or shoes). Refer to Chapter 2 in the text for more specific safety information.

3. Two or more pieces of 3-in. (76-mm) to 10-in. (254-mm) diameter schedule 40 mild steel pipe, prepared with single V-grooved joints.

4. Several Filler rods, having 1/16-in. (2-mm), 3/32-in. (2.4-mm), and 1/8-in. (3-mm) diameters.

INSTRUCTIONS

1. Bend the rods with the 1/16-in. (2-mm) and 3/32-in. (2.4-mm) diameters into a U-shape. These rods will be used to set the desired root opening.

2. Lay the pipe in an angle iron cradle. Slide the ends together so that the desired diameter wire is held between the ends.

3. Hold the torch nozzle against the beveled sides of the groove. The torch should be nearly parallel with the pipe.

4. Bring the filler rod end in close and rest it on the joint just ahead of the torch.

5. Lower your helmet and switch on the welding current by using the foot or hand control. Slowly straighten the torch so that the tungsten electrode is brought closer to the groove. This pivoting of the torch around the nozzle will keep the torch aligned with the joint and help prevent arc starts outside of the joint. When the tungsten electrode is close enough, an arc will be established.

6. Increase the current by depressing the foot or hand control until a molten weld pool is established on both root faces.

7. Dip the welding rod into the molten weld pool as the torch is pivoted from side to side.

8. Slowly move ahead and repeat this step until you have made a tack weld approximately 1/2-in. (13-mm) long. At the end of the tack weld, slowly reduce the current and fill the weld crater. If the tack weld crater is properly filled and it does not crack, grinding may not be required.

9. Roll the pipe 180° to the opposite side. Check the root opening and adjust the opening if needed. Make a tack weld 1/2-in. (13-mm) long.

10. Roll the pipe 90° to a spot halfway between the first two tack welds. Check and adjust the root opening if needed. Make a tack weld.

11. Roll the pipe 180° to the opposite side and make a tack weld halfway between the first two tack welds.

12. The completed joint should have four tack welds evenly spaced around the pipe (at 90° intervals). The root surfaces of the tack welds should be flat or slightly convex. The tack welds should have complete fusion into the base metal with no overlap at the start.

13. Continue making tack welds until this procedure is mastered.

14. Turn off the welding machine and shielding gas and clean up your work area when you are finished welding.

INSTRUCTOR'S COMMENTS _____

_____

_____

■ PRACTICE 18-2

Name _____   Date _____

Pipe Diameter _____   Filler Metal Type _____   Diameter _____

Class _____   Instructor _____   Grade _____

OBJECTIVE: After completing this practice, you should be able to weld root bead welds on carbon steel pipe with single V-groove joints using the GTAW process in the 1G position.

EQUIPMENT AND MATERIALS NEEDED FOR THIS PRACTICE

1. A properly set-up and adjusted GTA welding machine.

2. Proper safety protection (welding hood, safety glasses, wire brush, leather gloves, pliers, long-sleeved shirt, long pants, and leather boots or shoes). Refer to Chapter 2 in the text for more specific safety information.

3. Two or more pieces of 3-in. (76-mm) to 10-in. (254-mm) diameter schedule 40 mild steel pipe, prepared with single V-grooved joints.

4. Several filler rods, having 1/16-in. (2-mm), 3/32-in. (2.4-mm), and 1/8-in. (3-mm) diameters.

INSTRUCTIONS

1. Place the tacked pipe securely in an angle iron vee block. Using the same procedure as practiced for the tack welds, place the nozzle against the beveled sides of the joint with the torch at a steep angle.

2. Start the weld near the 3 o'clock position and weld toward the 12 o'clock position. Stop welding just past the 12 o'clock position and roll the pipe. Starting the weld above the 3 o'clock position is easier and starting the weld below this position is harder. Brace yourself against the pipe and try moving through the length of the weld to determine if you have full freedom of movement before welding.

3. Hold the welding rod close to the pipe so that it is protected by the shielding gas, thus preventing air from being drawn into the inert gas around the weld. The end of the rod must be far enough away from the starting point so that it will not be melted immediately by the arc. However, it must be close enough so that it can be seen by the light of the arc.

4. When you are comfortable and ready to start, lower your helmet. Switch on the welding current by depressing the foot or hand control. Slowly pivot the torch until the arc starts. Increase the current to establish molten weld pools on both sides of the root face. Add the filler rod to the molten weld pool as the torch is rocked from side to side and moved slowly ahead. The side-to-side motion can be used to walk the nozzle along the groove. The walking of the nozzle will ensure complete fusion in both root faces and help make a very uniform weld. When walking the nozzle along the groove, the tip of the torch cap will make a figure-8 motion.

   NOTE: For the best results with nozzle walking, you should use a high-temperature ceramic nozzle such as silicon nitride.

Nozzle walking the torch along a groove can be practiced without welding power so you can see how to move the torch. This technique can be used on most GTA groove welds in pipe or plate.

5. Watch the molten weld pool to make sure that it is melted into and of equal height along both beveled edges of the beveled pipe joint. If the molten weld pool is not balanced, then one side is probably not as hot nor is it penetrating as well.

6. Stop adding filler metal to the molten weld pool when it is within 1/16 in. (2 mm) of the tack weld. At this time, the keyhole will be much smaller or completely closed. Continue to rock the torch from side to side as you move ahead. Watch the molten weld pool. It should settle down or sink in when the weld has completely tied itself to the tack weld. Failure to do this will cause incomplete fusion or overlap at the root. If you are crossing a tack weld, carry the molten weld pool across the tack weld without adding filler metal and start adding filler rod at the other end. If you are at the closing end of the weld, slowly decrease the welding current as you add more filler rod to fill the weld crater.

7. After the weld is complete around the pipe, visually check the root for any defects. Repeat this practice as needed until it is mastered.

8. Turn off the welding machine and shielding gas and clean up your work area when you are finished welding.

INSTRUCTOR'S COMMENTS _____

_____

_____

■ PRACTICE 18-3

Name _____   Date _____

Pipe Diameter _____   Filler Metal Type _____   Diameter _____

Class _____   Instructor _____   Grade _____

OBJECTIVE: After completing this practice, you should be able to weld stringer beads on 3-in. (76-mm) to 10-in. (254-mm) diameter schedule 40 carbon steel pipe using the GTAW process in the 1G horizontal rolled position.

EQUIPMENT AND MATERIALS NEEDED FOR THIS PRACTICE

1. A properly set-up and adjusted GTA welding machine.

2. Proper safety protection (welding hood, safety glasses, wire brush, leather gloves, pliers, long-sleeved shirt, long pants, and leather boots or shoes). Refer to Chapter 2 in the text for more specific safety information.

3. Two or more pieces of 3-in. (76-mm) to 10-in. (254-mm) diameter schedule 40 mild steel pipe, prepared with single V-grooved joints.

4. Several filler rods having 1/16-in. (2-mm), 3/32-in. (2.4-mm), and 1/8-in. (3-mm) diameters.

INSTRUCTIONS

1. Start by cleaning a strip 1-in. (25-mm) wide around the pipe.

2. Brace yourself against the pipe so that you will not sway during welding. The torch should have a slight upward slope, about 5° to 10° from perpendicular to the pipe. The filler rod should be held so that it enters the molten weld pool from the top center at a right angle to the torch.

3. Start the weld near the 3 o'clock position and weld upward to a point just past the 12 o'clock position. In order to minimize stops and starts, try to make the complete weld without stopping.

4. With the power off and your helmet up, move the torch through the weld to be sure you have full freedom of movement.

5. With the torch held in position so that the electrode is just above the desired starting spot, lower your helmet. Start the welding current by depressing the foot or hand control switch. Establish a weld pool and dip the filler rod in it. Use a straight forward and backward movement with both the torch and rod. Never move farther ahead than the leading edge of the weld pool. Too large a movement will lead to inadequate weld pool coverage by the shielding gas. The short motion is to allow the filler rod to be dipped into the weld pool without the heat from the arc melting the rod back.

6. As the weld progresses from a vertical position to a flat position, the frequency of movement and pause times will change. Generally, the more vertical the weld, the faster the movement and the shorter the pause times. As the weld becomes flatter, the movement is slower and the pause times

are longer. In the vertical position, gravity tends to pull the weld toward the back center of the bead. This makes a high crown on the weld with the possibility of undercutting along the edge. Flatter positions cause the weld to be pulled out, resulting in a thinner weld with less apparent root reinforcement.

7. When the weld reaches the 12 o'clock position, stop and roll the pipe. To stop, slowly decrease the current and add filler rod to fill the weld crater. Keep the torch held over the weld until the post purge stops. This precaution will prevent the air from forming oxides on the hot weld bead.

8. After the weld is complete around the pipe, it should be visually inspected for straightness, uniformity, and defects. Repeat the process until you can consistently make the weld visually defect free.

9. Turn off the welding machine and shielding gas and clean up your work area when you are finished welding.

INSTRUCTOR'S COMMENTS _____

_____

_____

■ PRACTICE 18-4

Name _____ Date _____

Pipe Diameter _____ Filler Metal Type _____ Diameter _____

Class _____ Instructor _____ Grade _____

OBJECTIVE: After completing this practice, you should be able to make straight weave and lace beads on mild steel pipe in the horizontal rolled 1G position using the GTAW process.

EQUIPMENT AND MATERIALS NEEDED FOR THIS PRACTICE

1. A properly set-up and adjusted GTA welding machine.

2. Proper safety protection.

3. Filler rods, 36-in. (0.9-m) long by 1/16-in. (2-mm), 3/32-in. (2.4-mm), and 1/8-in. (3-mm) diameter.

4. One or more pieces of mild steel pipe, 3 in. (76 mm) to 10 in. (254 mm) in diameter.

INSTRUCTIONS

1. Clean a strip 1-in. (25-mm) wide around the pipe. Place the pipe securely in an angle iron vee block in the flat position.

2. Ensure that you have enough freedom of movement weld from the 3 o'clock to the 12 o'clock position. Hold the torch with a 5° to 10° upward angle. Start the weld in the 3 o'clock position.

3. To make a weave bead move the torch in a "C," "U," or zigzag pattern across the weld pool. Keep the pattern no wider than one-half the diameter of the cup.

4. To make a lace bead, move the torch in a zigzag pattern across the pipe. Keep the weld pool about the same size as for a stringer bead. The distance of the zigzag can be anything desired to produce a lace bead of any width.

5. Continue the weave or lace bead up the pipe to the top. Lower the current and fill the crater.

6. Rotate the pipe 90° and, beginning where you left off, repeat the steps above. Do this until you have welded 360° around the pipe.

7. After you have welded all the way around the pipe, visually inspect the bead. It must be straight, have good contour, and be visually defect free.

8. Turn off the welding machine and shielding gas and clean up your work area when you are finished welding.

INSTRUCTOR'S COMMENTS _____

_____

_____

■ PRACTICE 18-5

Name _____ Date _____

Pipe Diameter _____ Filler Metal Type _____ Diameter _____

Class _____ Instructor _____ Grade _____

OBJECTIVE: After completing this practice, you should be able to make a filler pass on mild steel pipe in the horizontal rolled 1G position using the GTAW process.

EQUIPMENT AND MATERIALS NEEDED FOR THIS PRACTICE

1. A properly set-up and adjusted GTA welding machine.

2. Proper safety protection.

3. Filler rods, 36-in. (0.9-m) long by 1/16-in. (2-mm), 3/32-in. (2.4-mm), and 1/8-in. (3-mm) diameter.

4. Two or more pieces of schedule 40 mild steel pipe, 3 in. (76 mm) to 10 in. (254 mm) in diameter with single V-groove prepared ends and already having a root pass in place.

INSTRUCTIONS

1. Place the pipe securely in an angle iron vee block in the flat position.

2. Place the cup against the beveled sides of the joint with a 5° to 10° upward angle. Be sure that you have full freedom of movement.

3. Establish the weld pool and use a forward and backward motion as you did in Practice 17-3. Start the weld near the 1 o'clock position and end just beyond the 12 o'clock position.

4. Add the filler metal at the top center of the molten weld pool until the bead surface is flat or slightly convex. Filler passes should have very little penetration so that maximum reinforcement can be added with each pass. Deep penetration will slow the joint fill-up rate.

5. Lower the current and fill the crater. Another method of ending the bead is to slowly decrease the current and pull the bead up on the beveled side of the joint.

6. Turn the pipe and, beginning where you left off, repeat the steps above. Do this until you have welded 360° around the pipe.

7. After you have welded all the way around the pipe, visually inspect the bead.

8. Turn off the welding machine and shielding gas and clean up your work area when you are finished welding.

INSTRUCTOR'S COMMENTS _____

_____

_____

■ PRACTICE 18-6

Name _____   Date _____

Pipe Diameter _____   Filler Metal Type _____   Diameter _____

Class _____   Instructor _____   Grade _____

OBJECTIVE: After completing this practice, you should be able to make a cover pass on mild steel pipe in the horizontal rolled 1G position using the GTAW process.

EQUIPMENT AND MATERIALS NEEDED FOR THIS PRACTICE

1. A properly set-up and adjusted GTA welding machine.

2. Proper safety protection.

3. Filler rods, 36-in. (0.9-m) long by 1/16 in. (2-mm), 3/32-in. (2.4-mm), and 1/8-in. (3-mm) diameter.

4. Two or more pieces of schedule 40 mild steel pipe, 3 in. (76 mm) to 10 in. (254 mm) in diameter with single V-groove prepared ends and already having a root pass and flush filler passes in place.

INSTRUCTIONS

1. Place the pipe securely in an angle iron vee block in the flat position.

2. Place the cup against the beveled sides of the joint with a 5° to 10° upward angle. Be sure that you have full freedom of movement. Start the weld near the 1 o'clock position and end just beyond the 12 o'clock position.

3. Using the same techniques and skill developed in Practice 17-5, make a cover pass.

4. The weld should not be more than 3/32 in. (2.4 mm) wider than the groove and should have a buildup of no more than 3/32 in. (2.4 mm). Fill the weld craters as you go.

5. Turn the pipe and, beginning where you left off, repeat the steps above. Do this until you have welded 360° around the pipe.

6. After you have welded all the way around the pipe, visually inspect the bead.

7. Turn off the welding machine and shielding gas and clean up your work area when you are finished welding.

INSTRUCTOR'S COMMENTS _____

_____

_____

■ PRACTICE 18-7

Name _____ Date _____

Pipe Diameter_____ Filler Metal Type_____ Diameter _____

Class _____ Instructor _____ Grade _____

OBJECTIVE: After completing this practice, you should be able to make a stringer bead on mild steel pipe in the horizontal fixed 5G position using the GTAW process.

EQUIPMENT AND MATERIALS NEEDED FOR THIS PRACTICE

1. A properly set-up and adjusted GTA welding machine.

2. Proper safety protection.

3. Filler rods, 36-in. (0.9-m) long by 1/16-in. (2-mm), 3/32-in. (2.4-mm), and 1/8-in. (3-mm) diameter.

4. One or more pieces of schedule 40 mild steel pipe, 3 in. (76 mm) to 10 in. (254 mm) in diameter.

INSTRUCTIONS

1. Clean a strip 1-in. (25-mm) wide around the pipe.

2. Mark the top for future reference and clamp the pipe at a comfortable work height. Be sure that you have complete freedom of movement.

3. Establish a weld pool at the 6 o'clock position. Keep the weld pool small for controllability and uniformity. Add filler metal at the front leading edge of the weld pool.

4. Move the torch forward and backward as the filler is added. The frequency of movement will increase as the weld becomes more vertical.

5. Continue the weld without stopping, if possible, to the 12 o'clock position.

6. Repeat the weld up the other side of the pipe in the same manner.

7. Inspect the weld for straightness, uniformity, and visual defects.

8. Repeat this practice until you can consistently make welds that are visually defect free.

9. Turn off the welding machine and shielding gas and clean up your work area when you are finished.

INSTRUCTOR'S COMMENTS _____

_____

_____

■ PRACTICE 18-8

Name _____ Date _____

Pipe Diameter _____ Filler Metal Type _____ Diameter _____

Class _____ Instructor _____ Grade _____

OBJECTIVE: After completing this practice, you should be able to make a stringer bead on mild steel pipe in the vertical fixed 2G position using the GTAW process.

EQUIPMENT AND MATERIALS NEEDED FOR THIS PRACTICE

1. A properly set-up and adjusted GTA welding machine.

2. Proper safety protection.

3. Filler rods, 36-in. (0.9-m) long by 1/16-in. (2-mm), 3/32-in. (2.4-mm), and 1/8-in. (3-mm) diameter.

4. One or more pieces of schedule 40 mild steel pipe, 3 in. (76 mm) to 10 in. (254 mm) in diameter.

INSTRUCTIONS

1. Clean a strip 1-in. (25-mm) wide around the pipe.

2. Hold the torch at a 5° angle from horizontal and 5° to 10° from perpendicular to the pipe. (Refer to Figure 17-46 in the textbook.)

3. Establish a small weld pool and add filler along the top front edge. Keep the weld pool small for controllability and uniformity.

4. A slight "J" pattern may help to control weld bead sag.

5. Continue the weld without stopping, if possible, all the way around the pipe.

6. Inspect the weld for straightness, uniformity, and visual defects.

7. Repeat this practice until you can consistently make welds that are visually defect free.

8. Turn off the welding machine and shielding gas and clean up your work area when you are finished.

INSTRUCTOR'S COMMENTS _____

_____

_____

■ PRACTICE 18-9

Name _____   Date _____

Pipe Diameter _____   Filler Metal Type _____   Diameter _____

Class _____   Instructor _____   Grade _____

OBJECTIVE: After completing this practice, you should be able to make a stringer bead on mild steel pipe at a 45° inclined angle (6G position) using the GTAW process.

EQUIPMENT AND MATERIALS NEEDED FOR THIS PRACTICE

1. A properly set-up and adjusted GTA welding machine.

2. Proper safety protection.

3. Filler rods, 36-in. (0.9-m) long by 1/16-in. (2-mm), 3/32-in. (2.4-mm), and 1/8-in. (3-mm) diameter.

4. One or more pieces of schedule 40 mild steel pipe, 3 in. (76 mm) to 10 in. (254 mm) in diameter.

INSTRUCTIONS

1. Clean a strip 1-in. (25-mm) wide around the pipe.

2. Clamp the pipe at a 45° inclined angle (6G position) to the table.

3. Starting at the 6:30 o'clock position with the torch at a slight downward angle, establish a small molten weld pool.

4. Add the filler metal at the upper leading edge of the molten weld pool. Keep the molten weld pool small. A slight "J" pattern may help to control weld bead shape. As the weld progresses around the pipe, the angle becomes more vertical and the rate of movement should increase.

5. Continue the weld without stopping, if possible, all the way around the pipe.

6. Inspect the weld for straightness, uniformity, and visual defects.

7. Repeat this practice until you can consistently make welds that are visually defect free.

8. Turn off the welding machine and shielding gas and clean up your work area when you are finished welding.

INSTRUCTOR'S COMMENTS _____

_____

_____

CHAPTER 18: QUIZ 1

Name _____   Date _____

Class _____   Instructor _____   Grade _____

INSTRUCTIONS

Carefully read Chapter 18 in the textbook and answer each question.

MATCHING

In the space provided to the left of Column A, write the letter from Column B that best answers or completes the statement in Column A.

Column A

Column B

_____ 1. Gas tungsten arc welding of pipe is used when the welded joint must have a high degree of _____.

a. bevel

_____ 2. Welders who are skilled in the GTA welding process have the ability to make consistently high-quality welds with a low _____ rate.

b. capped

_____ 3. The ends of pipe must be _____ before welding.

c. integrity

_____ 4. The _____ is the most commonly used pipe end preparation.

d. fatigue

_____ 5. The end of the pipe is prepared with a 37 1/2° _____, leaving a root face of 1/16 in. (2 mm) to 1/8 in. (3 mm).

e. burnthrough

_____ 6. The _____ is the small flat surface at the root of the groove.

f. heat

_____ 7. Before the joint is assembled, the welding surfaces must be cleaned and smoothed so that they are uniform and free of _____.

g. V-groove

_____ 8. The _____ of a weld is the deepest point into the joint where fusion between the base metal and filler metal occurs.

h. penetration

_____ 9. The root _____ is the distance measured between the original surface of the joint and the deepest point of fusion.

i. grooved

_____ 10. Sometimes when a weld does not completely penetrate the joint, or there may be a lack of fusion on one or both sides of the root, it is caused by not enough _____ penetrating the back side of the work.

j. pipe

_____ 11. Common causes for a concave root surface (suck back) are when insufficient _____ is added to the joint, or excessive heat is used in the overhead position.

k. leaks

_____ 12. Excessive root reinforcement or _____ is the excessive buildup of metal on the back side of a weld.

l. contaminants

——— 13. Root contamination can lead to faster corrosion, oxide flaking, weld brittleness, _____, stress points, or all of these.

m. consumable inserts

——— 14. When welding on alloy steel, stainless steel, aluminum, copper, and most other types of pipe, the easiest method of protecting the root from atmospheric contamination is to use a _____.

n. both

——— 15. On small diameters or short sections of pipe, the ends of the pipe are _____ to contain the backing gas in the pipe.

o. rejection

——— 16. For larger diameters, the pipe is plugged on _____ sides of the joint to be welded so that a smaller area can be purged.

p. preplacement

——— 17. The addition of filler metal to the joint can be done by dipping or by _____.

q. root

——— 18. _____ are preplaced filler metal, which is used for the root pass when consistent, high-quality welds are required.

r. root face

——— 19. Although the cup walking technique can be used on plate, it is most often used by _____ welders.

s. backing gas

——— 20. Cup walking reduces welder _____ because the welding torch nozzle and filler metal are both resting on the pipe.

t. filler metal

CHAPTER 18: QUIZ 2

Name _____   Date _____

Class _____   Instructor _____   Grade _____

INSTRUCTIONS

Carefully read Chapter 18 in the textbook and answer each question.

MATCHING

In the space provided to the left of Column A, write the letter from Column B that best answers or completes the statement in Column A.

Column A	Column B
_____ 1. The tungsten length for cup walking should be set so it extends out of the nozzle no more than the diameter of the _____ opening.	a. joint
_____ 2. Because the filler rod is rested in the welding joint, it is important that it should be slightly larger in diameter than the root opening so it does not _____ the root during welding.	b. insufficient
_____ 3. The outside edge of the nozzle must be _____ so it does not slip as it is stepped along the joint when the torch is moved in a figure-8 pattern.	c. figure-8
_____ 4. For the best results with cup walking, you should use a high-temperature _____ nozzle such as silicon nitride.	d. wrist
_____ 5. During cup walking, both the nozzle and the filler metal are rested on the pipe _____.	e. heat and temperature
_____ 6. The nozzle is held firmly against the pipe while only very _____ pressure is applied to the filler metal.	f. fall through
_____ 7. Too much _____ on the filler rod can cause it to push through the molten weld metal, causing root surface problems such as excessive root penetration or buildup.	g. shielding gas
_____ 8. The _____ movement is made during cup walking so that one edge of the nozzle is lifted off of the pipe joint edge.	h. penetration
_____ 9. When the torch nozzle begins touching the weld face as well as both sides of the groove, you should change to a _____ diameter nozzle to make it easier to walk the cup.	i. light
_____ 10. Moving your whole arm in the figure-8 pattern is less tiring than moving just your _____.	j. pressure
_____ 11. Hold the filler rod close to the pipe so it is protected by the _____, thus preventing air from being drawn into the inert gas around the weld.	k. metal and heat

———— 12. A _____ pass can be used to correct some of the problems caused by a poor root pass.

l. nozzle

———— 13. Normally, _____ is added during the hot pass, but if the root pass is overly filled and cannot be ground back, then the hot pass may be made without adding it all the time.

m. hot pass

———— 14. Concave root surfaces are generally caused by _____ filler metal for the joint.

n. filler pass

———— 15. To correct a concave root surface, a hot filler pass is used to add both the needed _____.

o. filler metal

———— 16. Incomplete fusion generally is caused by insufficient _____ for the joint.

p. sharp

———— 17. To correct incomplete fusion, a _____ is used with or without adding more filler metal.

q. 3/32 in. (2.4 mm)

———— 18. The _____ is the next weld layer(s) to be made after the hot pass.

r. larger

———— 19. The filler pass should have complete fusion but little _____.

s. ceramic

———— 20. The stringer bead can be continued to cap the weld; the bead should overlap the pipe surface no more than _____.

t. hot

CHAPTER 18: QUIZ 3

Name _____   Date _____

Class _____   Instructor _____   Grade _____

INSTRUCTIONS

Carefully read Chapter 18 in the textbook and answer each question.

IDENTIFICATION

Identify the numbered items on the drawing by writing the letter next to the identifying term in the space provided.

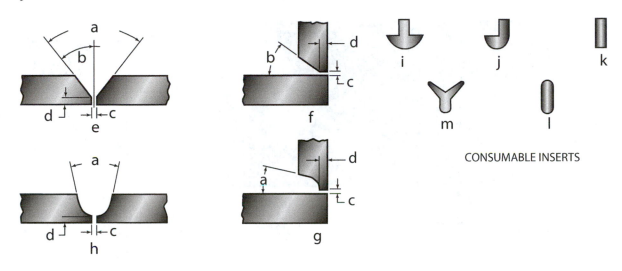

CONSUMABLE INSERTS

1. _____   A-SHAPE

2. _____   RECTANGULAR SHAPE

3. _____   ROOT OPENING

4. _____   J-SHAPE

5. _____   J-GROOVE

6. _____   ROOT FACE

7. _____   BEVEL

8. _____   V-GROOVE

9. _____   RECTANGULAR SHAPE (CONTOURED EDGE)

10. _____   Y-SHAPE

11. _____   BEVEL ANGLE

12. _____   GROOVE ANGLE

13. _____   U-GROOVE

CHAPTER 18: QUIZ 4

Name _____ Date _____

Class _____ Instructor _____ Grade _____

INSTRUCTIONS

Carefully read Chapter 18 in the textbook and answer each question.

IDENTIFICATION

Identify the numbered items on the drawing by writing the letter next to the identifying term in the space provided.

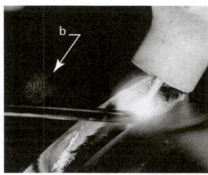

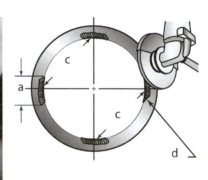

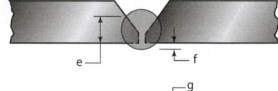

1. _____ TACK WELDS

2. _____ $\frac{1"}{8}$ (3 mm) MAX.

3. _____ WALL THICKNESS <u>LESS THAN</u> $\frac{3"}{8}$ (10 mm)

4. _____ WALL THICKNESS <u>MORE THAN</u> $\frac{3"}{8}$ (10 mm) OR THICKER

5. _____ 1" (25 mm)

6. _____ $\frac{3"}{32}$ (2.4 mm) MAX.

7. _____ DEPTH OF PENETRATION

8. _____ ROOT REINFORCEMENT

9. _____ OILY FINGERPRINT CONTAMINATION BURNED INTO PIPE SURFACE BY WELDING HEAT

10. _____ GRIND THE ENDS OF THE TACK WELDS TO A FEATHEREDGE

CHAPTER 18: QUIZ 5

Name _____   Date _____

Class _____   Instructor _____   Grade _____

INSTRUCTIONS

Carefully read Chapter 18 in the textbook and answer each question.

IDENTIFICATION

Identify the numbered items on the drawing by writing the letter next to the identifying term in the space provided.

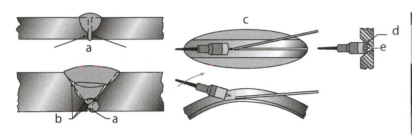

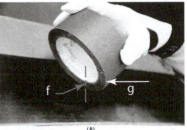

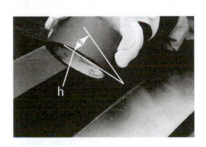

(A)

1. _____   CUP TOUCHES GROOVE FACE

2. _____   5° TO 10°

3. _____   TOP VIEW

4. _____   TUNGSTEN TOO SHORT TO TOUCH

5. _____   5°

6. _____   BRING THIS SIDE FORWARD

7. _____   INCOMPLETE PENETRATION AND FUSION

8. _____   PIVOT THE TAPE ON THE CENTER LINE

9. _____   CHANGING THE ANGLE WILL CHANGE
              THE WAY THE TAPE WALKS

10. _____   INCOMPLETE FUSION

11. _____   MARKS LEFT FROM CUP WALKING

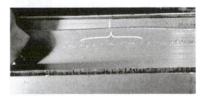

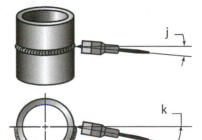

Gas Tungsten Arc Welding AWS SENSE Plate and Pipe Certification

AWS SENSE CERTIFICATION

Metal Specifications		Gas Flow			Nozzle Size in. (mm)	Amperage Min. to Max.
Thickness	Diameter of E70S-3*	Rates cfm (L/min)	Preflow Times	Postflow Times		
18 ga	1/16 in. (2 mm)	15 to 20 (7 to 9)	10 to 15 sec	10 to 25 sec	1/4 to 3/8 (6 to 10)	45 to 65
17 ga	1/16 in. (2 mm)	15 to 20 (7 to 9)	10 to 15 sec	10 to 25 sec	1/4 to 3/8 (6 to 10)	45 to 70
16 ga	1/16 in. (2 mm)	15 to 20 (7 to 9)	10 to 15 sec	10 to 25 sec	1/4 to 3/8 (6 to 10)	50 to 75
15 ga	1/16 in. (2 mm)	15 to 20 (7 to 9)	10 to 15 sec	10 to 25 sec	1/4 to 3/8 (6 to 10)	55 to 80
14 ga	3.32 in. (2.4 mm)	20 to 25 (9 to 12)	10 to 20 sec	10 to 30 sec	3/8 to 5/8 (10 to 16)	60 to 90
13 ga	3.32 in. (2.4 mm)	20 to 25 (9 to 12)	10 to 20 sec	10 to 30 sec	3/8 to 5/8 (10 to 16)	60 to 100
12 ga	3.32 in. (2.4 mm)	20 to 25 (9 to 12)	10 to 20 sec	10 to 30 sec	3/8 to 5/8 (10 to 16)	60 to 110
11 ga	3.32 in. (2.4 mm)	20 to 25 (9 to 12)	10 to 20 sec	10 to 30 sec	3/8 to 5/8 (10 to 16)	65 to 120
10 ga	3.32 in. (2.4 mm)	20 to 25 (9 to 12)	10 to 20 sec	10 to 30 sec	3/8 to 5/8 (10 to 16)	70 to 130

*Other E70S-X filler metal may be used.

Metal Specifications		Gas Flow			Nozzle Size in. (mm)	Amperage Min. to Max.
Thickness	Diameter of ER3XX* (in.)	Rates cfm (L/min)	Preflow Times	Postflow Times		
18 ga	1/16 in. (2 mm)	15 to 20 (7 to 9)	10 to 15 sec	10 to 25 sec	1/4 to 3/8 (6 to 10)	35 to 60
17 ga	1/16 in. (2 mm)	15 to 20 (7 to 9)	10 to 15 sec	10 to 25 sec	1/4 to 3/8 (6 to 10)	40 to 65
16 ga	1/16 in. (2 mm)	15 to 20 (7 to 9)	10 to 15 sec	10 to 25 sec	1/4 to 3/8 (6 to 10)	40 to 75
15 ga	1/16 in. (2 mm)	15 to 20 (7 to 9)	10 to 15 sec	10 to 25 sec	1/4 to 3/8 (6 to 10)	50 to 80
14 ga	3.32 in. (2.4 mm)	20 to 25 (9 to 12)	10 to 20 sec	10 to 30 sec	3/8 to 5/8 (10 to 16)	50 to 90
13 ga	3.32 in. (2.4 mm)	20 to 25 (9 to 12)	10 to 20 sec	10 to 30 sec	3/8 to 5/8 (10 to 16)	55 to 100
12 ga	3.32 in. (2.4 mm)	20 to 25 (9 to 12)	10 to 20 sec	10 to 30 sec	3/8 to 5/8 (10 to 16)	60 to 110
11 ga	3.32 in. (2.4 mm)	20 to 25 (9 to 12)	10 to 20 sec	10 to 30 sec	3/8 to 5/8 (10 to 16)	65 to 120
10 ga	3.32 in. (2.4 mm)	20 to 25 (9 to 12)	10 to 20 sec	10 to 30 sec	3/8 to 5/8 (10 to 16)	70 to 130

*Other ER3XX stainless steel A5.9 filler metal may be used.

Metal Specifications		Gas Flow			Nozzle Size in. (mm)	Amperage Min. to Max.
Thickness	Diameter of ER4043*	Rates cfm (L/min)	Preflow Times	Postflow Times		
18 ga	3/32 in. (2.4 mm)	20 to 30 (9 to 14)	10 to 15 sec	10 to 25 sec	1/4 to 3/8 (6 to 10)	40 to 60
17 ga	3/32 in. (2.4 mm)	20 to 30 (9 to 14)	10 to 15 sec	10 to 25 sec	1/4 to 3/8 (6 to 10)	50 to 70
16 ga	3/32 in. (2.4 mm)	20 to 30 (9 to 14)	10 to 15 sec	10 to 25 sec	1/4 to 3/8 (6 to 10)	60 to 75
15 ga	3/32 in. (2.4 mm)	20 to 30 (9 to 14)	10 to 15 sec	10 to 25 sec	1/4 to 3/8 (6 to 10)	65 to 85
14 ga	3/32 in. (2.4 mm)	20 to 30 (9 to 14)	10 to 15 sec	10 to 25 sec	1/4 to 3/8 (6 to 10)	75 to 90
13 ga	1/8 in. (3 mm)	25 to 40 (12 to 19)	10 to 20 sec	10 to 30 sec	3/8 to 5/8 (10 to 16)	85 to 100
12 ga	1/8 in. (3 mm)	25 to 40 (12 to 19)	10 to 20 sec	10 to 30 sec	3/8 to 5/8 (10 to 16)	90 to 110
11 ga	1/8 in. (3 mm)	25 to 40 (12 to 19)	10 to 20 sec	10 to 30 sec	3/8 to 5/8 (10 to 16)	100 to 115
10 ga	1/8 in. (3 mm)	25 to 40 (12 to 19)	10 to 20 sec	10 to 30 sec	3/8 to 5/8 (10 to 16)	100 to 125

*Other aluminum AWS A5.10 filler metal may be used if needed.

■ PRACTICE 19-1

Name _____   Date _____

Class _____   Instructor _____   Grade _____

OBJECTIVE: After completing this practice, you will make GTA welded butt joints, all four positions on 11 gauge to 14 gauge thickness mild steel with 100% joint penetration to be bend tested.

EQUIPMENT AND MATERIALS NEEDED FOR THIS PRACTICE

1. A properly set-up and adjusted GTA welding machine.

2. Assorted hand tools, vice, ball-peen hammer, saw or PAC torch, spare parts, and any other required materials.

3. Proper PPE.

4. ER70S-2 and/or ER70S-3 filler rod with 1/16 in. and/or 3/32 in. (2 mm and/or 2.4 mm) in diameter.

5. 100% Ar shielding gas.

6. Tungsten electrode ranging from 1/16 in. to 1/8 in. (2 mm to 3 mm) in diameter.

7. Two or more pieces of 6 in. long 1-1/2 in. wide (152 mm by 38 mm) 11 to 14 gauge mild steel.

INSTRUCTIONS

1. One at a time tack weld the sheets together to form the joints and place them in the desired position on a welding stand.

2. Starting at one end, run a bead along the joint. Watch the molten weld pool and bead for signs it is increasing or decreasing in size. Make any necessary changes in your technique to keep the weld consistent.

3. When the welds have cooled, use a hack saw or thermal cutting torch to cut out 1-in. (25-mm) wide strips as shown in Figure 19-1.

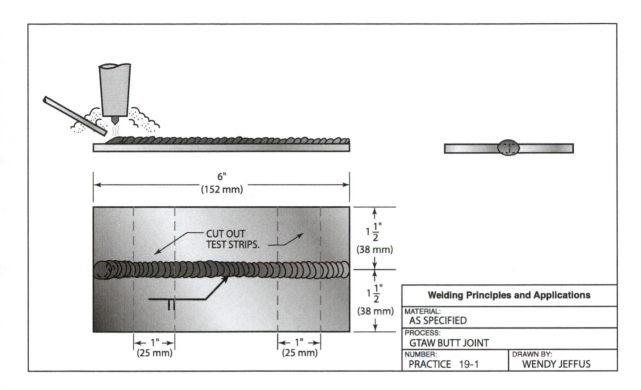

6"
(152 mm)

CUT OUT
TEST STRIPS.

$1\frac{1}{2}$"
(38 mm)

$1\frac{1}{2}$"
(38 mm)

1"
(25 mm)

1"
(25 mm)

Welding Principles and Applications	
MATERIAL: AS SPECIFIED	
PROCESS: GTAW BUTT JOINT	
NUMBER: PRACTICE 19-1	DRAWN BY: WENDY JEFFUS

4. Using a vice and a hammer, bend the strip of welded joint as shown in Figure 19-1.

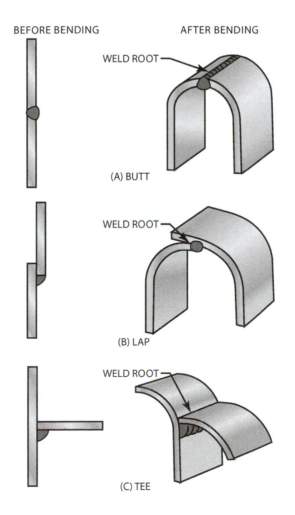

BEFORE BENDING   AFTER BENDING

WELD ROOT

(A) BUTT

WELD ROOT

(B) LAP

WELD ROOT

(C) TEE

5. Check for complete root fusion on the lap and tee joints and 100% penetration on the butt joint.

6. Repeat each type of joint in each position as needed until consistently good beads are obtained. Turn off the welding machine and shielding gas and clean up your work area when you are finished welding.

INSTRUCTOR'S COMMENTS _____

_____

_____

■ PRACTICE 19-2

Name _____   Date _____

Class _____   Instructor _____   Grade _____

OBJECTIVE: After completing this practice, you will make GTA welded lap and tee joints, all four positions on 11 gauge to 14 gauge thickness mild steel with 100% joint penetration to be bend tested.

EQUIPMENT AND MATERIALS NEEDED FOR THIS PRACTICE

1. A properly set-up and adjusted GTA welding machine.

2. Assorted hand tools, vice, ball-peen hammer, saw or PAC torch, spare parts, and any other required materials.

3. Proper PPE.

4. ER70S-2 and/or ER70S-3 filler rod with 1/16 in. and/or 3/32 in. (2 mm and/or 2.4 mm) in diameter.

5. 100% Ar shielding gas.

6. Tungsten electrode ranging from 1/16 in. to 1/8 in. (2 mm to 3 mm) in diameter.

7. Two or more pieces of 6 in. long 1-1/2 in. wide (152 mm by 38 mm) 11 to 14 gauge mild steel.

INSTRUCTIONS

1. One at a time tack weld the sheets together to form the joints and place them in the desired position on a welding stand.

2. Starting at one end, run a bead along the joint. Watch the molten weld pool and bead for signs it is increasing or decreasing in size. Make any necessary changes in your technique to keep the weld consistent.

3. When the welds have cooled, use a hack saw or thermal cutting torch to cut out 1-in. (25-mm) wide strips as shown in Figure 19-3 and Figure 19-4.

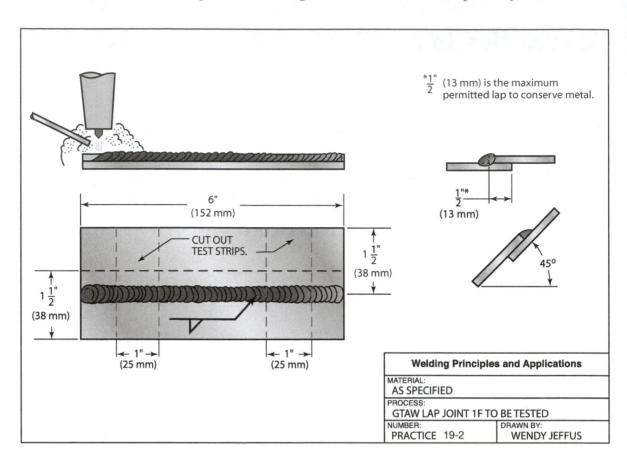

$*\frac{1}{2}"$ (13 mm) is the maximum permitted lap to conserve metal.

6"
(152 mm)

CUT OUT
TEST STRIPS.

$1\frac{1}{2}"$
(38 mm)

$1\frac{1}{2}"$
(38 mm)

$\frac{1}{2}"*$
(13 mm)

45°

1"
(25 mm)

1"
(25 mm)

Welding Principles and Applications

MATERIAL:	
AS SPECIFIED	
PROCESS:	
GTAW LAP JOINT 1F TO BE TESTED	
NUMBER:	DRAWN BY:
PRACTICE 19-2	WENDY JEFFUS

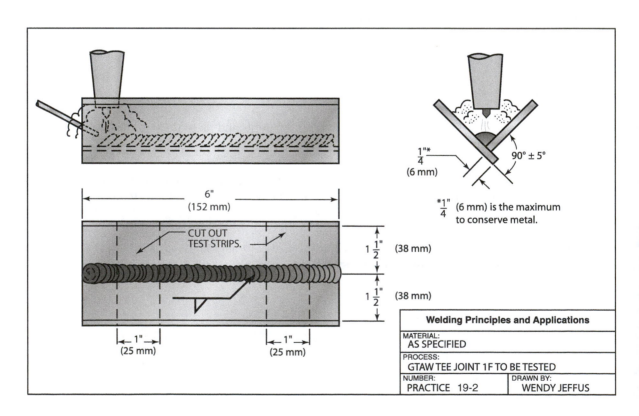

6"
(152 mm)

CUT OUT
TEST STRIPS.

$1\frac{1}{2}"$ (38 mm)

$1\frac{1}{2}"$ (38 mm)

$\frac{1}{4}"*$
(6 mm)

90° ± 5°

$*\frac{1}{4}"$ (6 mm) is the maximum to conserve metal.

1"
(25 mm)

1"
(25 mm)

Welding Principles and Applications

MATERIAL:	
AS SPECIFIED	
PROCESS:	
GTAW TEE JOINT 1F TO BE TESTED	
NUMBER:	DRAWN BY:
PRACTICE 19-2	WENDY JEFFUS

4. Using a vice and a hammer bend the strip of welded joint as shown in Figure 19-2.

5. Check for complete root fusion on the lap and tee joints and 100% penetration on the butt joint.

6. Repeat each type of joint in each position as needed until consistently good beads are obtained. Turn off the welding machine and shielding gas and clean up your work area when you are finished welding.

INSTRUCTOR'S COMMENTS _____

_____

_____

■ PRACTICE 19-3

Name _____   Date _____

Class _____   Instructor _____   Grade _____

OBJECTIVE: After completing this practice, you will make GTA welded butt joints, all four positions on 11 gauge to 14 gauge thickness stainless steel with 100% joint penetration to be bend tested.

EQUIPMENT AND MATERIALS NEEDED FOR THIS PRACTICE

1. A properly set-up and adjusted GTA welding machine.

2. Assorted hand tools, vice, ball-peen hammer, saw or PAC torch, spare parts, and any other required materials.

3. Proper PPE.

4. ER309 stainless steel filler rod with 1/16 in. and/or 3/32 in. (2 mm and/or 2.4 mm) in diameter.

5. 100% Ar shielding gas.

6. Tungsten electrode ranging from 1/16 in. to 1/8 in. (2 mm to 3 mm) in diameter.

7. Two or more pieces of 6 in. long 1-1/2 in. wide (152 mm by 38 mm) 11 to 14 gauge stainless steel.

   NOTE: Mild steel can be used in place of stainless steel as long as stainless steel filler metal is used.

INSTRUCTIONS

1. One at a time tack weld the sheets together to form the joints and place them in the desired position on a welding stand.

2. Starting at one end, run a bead along the joint. Watch the molten weld pool and bead for signs it is increasing or decreasing in size. Make any necessary changes in your technique to keep the weld consistent.

3. When the welds have cooled, use a hack saw or thermal cutting torch to cut out 1-in. (25-mm) wide strips.

4. Using a vice and a hammer, bend the strip of welded joint.

5. Check for complete root fusion on the lap and tee joints and 100% penetration on the butt joint.

6. Repeat each type of joint in each position as needed until consistently good beads are obtained. Turn off the welding machine and shielding gas and clean up your work area when you are finished welding.

INSTRUCTOR'S COMMENTS _____

_____

_____

■ PRACTICE 19-4

Name _____   Date _____

Class _____   Instructor _____   Grade _____

OBJECTIVE: After completing this practice, you will make GTA welded lap and tee joints, all four positions on 11 gauge to 14 gauge thickness stainless steel with 100% joint penetration to be bend tested.

EQUIPMENT AND MATERIALS NEEDED FOR THIS PRACTICE

1. A properly set-up and adjusted GTA welding machine.

2. Assorted hand tools, vice, ball-peen hammer, saw or PAC torch, spare parts, and any other required materials.

3. Proper PPE.

4. ER309 stainless steel filler rod with 1/16 in. and/or 3/32 in. (2 mm and/or 2.4 mm) in diameter.

5. 100% Ar shielding gas.

6. Tungsten electrode ranging from 1/16 in. to 1/8 in. (2 mm to 3 mm) in diameter.

7. Two or more pieces of 6 in. long 1-1/2 in. wide (152 mm by 38 mm) 11 to 14 gauge stainless steel.

   NOTE: Mild steel can be used in place of stainless steel as long as stainless steel filler metal is used.

INSTRUCTIONS

1. One at a time tack weld the sheets together to form the joints and place them in the desired position on a welding stand.

2. Starting at one end, run a bead along the joint. Watch the molten weld pool and bead for signs it is increasing or decreasing in size. Make any necessary changes in your technique to keep the weld consistent.

3. When the welds have cooled, use a hack saw or thermal cutting torch to cut out 1-in. (25-mm) wide strips.

4. Using a vice and a hammer, bend the strip of welded joint.

5. Check for complete root fusion on the lap and tee joints and 100% penetration on the butt joint.

6. Repeat each type of joint in each position as needed until consistently good beads are obtained. Turn off the welding machine and shielding gas and clean up your work area when you are finished welding.

INSTRUCTOR'S COMMENTS _____

_____

_____

■ PRACTICE 19-5

Name _____   Date _____

Class _____   Instructor _____   Grade _____

OBJECTIVE: After completing this practice, you will make GTA welded butt joints, all four positions on 11 gauge to 14 gauge thickness aluminum with 100% joint penetration to be bend tested.

EQUIPMENT AND MATERIALS NEEDED FOR THIS PRACTICE

1. A properly set-up and adjusted GTA welding machine.

2. Assorted hand tools, vice, ball-peen hammer, saw or PAC torch, spare parts, and any other required materials.

3. Proper PPE.

4. ER4043 filler rod with 1/16 in. and/or 3/32 in. (2 mm and/or 2.4 mm) in diameter.

5. 100% Ar shielding gas.

6. Tungsten electrode ranging from 1/16 in. to 1/8 in. (2 mm to 3 mm) in diameter.

7. Two or more pieces of 6 in. long 1-1/2 in. wide (152 mm by 38 mm) 11 to 14 aluminum.

INSTRUCTIONS

1. One at a time tack weld the sheets together to form the joints and place them in the desired position on a welding stand.

2. Starting at one end, run a bead along the joint. Watch the molten weld pool and bead for signs it is increasing or decreasing in size. Make any necessary changes in your technique to keep the weld consistent.

3. When the welds have cooled, use a hack saw or thermal cutting torch to cut out 1-in. (25-mm) wide strips.

4. Using a vice and a hammer, bend the strip of welded joint.

5. Check for complete root fusion on the lap and tee joints and 100% penetration on the butt joint.

6. Repeat each type of joint in each position as needed until consistently good beads are obtained. Turn off the welding machine and shielding gas and clean up your work area when you are finished welding.

INSTRUCTOR'S COMMENTS _____

_____

_____

■ PRACTICE 19-6

Name _____   Date _____

Class _____   Instructor _____   Grade _____

OBJECTIVE: After completing this practice, you will make GTA welded lap and tee joints, all four positions on 11 gauge to 14 gauge thickness aluminum with 100% joint penetration to be bend tested.

EQUIPMENT AND MATERIALS NEEDED FOR THIS PRACTICE

1. A properly set-up and adjusted GTA welding machine.

2. Assorted hand tools, vice, ball-peen hammer, saw or PAC torch, spare parts, and any other required materials.

3. Proper PPE.

4. ER4043 filler rod with 1/16 in. and/or 3/32 in. (2 mm and/or 2.4 mm) diameter.

5. 100% Ar shielding gas.

6. Tungsten electrode ranging from 1/16 in. to 1/8 in. (2 mm to 3 mm) in diameter.

7. Two or more pieces of 6 in. long 1-1/2 in. wide (152 mm by 38 mm) 11 to 14 aluminum.

INSTRUCTIONS

1. One at a time tack weld the sheets together to form the joints and place them in the desired position on a welding stand.

2. Starting at one end, run a bead along the joint. Watch the molten weld pool and bead for signs it is increasing or decreasing in size. Make any necessary changes in your technique to keep the weld consistent.

3. When the welds have cooled, use a hack saw or thermal cutting torch to cut out 1-in. (25-mm) wide strips.

4. Using a vice and a hammer, bend the strip of welded joint.

5. Check for complete root fusion on the lap and tee joints and 100% penetration on the butt joint.

6. Repeat each type of joint in each position as needed until consistently good beads are obtained. Turn off the welding machine and shielding gas and clean up your work area when you are finished welding.

INSTRUCTOR'S COMMENTS _____

_____

_____

■ PRACTICE 19-7 AWS SENSE

GAS TUNGSTEN ARC WELDING (GTAW) ON PLAIN CARBON STEEL WORKMANSHIP SAMPLE

Name _____   Date _____

Class _____   Instructor _____   Grade _____

Welding Procedure Specification (WPS) No.: Practice 19-7.

TITLE:
Welding GTAW of sheet to sheet.

SCOPE:
This procedure is applicable for square groove and fillet welds within the range of 18 gauge through 10 gauge.

WELDING MAY BE PERFORMED IN THE FOLLOWING POSITIONS:
2G, 3G, 2F, 3F, and 4F.

BASE METAL:
The base metal shall conform to carbon steel M-1, Group 1.

BACKING MATERIAL SPECIFICATION:
None.

FILLER METAL:
The filler metal shall conform to AWS specification no. #70S-3 for 1/16 in. (2 mm) to 3/32 in. (2.4 mm) diameter as listed in AWS specification A5.18. This filler metal falls into F-number F-6 and A-number A-1.

ELECTRODE:
The tungsten electrode shall conform to AWS specification no. EWTh-2, EWCe-2, or EWLa from AWS specification A5.12. The tungsten diameter shall be 1/8 in. (3.2 mm) maximum.

The tungsten end shape shall be tapered at two to three times its length to its diameter.

SHIELDING GAS:
The shielding gas, or gases, shall conform to the following compositions and purity: welding grade argon.

JOINT DESIGN AND TOLERANCES:
Refer to the drawing and specifications in Figure 19-9 for the workmanship sample layout.

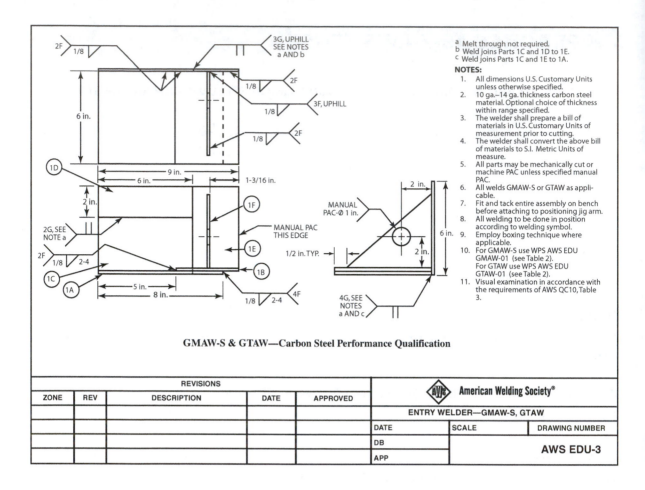

GMAW-S & GTAW—Carbon Steel Performance Qualification

		REVISIONS					
ZONE	REV	DESCRIPTION	DATE	APPROVED			

American Welding Society®

ENTRY WELDER—GMAW-S, GTAW

DATE		SCALE	DRAWING NUMBER
DB			AWS EDU-3
APP			

PREPARATION OF BASE METAL:

All hydrocarbons and other contaminants, such as cutting fluids, grease, oil, and primers, must be cleaned off all parts and filler metals before welding. This cleaning can be done with any suitable solvents or detergents. The joint face and inside and outside plate surface within 1 in. (25 mm) of the joint must be mechanically cleaned of slag, rust, and mill scale. Cleaning must be done with a wire brush or grinder down to bright metal.

ELECTRICAL CHARACTERISTICS:

Set the welding current to DCEN and the amperage and shielding gas flow according to Table 19-1.

PREHEAT:

The parts must be heated to a temperature higher than 50°F (10°C) before any welding is started.

BACKING GAS:

None.

SAFETY:

Proper protective clothing and equipment must be used. The area must be free of all hazards that may affect the welder or others in the area. The welding machine, welding leads, work clamp, electrode holder, and other equipment must be in safe working order.

WELDING TECHNIQUE:

Tack Welds: With the parts securely clamped in place with the correct root gap, the tack welds are to be performed. Holding the electrode so that it is very close to the root face but not touching, slowly increase the current until the arc starts and a molten weld pool is formed. Add filler metal as required to maintain a slightly convex weld face and a flat or slightly concave root face. When it is time to end the tack weld, lower the current slowly so that the molten weld pool can be tapered down in size. When all tack welds are complete, allow the parts to cool as needed before assembling the remaining parts. Repeat the tack welding procedure until the entire part is assembled.

Square Groove and Fillet Welds: Holding the electrode so that it is very close to the metal surface but not touching, slowly increase the current until the arc starts and a molten weld pool is formed. As the weld progresses, add filler metal as required to maintain a flat or slightly convex weld face. If it is necessary to stop the weld or to reposition yourself or if the weld is completed, the current must be lowered slowly so that the molten weld pool can be tapered down in size.

INTERPASS TEMPERATURE:

The plate should not be heated to a temperature higher than 120°F (49°C) during the welding process. After each weld pass is completed, allow it to cool but never to a temperature below 50°F (10°C). The weldment must not be quenched in water.

CLEANING:

Recleaning may be required if the parts or filler metal becomes contaminated or reoxidized to a degree that the weld quality will be affected. Reclean using the same procedure used for the original metal preparation.

VISUAL INSPECTION:

Visual inspection criteria for entry welders:

- There shall be no cracks and no incomplete fusion.
- There shall be no incomplete joint penetration in groove welds except as permitted for partial joint penetration groove welds.
- The Test Supervisor shall examine the weld for acceptable appearance and shall be satisfied that the welder is skilled in using the process and procedure specified for the test.
- Undercut shall not exceed the lesser of 10% of the base metal thickness or 1/32 in. (0.8 mm).
- Where visual examination is the only criterion for acceptance, all weld passes are subject to visual examination at the discretion of the Test Supervisor.
- The frequency of porosity shall not exceed one in each 4 in. (100 mm) of weld length, and the maximum diameter shall not exceed 3/32 in. (2.4 mm).
- Welds shall be free from overlap.

SKETCHES:

Gas Tungsten Arc Welding (GTAW) Workmanship Sample drawing for Carbon Steel.

INSTRUCTOR'S COMMENTS _____

_____

_____

■ PRACTICE 19-8 AWS SENSE

GAS TUNGSTEN ARC WELDING (GTAW) ON STAINLESS STEEL WORKMANSHIP SAMPLE

Name _____ Date _____

Class _____ Instructor _____ Grade _____

Welding Procedure Specification (WPS) No.: Practice 19-8.

TITLE:

Welding GTAW of sheet to sheet.

SCOPE:

This procedure is applicable for square groove and fillet welds within the range of 10 gauge through 14 gauge.

WELDING MAY BE PERFORMED IN THE FOLLOWING POSITIONS:

1G and 2F.

BASE METAL:

The base metal shall conform to austenitic stainless steel M-8 or P-8.

BACKING MATERIAL SPECIFICATION:

None.

FILLER METAL:

The filler metal shall conform to AWS specification no. ER308 or ER308L from AWS specification A5.9. This filler metal falls into F-number F-6 and A-number A-8.

ELECTRODE:

The tungsten electrode shall conform to AWS specification no. EWTh-2, EWCe-2, or EWLa from AWS specification A5.12. The tungsten diameter shall be 1/8 in. (3.2 mm) maximum. The tungsten end shape shall be tapered at two to three times its length to its diameter.

SHIELDING GAS:

The shielding gas, or gases, shall conform to the following compositions and purity: welding grade argon.

JOINT DESIGN AND TOLERANCES:

Refer to the drawing and specifications in Figure 19-11 for the workmanship sample layout.

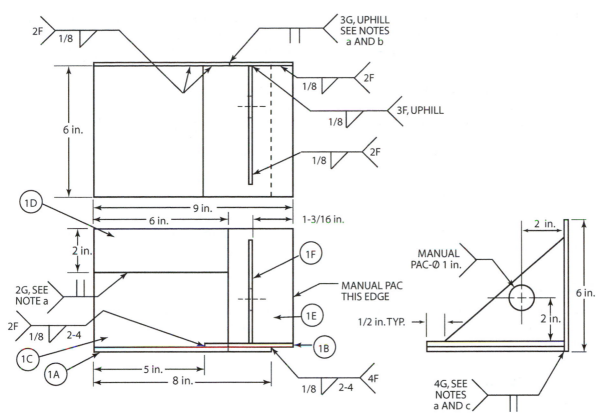

PREPARATION OF BASE METAL:

All hydrocarbons and other contaminants, such as cutting fluids, grease, oil, and primers, must be cleaned off all parts and filler metals before welding. This cleaning can be done with any suitable solvents or detergents. The joint face and inside and outside plate surface within 1 in. (25 mm) of the joint must be cleaned of slag, oxide, and scale. Cleaning can be mechanical or chemical. Mechanical metal cleaning can be done by grinding, stainless steel wire brushing, scraping, machining, or filing. Chemical cleaning can be done by using acids, alkalis, solvents, or detergents. Cleaning must be done down to bright metal.

ELECTRICAL CHARACTERISTICS:

Set the welding current to DCEN and the amperage and shielding gas flow according to Table 19-2.

PREHEAT:

The parts must be heated to a temperature higher than 50°F (10°C) before any welding is started.

BACKING GAS:

None.

SAFETY:

Proper protective clothing and equipment must be used. The area must be free of all hazards that may affect the welder or others in the area. The welding machine, welding leads, work clamp, electrode holder, and other equipment must be in safe working order.

WELDING TECHNIQUE:

Tack Welds: With the parts securely clamped in place with the correct root gap, the tack welds are to be performed. Holding the electrode so that it is very close to the root face but not touching, slowly increase the current until the arc starts and a molten weld pool is formed. Add filler metal as required to maintain a slightly convex weld face and a flat or slightly concave root face. When it is time to end the tack weld, lower the current slowly so that the molten weld pool can be tapered down in size. When all tack welds are complete, allow the parts to cool as needed before assembling the remaining parts. Repeat the tack welding procedure until the entire part is assembled.

Square Groove and Fillet Welds: Holding the electrode so that it is very close to the metal surface but not touching, slowly increase the current until the arc starts and a molten weld pool is formed. As the weld progresses, add filler metal as required to maintain a flat or slightly convex weld face. If it is necessary to stop the weld or to reposition yourself or if the weld is completed, the current must be lowered slowly so that the molten weld pool can be tapered down in size.

INTERPASS TEMPERATURE:

The plate should not be heated to a temperature higher than 350°F (180°C) during the welding process. After each weld pass is completed, allow it to cool but never to a temperature below 50°F (10°C). The weldment must not be quenched in water.

CLEANING:

Recleaning may be required if the parts or filler metal become contaminated or oxidized to a degree that the weld quality will be affected. Reclean using the same procedure used for the original metal preparation.

VISUAL INSPECTION:

Visual inspection criteria for entry welders:

- There shall be no cracks and no incomplete fusion.
- There shall be no incomplete joint penetration in groove welds except as permitted for partial joint penetration groove welds.
- The Test Supervisor shall examine the weld for acceptable appearance and shall be satisfied that the welder is skilled in using the process and procedure specified for the test.
- Undercut shall not exceed the lesser of 10% of the base metal thickness or 1/32 in. (0.8 mm).
- Where visual examination is the only criterion for acceptance, all weld passes are subject to visual examination at the discretion of the Test Supervisor.
- The frequency of porosity shall not exceed one in each 4 in. (100 mm) of weld length and the maximum diameter shall not exceed 3/32 in. (2.4 mm).
- Welds shall be free from overlap.

SKETCHES:

Gas Tungsten Arc Welding (GTAW) Workmanship Sample drawing for Stainless Steel.

INSTRUCTOR'S COMMENTS _____

_____

_____

■ PRACTICE 19-9 AWS SENSE

GAS TUNGSTEN ARC WELDING (GTAW) ON ALUMINUM WORKMANSHIP SAMPLE

Name _____   Date _____

Class _____   Instructor _____   Grade _____

Welding Procedure Specification (WPS) No.: Practice 19-9.

TITLE:
Welding GTAW of sheet to sheet.

SCOPE:
This procedure is applicable for square groove and fillet welds within the range of 18 gauge through 10 gauge.

WELDING MAY BE PERFORMED IN THE FOLLOWING POSITIONS:
1G and 2F.

BASE METAL:
The base metal shall conform to aluminum M-22 or P-22.

BACKING MATERIAL SPECIFICATION:
None.

FILLER METAL:
The filler metal shall conform to AWS specification no. ER4043 from AWS specification A5.10. This filler metal falls into F-number F-22 and A-number A-5.10.

ELECTRODE:
The tungsten electrode shall conform to AWS specification no. EWCe-2, EWZr, EWLa, or EWP from AWS specification A5.12. The tungsten diameter shall be 1/8 in. (3.2 mm) maximum. The tungsten end shape shall be rounded.

SHIELDING GAS:
The shielding gas, or gases, shall conform to the following compositions and purity: welding grade argon.

JOINT DESIGN AND TOLERANCES:
Refer to the drawing and specifications in Figure 19-13 for the workmanship sample layout.

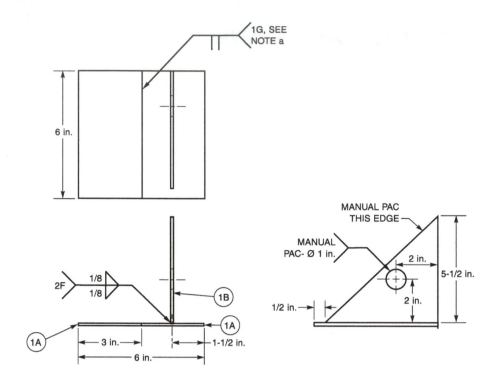

PREPARATION OF BASE METAL:

All hydrocarbons and other contaminants, such as cutting fluids, grease, oil, and primers, must be cleaned off all parts and filler metals before welding. This cleaning can be done with any suitable solvents or detergents. The joint face and inside and outside plate surface within 1 in. (25 mm) of the joint must be mechanically or chemically cleaned of oxides. Mechanical cleaning may be done by stainless steel wire brushing, scraping, machining, or filing. Chemical cleaning may be done by using acids, alkalis, solvents, or detergents. Because the oxide layer may reform quickly and affect the weld, welding should be started within 10 minutes of cleaning.

ELECTRICAL CHARACTERISTICS:

Set the welding current to AC high-frequency stabilized and the amperage and shielding gas flow according to Table 19-3.

PREHEAT:

The parts must be heated to a temperature higher than 50°F (10°C) before any welding is started.

BACKING GAS:

N/A.

SAFETY:

Proper protective clothing and equipment must be used. The area must be free of all hazards that may affect the welder or others in the area. The welding machine, welding leads, work clamp, electrode holder, and other equipment must be in safe working order.

WELDING TECHNIQUE:

The welder's hands or gloves must be clean and oil free to prevent contamination of the metal or filler rods.

Tack Welds: With the parts securely clamped in place with the correct root gap, the tack welds are to be performed. Holding the electrode so that it is very close to the root face but not touching, slowly increase the current until the arc starts and a molten weld pool is formed. Add filler metal as required to maintain a slightly convex weld face and a flat or slightly concave root face. When it is time to end the tack weld, lower the current slowly so that the molten weld pool can be tapered down in size. When all tack welds are complete, allow the parts to cool as needed before assembling the remaining parts. Repeat the tack welding procedure until the entire part is assembled.

Square Groove and Fillet Welds: Holding the electrode so that it is very close to the metal surface but not touching, slowly increase the current until the arc starts and a molten weld pool is formed. As the weld progresses, add filler metal as required to maintain a flat or slightly convex weld face. If it is necessary to stop the weld or to reposition yourself or the weld is completed, the current must be lowered slowly so that the molten weld pool can be tapered down in size.

INTERPASS TEMPERATURE:

The plate should not be heated to a temperature higher than 120°F (49°C) during the welding process. After each weld pass is completed, allow it to cool but never to a temperature below 50°F (10°C). The weldment must not be quenched in water.

CLEANING:

Recleaning may be required if the parts or filler metal become contaminated or oxidized to a degree that the weld quality will be affected. Reclean using the same procedure used for the original metal preparation.

VISUAL INSPECTION:

Visual inspection criteria for entry welders:

- There shall be no cracks and no incomplete fusion.
- There shall be no incomplete joint penetration in groove welds except as permitted for partial joint penetration groove welds.
- The Test Supervisor shall examine the weld for acceptable appearance and shall be satisfied that the welder is skilled in using the process and procedure specified for the test.
- Undercut shall not exceed the lesser of 10% of the base metal thickness or 1/32 in. (0.8 mm).
- Where visual examination is the only criterion for acceptance, all weld passes are subject to visual examination at the discretion of the Test Supervisor.
- The frequency of porosity shall not exceed one in each 4 in. (100 mm) of weld length, and the maximum diameter shall not exceed 3/32 in. (2.4 mm).
- Welds shall be free from overlap.

SKETCHES:

Gas Tungsten Arc Welding (GTAW) Workmanship Sample drawing for Aluminum.

INSTRUCTOR'S COMMENTS _____

_____

_____

■ PRACTICE 19-10 AWS SENSE

Name _____ Date _____

Class _____ Instructor _____ Grade _____

OBJECTIVE: After completing this practice, you will make GTA welded butt pipe joints, in 2G and 5G positions on 1 in. to 2-7/8 in. (25 mm to 70 mm) 10 gauge to 18 gauge thickness mild steel tubing with 100% joint penetration.

EQUIPMENT AND MATERIALS NEEDED FOR THIS PRACTICE

1. A properly set-up and adjusted GTA welding machine.

2. Assorted hand tools, spare parts, and any other required materials.

3. Proper PPE.

4. ER70S-2 and/or ER70S-3 filler rod with 1/16 in. and/or 3/32 in. (2 mm and/or 2.4 mm) in diameter.

5. 100% Ar shielding gas.

6. Tungsten electrode ranging from 1/16 in. to 1/8 in. (2 mm to 3 mm) in diameter.

7. Two or more pieces of 3 in. long 1 in. to 2-7/8 in. (25 mm to 70 mm) 10 gauge to 18 gauge thickness mild steel tubing.

8. One or more backing rings for the size of tubing being welded.

INSTRUCTIONS

1. Tack weld the tubing together with the proper root opening, Figure 19-8. Make the tack welds at the 12:00, 3:00, 6:00, and 9:00 o'clock positions.

2. Grind the ends of the tack welds to a featheredge.

3. Attach the pipe section in the proper position to the welding stand at a comfortable height.

   NOTE: You may use the cup walking or freehand technique for making these welds.

2G Position (Pipe is Vertical and the Weld is Horizontal)

4. Start welding on one of the tack welds by resting your gloved hand with the torch on the pipe or pipe stand.

5. Once the tack weld is melted, slowly add filler metal to the leading edge so that the keyhole stays about the same size as the root opening. If the keyhole increases in size reduce the power setting or add filler metal faster.

6. Stop your weld when you reach the next tack weld and reposition to continue the weld.

5G Position (Pipe is Horizontal and the Weld is Vertical)

7. Starting your weld at the 3:00 or 9:00 o'clock positions and welding upward is an easier way of developing the skills of watching the keyhole and adding filler metal than starting at the 6:00 o'clock position.

8. Start welding on one of the tack welds by resting your gloved hand with the torch on the pipe or pipe stand.

9. Once the tack weld is melted, slowly add filler metal to the leading edge so that the keyhole stays about the same size as the root opening. If the keyhole increases in size reduce the power setting or add filler metal faster.

10. Stop your weld when you reach the next tack weld and reposition to continue the weld.

11. Repeat each of joint in each position as needed until consistently good beads are obtained. Turn off the welding machine and shielding gas and clean up your work area when you are finished welding.

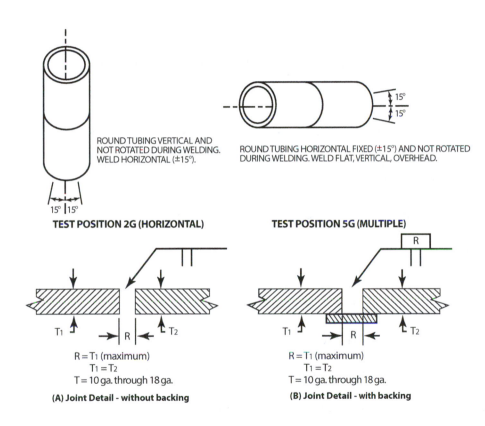

ROUND TUBING VERTICAL AND NOT ROTATED DURING WELDING. WELD HORIZONTAL (±15°).

ROUND TUBING HORIZONTAL FIXED (±15°) AND NOT ROTATED DURING WELDING. WELD FLAT, VERTICAL, OVERHEAD.

TEST POSITION 2G (HORIZONTAL)

TEST POSITION 5G (MULTIPLE)

$R = T_1$ (maximum)
$T_1 = T_2$
$T = 10$ ga. through 18 ga.

(A) Joint Detail - without backing

$R = T_1$ (maximum)
$T_1 = T_2$
$T = 10$ ga. through 18 ga.

(B) Joint Detail - with backing

INSTRUCTOR'S COMMENTS _____

_____

_____

■ PRACTICE 19-11 AWS SENSE

Name _____   Date _____

Class _____   Instructor _____   Grade _____

OBJECTIVE: After completing this practice, you will make GTA welded butt pipe joints, in 2G and 5G positions on 1 in. to 2-7/8 in. (25 mm to 70 mm) 10 gauge to 18 gauge thickness stainless steel tubing with 100% joint penetration.

EQUIPMENT AND MATERIALS NEEDED FOR THIS PRACTICE

1.   A properly set-up and adjusted GTA welding machine.

2.   Assorted hand tools, spare parts, and any other required materials.

3.   Proper PPE.

4.   ER3XX filler rod with 1/16 in. and/or 3/32 in. (2 mm and/or 2.4 mm) in diameter.

5.   100% Ar shielding gas.

6.   Tungsten electrode ranging from 1/16 in. to 1/8 in. (2 mm to 3 mm) in diameter.

7.   Two or more pieces of 3 in. long 1 in. to 2-7/8 in. (25 mm to 70 mm) 10 gauge to 18 gauge thickness stainless steel tubing.

8.   One or more backing rings for the size of tubing being welded.

INSTRUCTIONS

1.   Tack weld the tubing together with the proper root opening, Figure 19-3. Make the tack welds at the 12:00, 3:00, 6:00, and 9:00 o'clock positions.

2.   Grind the ends of the tack welds to a featheredge.

3.   Attach the pipe section in the proper position to the welding stand at a comfortable height.

     NOTE: You may use the cup walking or freehand technique for making these welds.

2G Position (Pipe is Vertical and the Weld is Horizontal)

4.   Start welding on one of the tack welds by resting your gloved hand with the torch on the pipe or pipe stand.

5.   Once the tack weld is melted, slowly add filler metal to the leading edge so that the keyhole stays about the same size as the root opening. If the keyhole increases in size reduce the power setting or add filler metal faster.

6.   Stop your weld when you reach the next tack weld and reposition to continue the weld.

5G Position (Pipe is Horizontal and the Weld is Vertical)

7. Starting your weld at the 3:00 or 9:00 o'clock positions and welding upward is an easier way of developing the skills of watching the keyhole and adding filler metal than starting at the 6:00 o'clock position.

8. Start welding on one of the tack welds by resting your gloved hand with the torch on the pipe or pipe stand.

9. Once the tack weld is melted, slowly add filler metal to the leading edge so that the keyhole stays about the same size as the root opening. If the keyhole increases in size reduce the power setting or add filler metal faster.

10. Stop your weld when you reach the next tack weld and reposition to continue the weld.

11. Repeat each of joint in each position as needed until consistently good beads are obtained. Turn off the welding machine and shielding gas and clean up your work area when you are finished welding.

INSTRUCTOR'S COMMENTS _____

_____

_____

■ PRACTICE 19-12

Name _____ Date _____

Class _____ Instructor _____ Grade _____

OBJECTIVE: After completing this practice, you will make GTA welded butt pipe joints, in 2G and 5G positions on 1 in. to 2-7/8 in. (25 mm to 70 mm) 10 gauge to 18 gauge thickness aluminum tubing with 100% joint penetration.

EQUIPMENT AND MATERIALS NEEDED FOR THIS PRACTICE

1. A properly set-up and adjusted GTA welding machine.

2. Assorted hand tools, spare parts, and any other required materials.

3. Proper PPE.

4. ER4042 or ER5XXX filler rod with 1/16 in. and/or 3/32 in. (2 mm and/or 2.4 mm) in diameter.

5. 100% Ar shielding gas.

6. Tungsten electrode ranging from 1/16 in. to 1/8 in. (2 mm to 3 mm) in diameter.

7. Two or more pieces of 3 in. long 1 in. to 2-7/8 in. (25 mm to 70 mm) 10 gauge to 18 gauge thickness aluminum tubing.

8. One or more backing rings for the size of tubing being welded.

INSTRUCTIONS

1. Tack weld the tubing together with the proper root opening, Figure 19-3. Make the tack welds at the 12:00, 3:00, 6:00, and 9:00 o'clock positions.

2. Grind the ends of the tack welds to a featheredge.

3. Attach the pipe section in the proper position to the welding stand at a comfortable height.

   NOTE: You may use the cup walking or freehand technique for making these welds.

2G Position (Pipe is Vertical and the Weld is Horizontal)

4. Start welding on one of the tack welds by resting your gloved hand with the torch on the pipe or pipe stand.

5. Once the tack weld is melted, slowly add filler metal to the leading edge so that the keyhole stays about the same size as the root opening. If the keyhole increases in size reduce the power setting or add filler metal faster.

6. Stop your weld when you reach the next tack weld and reposition to continue the weld.

5G Position (Pipe is Horizontal and the Weld is Vertical)

7. Starting your weld at the 3:00 or 9:00 o'clock positions and welding upward is an easier way of developing the skills of watching the keyhole and adding filler metal than starting at the 6:00 o'clock position.

8. Start welding on one of the tack welds by resting your gloved hand with the torch on the pipe or pipe stand.

9. Once the tack weld is melted, slowly add filler metal to the leading edge so that the keyhole stays about the same size as the root opening. If the keyhole increases in size, reduce the power setting or add filler metal faster.

10. Stop your weld when you reach the next tack weld and reposition to continue the weld.

11. Repeat each of joint in each position as needed until consistently good beads are obtained. Turn off the welding machine and shielding gas and clean up your work area when you are finished welding.

INSTRUCTOR'S COMMENTS _____

_____

_____

■ PRACTICE 19-13

Name _____ Date _____

Pipe Diameter _____ Filler Metal Type _____ Diameter _____

Class _____ Instructor _____ Grade _____

OBJECTIVE: After completing this practice, you should be able to make a pipe weld on mild steel pipe in the horizontal rolled 1G position using the GTAW process. The weld is to be tested.

EQUIPMENT AND MATERIALS NEEDED FOR THIS PRACTICE

1. A properly set-up and adjusted GTA welding machine.

2. Proper safety protection.

3. Filler rods, 36-in. (0.9-m) long by 1/16-in. (2-mm), 3/32-in. (2.4-mm), and 1/8-in. (3-mm) diameter.

4. Two or more pieces of schedule 40 mild steel pipe, 3 in. (76 mm) to 10 in. (254 mm) in diameter with single V-groove prepared ends.

INSTRUCTIONS

1. Using the skills you have developed so far, tack weld two pieces of pipe together, then do a root pass, filler passes (until flush), and a cover pass.

2. After you have completed all passes all the way around the pipe, visually inspect the bead.

3. If it passes the visual inspection, cut out guided-bend test specimens.

4. Subject the specimens to the bend test.

5. Repeat this practice until all the specimens consistently pass the bend test.

6. Turn off the welding machine and shielding gas and clean up your work area when you are finished welding.

INSTRUCTOR'S COMMENTS _____

_____

_____

■ PRACTICE 19-14

Name _____ Date _____

Pipe Diameter _____ Filler Metal Type _____ Diameter _____

Class _____ Instructor _____ Grade _____

OBJECTIVE: After completing this practice, you should be able to make a single-V butt joint bead on mild steel pipe in the horizontal fixed 5G position using the GTAW process, 100% root penetration to be tested.

EQUIPMENT AND MATERIALS NEEDED FOR THIS PRACTICE

1. A properly set-up and adjusted GTA welding machine.

2. Proper safety protection.

3. Filler rods, 36-in. (0.9-m) long by 1/16-in. (2-mm), 3/32-in. (2.4-mm), and 1/8-in. (3-mm) diameter.

4. Two or more pieces of schedule 40 mild steel pipe, 3 in. (76 mm) to 10 in. (254 mm) in diameter having the ends prepared with a single V-groove.

INSTRUCTIONS

1. Tack weld the pipes together in the 12:00, 3:00, 6:00, and 9:00 o'clock positions. Clamp the pipe in the 5G position.

2. Start the weld at the 6:30 o'clock position and weld uphill around the pipe to the 12:30 o'clock position. Place the cup against the pipe bevels and add the filler rod at the leading edge of the molten weld pool.

3. Repeat this process up the other side of the pipe. The starts and stops should overlap slightly to ensure a good tie-in.

4. The filler passes should also start at the 6:30 o'clock position and go up to the 12:30 o'clock position. Stagger the bead locations to prevent defects arising from discontinuities at the start and stop points.

5. Use a lace or weave bead and put on a cover pass.

6. Inspect the weld for straightness, uniformity, and visual defects. Repeat this practice until you can consistently make welds that are visually defect free.

7. Cut out guided-bend test specimens and test them. Repeat this practice until you can consistently make welds that pass the bend test.

8. Turn off the welding machine and shielding gas and clean up your work area when you are finished welding.

INSTRUCTOR'S COMMENTS _____

_____

_____

■ PRACTICE 19-15

Name _____ Date _____

Pipe Diameter _____ Filler Metal Type _____ Diameter _____

Class _____ Instructor _____ Grade _____

OBJECTIVE: After completing this practice, you should be able to make a single-V butt joint bead on mild steel pipe in the vertical fixed 2G position using the GTAW process, 100% root penetration to be tested.

EQUIPMENT AND MATERIALS NEEDED FOR THIS PRACTICE

1. A properly set-up and adjusted GTA welding machine.

2. Proper safety protection.

3. Filler rods, 36-in. (0.9-m) long by 1/16-in. (2-mm), 3/32-in. (2.4-mm), and 1/8-in. (3-mm) diameter.

4. Two or more pieces of schedule 40 mild steel pipe, 3 in. (76 mm) to 10 in. (254 mm) in diameter having the ends prepared with a single V-groove.

INSTRUCTIONS

1. Tack weld the pipes together in the 12:00, 3:00, 6:00, and 9:00 o'clock positions. Clamp the pipe in the 2G position.

2. Start the root weld at a point between the tack welds. The root pass should be small enough that surface tension will hold it in place.

3. The filler passes can be larger if they are made along the lower beveled surface which will support the bead. The next pass goes on top side of the first, and so forth.

4. The cover pass is started around the lower side of the joint, overlapping the pipe surface by no more than 1/8 in. (3 mm). The next pass covers about one-half of the first pass and should be slightly larger. This process of making each pass larger continues until the center of the weld is reached, then each weld is made successively smaller.

5. Inspect the weld for uniformity and visual defects. Repeat this practice until you can consistently make welds that are visually defect free.

6. Cut out guided-bend test specimens and test them. Repeat this practice until you can consistently make welds that pass the bend test.

7. Turn off the welding machine and shielding gas and clean up your work area when you are finished.

INSTRUCTOR'S COMMENTS _____

_____

_____

■ PRACTICE 19-16

Name _____   Date _____

Pipe Diameter _____   Filler Metal Type _____   Diameter _____

Class _____   Instructor _____   Grade _____

OBJECTIVE: After completing this practice, you should be able to make a single-V butt joint bead on mild steel pipe at a 45° fixed inclined angle (6G position) using the GTAW process, 100% root penetration to be tested.

EQUIPMENT AND MATERIALS NEEDED FOR THIS PRACTICE

1. A properly set-up and adjusted GTA welding machine.

2. Proper safety protection.

3. Filler rods, 36-in. (0.9-m) long by 1/16-in. (2-mm), 3/32-in. (2.4-mm), and 1/8-in. (3-mm) diameter.

4. Two or more pieces of schedule 40 mild steel pipe, 3 in. (76 mm) to 10 in. (254 mm) in diameter having the ends prepared with a single V-groove.

INSTRUCTIONS

1. Tack weld the pipes together in the 12:00, 3:00, 6:00, and 9:00 o'clock positions. Clamp the pipe in the 6G position.

2. Start the root weld at the bottom and at a point between the tack welds. The root pass should be small enough that surface tension will hold it in place. The root pass may be off to one side.

3. The filler passes are applied to the downhill side first so that the upper side will be supported. Stagger the starts and stops to ensure against proximity of defects.

4. The cover pass is easy to control if it is a series of stringer beads. Start on the lower side and build up the cover pass as you did in the 2G position. Each weld should overlap the preceding weld for continuity.

5. Inspect the weld for uniformity and visual defects. Repeat this practice until you can consistently make welds that are visually defect free.

6. Cut out guided-bend test specimens and test them. Repeat this practice until you can consistently make welds that pass the bend test.

7. Turn off the welding machine and shielding gas and clean up your work area when you are finished.

INSTRUCTOR'S COMMENTS _____

_____

_____

■ PRACTICE 19-17

Name _____   Date _____

Class _____   Instructor _____   Grade _____

OBJECTIVE: After completing this practice, you should be able to make single pass fillet welds in the 2F, 4F, and 5F positions.

EQUIPMENT AND MATERIALS NEEDED FOR THIS PRACTICE

Use a properly setup and adjusted GTA welding machine, E70S-2 and/or E70S-3, 3/32 in. or 1/8 in. (2.4 mm or 3 mm) filler metal, 4 in.to 8 in. Sch. 40 or 80 Pipe 6 in. or 10 in. square plate (100 mm to 200 mm Sch. 40 or 80 Pipe 150 mm or 10 mm square plate), layout and assembly tools, and any required PPE, you will safely assemble the pipe to the plate and make the required single pass fillet welds.

INSTRUCTIONS

1. Make four to six tack welds to join the pipe to the plate, depending on the diameter of the pipe.

2. Use a thin grinding disk and feather the start and stops of the tack welds.

3. Start the weld on top of one of the tack welds. Using the cup walking technique, place the filler metal in the root of the joint.

4. Hold the arc in one place and allow the molten weld pool to form and gently slide the end of the welding rod into the weld pool.

5. Using the figure-8 motion, walk the cup around the weldment.

6. The fillet weld should be made as an equal legged fillet weld with a flat or nearly flat weld face.

7. Try to end the weld on the next tack weld. If you did not stop on a tack weld, then use the thin grinding disk to feather the end of the weld.

8. When you get to a point where you have to stop to reposition, gradually reduce the current.

9. Once you are ready, repeat the starting process as above until the weld is made all the way around the pipe.

Repeat each weld position until you can pass the required visual inspection.

Turn off the welding machine, shielding gas, and cooling water, and clean your work area when you are finished welding.

Complete a copy of the "Student Welding Report" listed in Appendix I or provided by your instructor.

INSTRUCTOR'S COMMENTS _____

_____

_____

■ PRACTICE 19-18

Name _____   Date _____

Class _____   Instructor _____   Grade _____

OBJECTIVE: After completing this practice, you should be able to make multiple pass fillet welds in the 2F, 5F, and 6F positions.

EQUIPMENT AND MATERIALS NEEDED FOR THIS PRACTICE

Use a properly setup and adjusted GTA welding machine, E70S-2 and/or E70S-3, 3/32 in. or 1/8 in. (2.4 mm or 3 mm) filler metal, 4 in.to 8 in. Sch. 40 or 80 Pipe 6 in. or 10 in. square plate (100 mm to 200 mm Sch. 40 or 80 Pipe 150 mm or 10 mm square plate), layout and assembly tools, and any required PPE, you will safely assemble the pipe to the plate and make the required single pass fillet welds.

INSTRUCTIONS

1. Make four to six tack welds to join the pipe to the plate, depending on the diameter of the pipe.

2. To make the first weld pass on this multiple pass fillet, follow the same steps listed in Practice 19-17.

3. You may need a larger cut to help with your cup walking technique.

    NOTE: **The first weld pass, the root pass, should be made as an equal legged fillet weld with a flat or nearly flat weld face. That will make it easier to do the next weld passes.**

4. When the first fillet weld pass is completed, you will start on the second filler pass.

5. The second filler pass will be made on the downward side of the first filler pass.

6. This pass can have a slightly convex surface that will help hold the third pass in place.

7. When it becomes necessary to stop the weld to reposition yourself, slowly reduce the current to taper the weld size down, and do not stop at the end of any of the first pass welds.

8. Once you are ready, repeat the starting process as above until the weld is made all the way around the pipe.

    NOTE: **The weld bead position changes as the weld progresses around the 6F weldment, so you will have to make changes in the placement of the filler metal in the molten weld pool.**

9. The third fillet weld pass will be made so it overlaps 1/2 to 2/3 of the second weld.

    Repeat each weld position until you can pass the required visual inspection.

    Turn off the welding machine, shielding gas, and cooling water, and clean your work area when you are finished welding.

    Complete a copy of the "Student Welding Report" listed in Appendix I or provided by your instructor.

INSTRUCTOR'S COMMENTS _____

_____

_____

CHAPTER 19: QUIZ 1

Name _____   Date _____

Class _____   Instructor _____   Grade _____

INSTRUCTIONS

Carefully read Chapter 19 in the textbook and answer each question.

MATCHING

In the space provided to the left of Column A, write the letter from Column B that best answers or completes the statement in Column A.

Column A	Column B
_____ 1. The AWS SENSE Level I Workmanship Qualifications test requires GTA welders to be able to make butt welds in the vertical up, horizontal, and _____ positions.	a. free bend
_____ 2. The AWS SENSE Level I Workmanship Qualifications test requires GTA welders to be able to make intermittent lap welds in the _____ and overhead positions.	b. face
_____ 3. The AWS SENSE Level I Workmanship Qualifications test requires GTA welders to be able to make tee welds in the horizontal and _____ positions.	c. root penetration
_____ 4. The AWS SENSE Level II Workmanship Qualification test requires GTA welders to be able to make square groove butt welds with and without a _____ in the 2G and 5G positions in thin wall small diameter mild steel, stainless steel, and aluminum tubing.	d. horizontal
_____ 5. All the welds must pass _____ (VT) based on the criteria for each type of weld.	e. 1/8 in. (3.2 mm)
_____ 6. The visual inspection acceptance criteria for butt joints requires that there be no _____.	f. 1/16 in. (2 mm)
_____ 7. Butt joint welds must have complete joint _____.	g. backing ring
_____ 8. Butt joint welds must have 100% _____.	h. contaminated
_____ 9. The visual inspection acceptance criteria for lap and tee joint fillet welds requires that the maximum weld size be _____.	i. pipe surface
_____ 10. For lap and tee joint fillet welds, the maximum porosity is one per _____ of weld that is no larger that 25% of the base metal thickness.	j. overhead
_____ 11. Lap and tee joint fillet welds must have no undercut greater in depth than 15% of the _____ thickness.	k. maximum

_____ 12. The visual inspection acceptance criteria for tubing welds requires that the _____ weld reinforcement be 1/8 in. (3.2 mm).

_____ 13. The visual inspection acceptance criteria for tubing welds requires that maximum root face concavity be _____.

_____ 14. For tubing welds, the weld surface must be at least flush with the outside of the _____.

_____ 15. For tubing welds, the weld face must have a smooth uniform transition from the sides of the weld to the weld _____.

_____ 16. For tubing welds, all weld craters must be _____.

_____ 17. Mechanical testing is not required for the GTA AWS SENSE qualification tests, but the _____ test can help you identify any root problems that might exist in your weld.

_____ 18. Always remember to wear all the required _____ when grinding tungsten.

_____ 19. It is very important that the pipe is mounted so that it does not move or vibrate as it is being welded because movement can increase the possibility that the tungsten will become _____.

_____ 20. _____ welds are used to join sections of pipe to pipe flanges.

l. PPE

m. base metal

n. fusion

o. fillet

p. vertical up

q. cracks

r. 1 in. (25 mm)

s. filled

t. visual inspection

CHAPTER 19: QUIZ 2

Name _____   Date _____

Class _____   Instructor _____   Grade _____

INSTRUCTIONS

Carefully read Chapter 19 in the textbook and answer each question.

IDENTIFICATION

Identify the numbered items on the drawing by writing the letter next to the identifying term in the space provided.

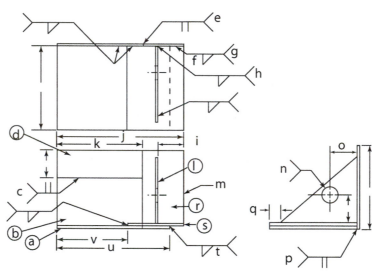

GMAW-S & GTAW—Carbon Steel Performance Qualification

1. _____ 1A

2. _____ 1/8

3. _____ 2-4

4. _____ MANUAL PAC THIS EDGE

5. _____ 1D

6. _____ 3G, UPHILL SEE NOTES a AND b

7. _____ 9 in.

8. _____ 1B

9. _____ 1-3/16 in.

10. _____ 5 in.

11. _____ 1F

12. _____ 2F

13. _____ MANUAL PAC-O/ 1 in.

14. _____ 8 in.

15. _____ 4G, SEE NOTES a AND c

16. _____ 1C

17. _____ 3F, UPHILL

18. _____ 1/2 in. TYP.

19. _____ 1E

20. _____ 6 in.

21. _____ 2G, SEE NOTE a

22. _____ 2 in.

CHAPTER 19: QUIZ 3

Name _____ Date _____

Class _____ Instructor _____ Grade _____

INSTRUCTIONS

Carefully read Chapter 19 in the textbook and answer each question.

IDENTIFICATION

Identify the numbered items on the drawing by writing the letter next to the identifying term in the space provided.

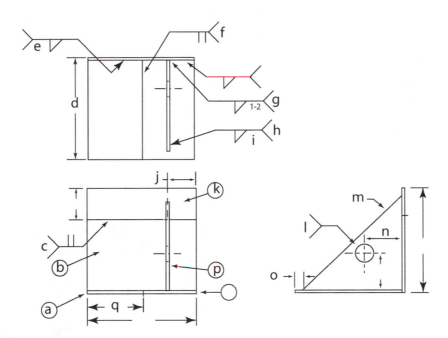

1. _____ 6 in.		10. _____ MANUAL PAC-Ø 1 in.
2. _____ 1/8		11. _____ 1C
3. _____ 1D		12. _____ 1B
4. _____ 2F		13. _____ MANUAL PAC THIS EDGE
5. _____ 1/2 in. TYP.		14. _____ 3F
6. _____ 1A		15. _____ 2 in.
7. _____ 1-2		16. _____ 3 in.
8. _____ 2G, SEE NOTE a		17. _____ 1G, SEE NOTE a
9. _____ 1-1/2 in.		

CHAPTER 19: QUIZ 4

Name _____   Date _____

Class _____   Instructor _____   Grade _____

INSTRUCTIONS

Carefully read Chapter 19 in the textbook and answer each question.

IDENTIFICATION

Identify the numbered items on the drawing by writing the letter next to the identifying term in the space provided.

1. _____ PIPE SURFACE GROUND CLEAN

2. _____ FACE BEND

3. _____ KEY HOLE

4. _____ 45°

5. _____ 1ST COVER PASS

6. _____ $1\frac{1}{2}"$ (38 mm)

7. _____ GROUND TACK WELD

8. _____ TACK WELD TO BE GROUND

9. _____ HOT PASS

10. _____ VERTICAL

11. _____ $\frac{3"}{32}$ (2 mm) MAX. REINFORCEMENT

12. _____ OVERHEAD

13. _____ FILLER PASSES

14. _____ ROOT WELD

15. _____ HORIZONTAL

16. _____ ROOT BEND

17. _____ CHANGING WELDING POSITIONS

18. _____ WELD

Shop Math and Weld Cost

■ PRACTICE 20-1

Name _____ Date _____

Class _____ Instructor _____ Grade _____

OBJECTIVE: After completing this practice, you will be able to determine the total volume of groove welds having various dimensions.

EQUIPMENT AND MATERIALS NEEDED FOR THIS PRACTICE

Piece of paper, pencil, calculator, and weld groove dimensions.

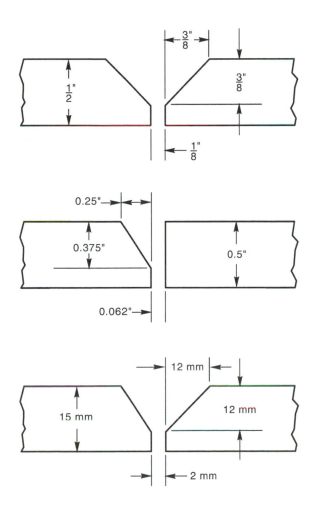

INSTRUCTIONS

1. Using Formulas 20-1, 20-2, 20-3, and 20-4 from the text, you will figure the total volume given the following groove weld dimensions.

2. Using Formula 20-1, determine the cross-sectional area for the given groove joints.

3. Using Formula 20-2, determine the cross-sectional area for the root opening.

4. Using Formula 20-3, determine the total cross-sectional area for each weld groove.

5. Plug the results from Formula 20-1 and Formula 20-2 into Formula 20-4. This will give you the total groove volume.

6. V-groove joint with the following dimensions:

   Width 3/8 in. _____

   Depth 3/8 in. _____

   Root opening 1/8 in. _____

   Thickness 1/2 in. _____

   Weld length 144 in. _____

7. Single bevel joint with the following dimensions:

   Width 0.25 in. _____

   Depth 0.375 in. _____

   Root opening 0.062 in. _____

   Thickness 0.5 in. _____

   Weld length 96 in. _____

8. V-groove joint with the following dimensions:

   Width 12 mm _____

   Depth 12 mm _____

   Root opening 2 mm _____

   Thickness 15 mm _____

   Weld length 3600 mm _____

INSTRUCTOR'S COMMENTS _____

_____

_____

■ PRACTICE 20-2

Name _____ Date _____

Class _____ Instructor _____ Grade _____

OBJECTIVE: After completing this practice, you will be able to calculate the weight of metal required for each of the welds described in Practice 20-1. Calculate the weight for both steel and aluminum base and filler metals.

EQUIPMENT AND MATERIALS NEEDED FOR THIS PRACTICE

Paper, pencil, calculator, and from the textbook, Table 20-7 Densities of Metals.

INSTRUCTIONS

Each example given is of a groove-type weld. Therefore, use Formula 20-4 and determine the GV (groove volume). Multiply the resulting groove volume by the appropriate metal density given in Table 20-7. Figure each weld groove for both aluminum and steel.

INSTRUCTOR'S COMMENTS _____

_____

_____

■ PRACTICE 20-3

Name _____ Date _____

Class _____ Instructor _____ Grade _____

OBJECTIVE: After completing this practice, you will be able to calculate the cost per pound of filler metal as it would be used to create a bill of materials.

EQUIPMENT AND MATERIALS NEEDED FOR THIS PRACTICE

Paper, pencil, and calculator.

INSTRUCTIONS

Use the data supplied below in A. and B. to fill out the information in the Total Cost per Pound of Weld Metal Deposited.

Total Cost per Pound of Weld Metal Deposited.
Welding Process: _____
Filler metal: AWS Number _____ Diameter _____
Welding Current: Amperage _____ Voltage _____

LABOR and OVERHEAD	(Labor & Overhead Cost/hr) ———————————————————————— (Deposition Rate [ib/hr]) X (Operating Factor)	= _____
ELECTRODE	(Electrode Cost/ib) ———————————————————————— (Deposition Efficiency)	= _____
GAS	(Gas Flow Rate [cfh]) X (Gas Cost/cu ft) ———————————————————————— (Deposition Rate [ib/hr])	= _____
Total Cost/ib of Weld Metal Deposited (Sum of the Above)		_____

A.

                        FCAW

Electrode Type	0.045-in. dia. E71T-2
Labor & Overhead	$25.00/hr
Welding Current	200 amperes
Deposition Rate	5.5 lb/hr
Operation Factor	45%
Electrode Cost	$1.55/lb.
Deposition Efficiency	85%
Gas Flow Rate	35 cfh
Gas Cost per Cu Ft	$0.04 CO_2

B.

	SMAW
Electrode Type	1/8-in. dia. E7018
Labor & Overhead	$23.00/hr
Welding Current	120 amperes
Deposition Rate	2.58 lb/hr
Operation Factor	30%
Electrode Cost	$0.97/lb
Deposition Efficiency	71.6%
Gas Flow Rate	Not Applicable
Gas Cost per Cu Ft	Not Applicable

INSTRUCTOR'S COMMENTS _____

_____

_____

CHAPTER 20: QUIZ 1

Name _____ Date _____

Class _____ Instructor _____ Grade _____

INSTRUCTIONS

Carefully read Chapter 20 in the textbook and answer each question.

MATCHING

In the space provided to the left of Column A, write the letter from Column B that best answers or completes the statement in Column A.

Column A

Column B

_____ 1. The most common use of math in welding shops is for dimensioning and _____.

a. parentheses

_____ 2. _____—the basic branch of mathematics that is primarily involved with combining numbers by addition, subtraction, multiplication, and division—is the math that most welders use on a daily basis.

b. reduced

_____ 3. _____ numbers are numbers used to express units in increments of 1, so they can be divided evenly by the number 1.

c. arithmetic

_____ 4. A decimal fraction is a number that uses a decimal point to denote a unit that is _____ than 1.

d. two

_____ 5. An example of a _____ is a linear measurement such as 2 ft 6 in., with part of the measurement expressed in feet (2) and the other part in inches (6).

e. common denominator

_____ 6. Examples of _____ are 1/2, 3/4, 5/16, 2 3/8, and 91/2.

f. pricing

_____ 7. In the expression 3^2, it means that the number 3 is squared, or multiplied by itself _____ times.

g. numerator

_____ 8. An equation is a mathematical statement in which both sides are _____ to each other; for example, 2X = 1Y.

h. type of number

_____ 9. When a formula has more than one mathematical operation, the operations within _____ must be performed first.

i. smaller

_____ 10. When adding mixed units such as feet and inches, you have to add each _____ together first; for example, you would add the inches to inches and feet to feet.

j. largest

_____ 11. When some mixed units are added, the sum can be _____.

k. measurements

_____ 12. Fractions are commonly used in welding fabrication for dimensioning a distance that is less than an _____.

l. whole

——— 13. When the denominators of two fractions to be added or subtracted are different, you must first find the _____.

m. vertex

——— 14. To reduce a fraction, find the _____ number that can be divided into both the denominator and numerator.

n. equal

——— 15. By converting a fraction to a decimal fraction, you can easily use a _____ to multiply and divide.

o. conversion chart

——— 16. Sometimes you will be working from the manufacturer's drawing, and then you may have to make conversions of _____.

p. fractions

——— 17. To convert a fraction to a decimal, divide the _____ (top number in the fraction) by the denominator (bottom number in the fraction).

q. tolerance

——— 18. Dimensioning _____ is the difference between the exact dimension as shown on a drawing and the actual acceptable size of a part.

r. calculator

——— 19. An easy way to convert between fractions, decimals, and metric units is to use a _____.

s. inch

——— 20. An angle is formed by two lines called sides that radiate outward from a single point called the _____.

t. mixed unit

CHAPTER 20: QUIZ 2

Name _____   Date _____

Class _____   Instructor _____   Grade _____

INSTRUCTIONS

Carefully read Chapter 20 in the textbook and answer each question.

MATCHING

In the space provided to the left of Column A, write the letter from Column B that best answers or completes the statement in Column A.

Column A	Column B
_____ 1. A right triangle is a triangle that has one of its angles at _____ between two of its sides.	a. isosceles
_____ 2. All three sides of an equilateral triangle are the same length, and each of the angles is _____.	b. cost estimation
_____ 3. An _____ triangle has two sides that are the same length and two angles that are the same.	c. smallest
_____ 4. The distance around the outside edge of a weldment is its _____.	d. widely varying
_____ 5. The perimeter is calculated by finding the _____ of all of the individual sides of an object.	e. 60°
_____ 6. Area is always expressed as _____ inches (in²), feet (ft²), or yards (yd²) (mm², cm², or m²).	f. 90°
_____ 7. Volume is expressed in _____ measurements such as cubic in. (in.³), cubic feet (ft³), or cubic yards (yd³) (mm³, cm³, or m³).	g. intangible
_____ 8. Measuring for most welded fabrications does not require accuracies greater than what can be obtained with a _____.	h. metal is thick
_____ 9. To remain profitable and competitive, some _____ is required.	i. benefits
_____ 10. _____ costs are those expenses that must be paid each and every day, week, month, or year, regardless of work or production.	j. bids
_____ 11. _____ costs are those expenses that change with the quantity of work being produced.	k. only a small part
_____ 12. The cost of new stock required to produce the weldment is fixed by the supplier, but it is often possible to help control these costs by getting _____ from several suppliers.	l. sum

_____ 13. Scrap costs a company in two ways: by wasting expensive resources and by requiring _____.

m. cubic

_____ 14. The cost of filler metal per pound is _____ of its actual cost.

n. square

_____ 15. The major welding processes (SMAW, GMAW, FCAW, and GTAW) have _____ deposition and efficiency rates.

o. production cost

_____ 16. Total labor cost includes wages and _____.

p. perimeter

_____ 17. Overhead costs are often _____ costs related to doing business.

q. cleanup and removal

_____ 18. Postwelding cleanup, grinding, painting, or other finishing adds to the weldment's final _____.

r. variable

_____ 19. To keep welding costs down, the joints should have the _____ possible root opening and the smallest reasonable bevel angle.

s. fixed

_____ 20. The weld should be approximately the same size as the _____.

t. steel rule or a steel tape

CHAPTER 20: QUIZ 3

Name _____ Date _____

Class _____ Instructor _____ Grade _____

INSTRUCTIONS

Carefully read Chapter 20 in the textbook and answer each question.

MATCHING

In the space provided to the left of Column A, write the letter from Column B that best answers or completes the statement in Column A.

	Column A	Column B
_____	1. Reducing the amount of filler metal needed cuts the _____ needed to make the welds.	a. operating factor
_____	2. For groove welds, the bevel angle greatly affects the _____ volume.	b. Bill of Materials
_____	3. The cross-sectional area of a fillet weld is equal to one-half of the weld leg height times the _____.	c. deposition efficiency
_____	4. The weight of weld metal is determined by multiplying the weld volume times the _____ of the metal.	d. pounds per hour
_____	5. To determine how much shielding gas will be used, you must know the _____.	e. continuous
_____	6. Some portion of every electrode is lost as _____.	f. labor
_____	7. The deposition rate is the rate at which weld metal can be deposited by a given electrode or welding wire, expressed in _____.	g. filler metal
_____	8. The _____ of a welding process refers to the percentage of the welding filler material that actually becomes part of the weld deposit.	h. 1%
_____	9. The deposition efficiency of _____ electrodes does not subtract the unused electrode stub that is discarded.	i. 99%
_____	10. Stub loss need not be considered in the efficiency of flux cored wires because the wire is _____.	j. time
_____	11. The efficiency of solid wires in GMAW is very high and will vary with the _____ or gas mixture used.	k. coated
_____	12. In submerged arc welding, there is no spatter loss, and an efficiency of _____ may be assumed.	l. invoice
_____	13. _____ is the percentage of a welder's working day actually spent on welding.	m. finished part

——— 14. Knowing the productivity of welders is necessary when determining the cost of the _____.

——— 15. _____ is the welder's hourly rate of pay including wages and benefits.

——— 16. _____ includes allocated portions of plant operating and maintenance costs.

——— 17. Cost of electrical power is a very small part of the cost of depositing weld metal and in most cases is less than _____ of the total.

——— 18. A _____ is a list of the materials, supplies, and consumables required to fabricate a project for a customer.

——— 19. The term _____ is usually used to indicate an estimate of the time required for the fabrication; it may also be used on an invoice.

——— 20. An _____ is a list of all of the items provided to a customer along with the total sum due.

n. flow rate

o. shielding gas

p. weld leg width

q. slag, spatter, and/or fume

r. shop time

s. density

t. overhead

CHAPTER 20: QUIZ 4

Name _____   Date _____

Class _____   Instructor _____   Grade _____

INSTRUCTIONS

Carefully read Chapter 20 in the textbook and answer each question.

IDENTIFICATION

Identify the numbered items on the drawing by writing the letter next to the identifying term in the space provided.

1. _____ .5

2. _____ 1 FOOT

3. _____ 1

4. _____ $\frac{3}{8}$

5. _____ 1 CENTIMETER

6. _____ $\frac{7}{8}$

7. _____ $\frac{11}{16}$

8. _____ $\frac{1}{4}$

9. _____ 3 FEET = 1 YARD

10. _____ $\frac{15}{16}$

11. _____ 1 MILLIMETER

12. _____ $\frac{9}{16}$

13. _____ $\frac{3}{4}$

14. _____ $\frac{3}{16}$

15. _____ 12 INCHES = 1 FOOT

16. _____ $\frac{7}{16}$

17. _____ $\frac{1}{2}$

18. _____ $\frac{1}{8}$

19. _____ $\frac{5}{16}$

20. _____ 10 MILLIMETERS = 1 CENTIMETER

21. _____ $\frac{5}{8}$

22. _____ $\frac{13}{16}$

23. _____ $\frac{1}{16}$

24. _____ 100 CENTIMETERS = 1 METER

CHAPTER 20: QUIZ 5

Name _____ Date _____

Class _____ Instructor _____ Grade _____

INSTRUCTIONS

Carefully read Chapter 20 in the textbook and answer each question.

IDENTIFICATION

Identify the numbered items on the drawing by writing the letter next to the identifying term in the space provided.

1. _____ BOX

2. _____ SQUARE

3. _____ PARALLELOGRAM

4. _____ OCTAGON

5. _____ HEXAGON

6. _____ CYLINDER

7. _____ CUBE

8. _____ ELLIPSE

9. _____ RECTANGLE

10. _____ TRIANGLE

11. _____ CIRCLE

a

b

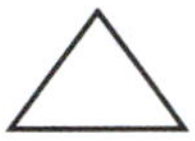

c

d

e

f

g

h

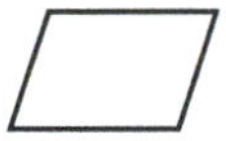

i

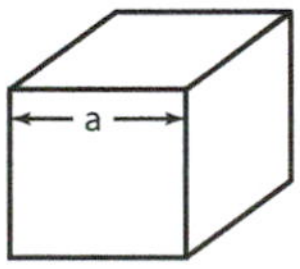

j

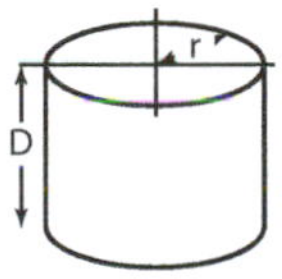

k

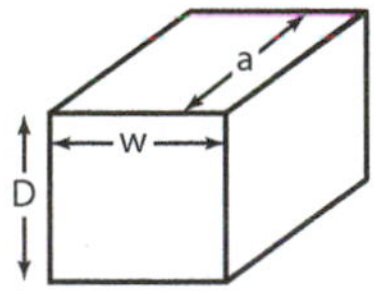

CHAPTER 20: QUIZ 6

Name _____   Date _____

Class _____   Instructor _____   Grade _____

INSTRUCTIONS

Carefully read Chapter 20 in the textbook and answer each question.

MATH

Calculate the area of the shapes below. 1st write the formula, 2nd ll in the values, 3rd calculate the area. Show your work.

1. _____
   Write Area Formula for Squares

$\uparrow$ a $\downarrow$

4″

Area = a²

1. _____
   Write Answer Here

---

2. _____
   Write Area Formula for Parallelograms

h

← b →

3″

7″

Area = b × h

2. _____
   Write Answer Here

CHAPTER 20: QUIZ 7

Name _____ Date _____

Class _____ Instructor _____ Grade _____

INSTRUCTIONS

Carefully read Chapter 20 in the textbook and answer each question.

MATH

Calculate the area of the shapes below. 1st write the formula, 2nd ll in the values, 3rd calculate the area. Show your work.

3. _____
    Write Area Formula for Circles

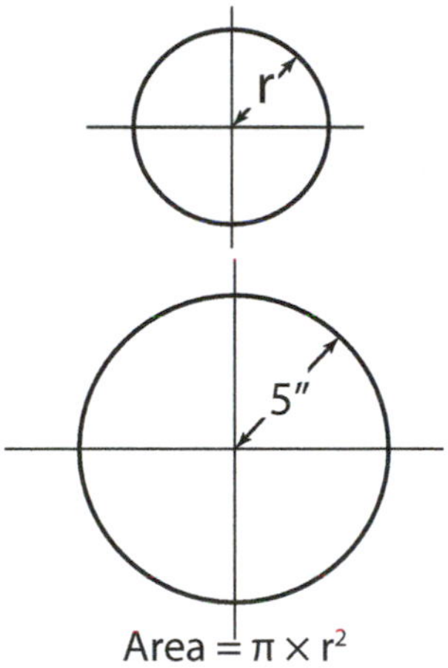

Area $= \pi \times r^2$

3. _____
    Write Answer Here

_____

4. _____
    Write Area Formula for Triangles

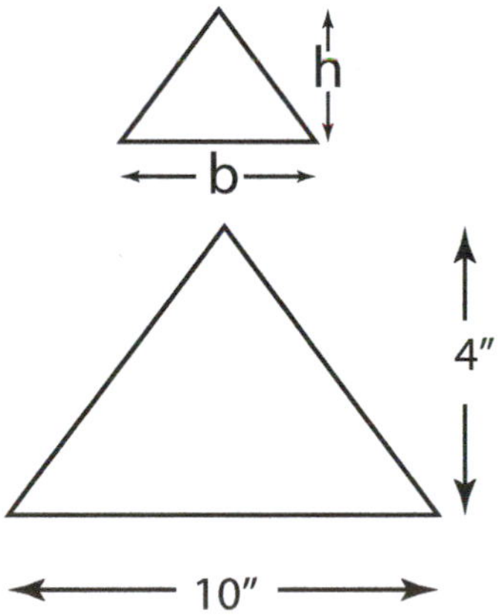

Area $= 1/2 \times b \times h$

4. _____
    Write Answer Here

CHAPTER 20: QUIZ 8

Name _____   Date _____

Class _____   Instructor _____   Grade _____

INSTRUCTIONS

Carefully read Chapter 20 in the textbook and answer each question.

MATH

Calculate the area of the shapes below. 1st write the formula, 2nd ll in the values, 3rd calculate the area. Show your work.

5. _____
    Write Area Formula for Rectangles

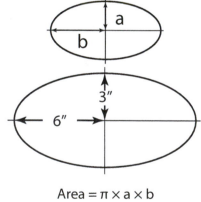

Area = w × h

5. _____
    Write Answer Here

---

6. _____
    Write Area Formula for Ellipses

Area = π × a × b

6. _____
    Write Answer Here

Reading Technical Drawings

Name _____ Date _____

Class _____ Instructor _____ Grade _____

OBJECTIVE: After completing this practice, you will be able to sketch a series of 6-in.-long straight lines.

EQUIPMENT AND MATERIALS NEEDED FOR THIS PRACTICE

You will need a pencil and unlined paper.

INSTRUCTIONS

Using a pencil and unlined paper, you are going to sketch a series of 6-in.-long horizontal straight lines spaced approximately 1/2 in. to 3/4 in. apart. Practice sketching from the right to the left and left to the right. The direction you sketch is not as important as your ability to make straight lines.

Once you have completed six or eight lines, lay a straightedge next to the lines and see how straight you were able to make them. Keep practicing sketching straight lines until you are able to make 6-in.-long lines that are within ±1/8 in. of being straight.

INSTRUCTOR'S COMMENTS _____

_____

_____

■ PRACTICE 21-2

Name _____   Date _____

Class _____   Instructor _____   Grade _____

OBJECTIVE: After completing this practice, you will be able to sketch a series of circles.

EQUIPMENT AND MATERIALS NEEDED FOR THIS PRACTICE

You will need a pencil and unlined paper.

INSTRUCTIONS

1. Start by sketching two light construction lines that cross at right angles.

2. Make two marks on each of the construction lines about 1/2 in. from the center point. These points will serve as your aiming points as you sketch the circle. The circle you sketch will be tangent to these points. A tangent straight line is one that meets a circle at a point where the circular line and straight line are going in the same direction, much like placing a 12-in. ruler on a round pipe; where the ruler and pipe meet is the tangent point. If you were to make a short, straight line at the tangent point and keep doing this all the way around the pipe, you would wind up with a circle drawn from a series of short, straight lines.

3. Sketching a tangent line starting at the top mark, keep sketching and gradually turn the line toward the mark on the next construction line. Once you have completed the first quarter of the circle, you may find it easier to continue if you turn the paper.

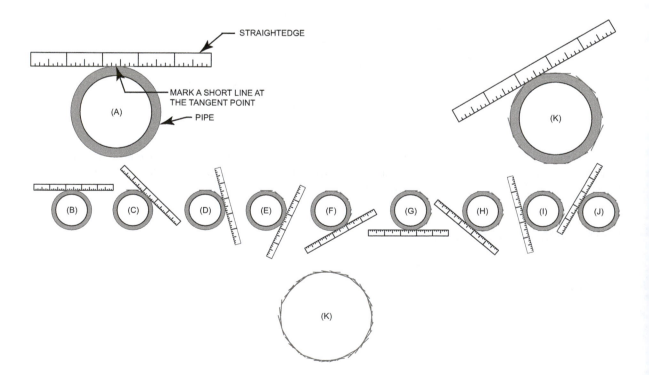

4. Repeat the sketching process until you have completed sketching the circle.

5. Repeat this process making several different size circles. Use a circle template to check your circles for accuracy.

INSTRUCTOR'S COMMENTS _____

_____

_____

■ PRACTICE 21-3

Name _____   Date _____

Class _____   Instructor _____   Grade _____

OBJECTIVE: After completing this practice, you will be able to sketch a mechanical drawing showing three views of a block, as shown in Figure 21-21 in the text.

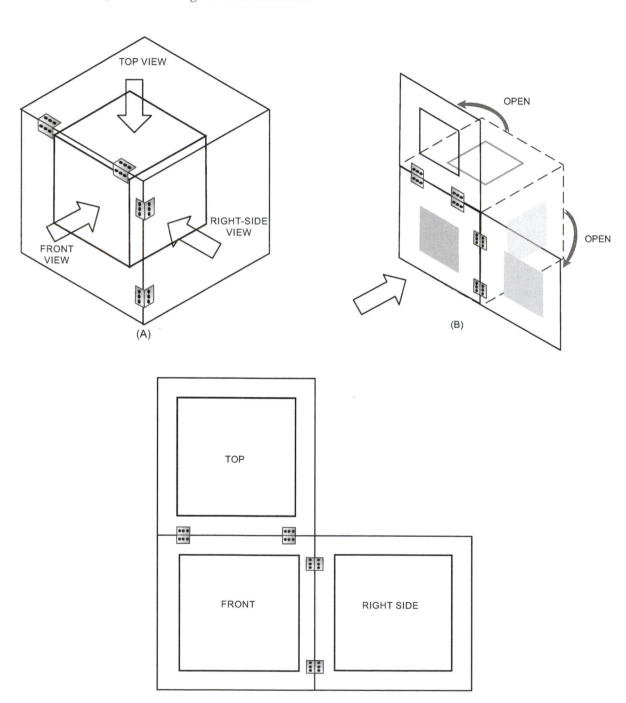

EQUIPMENT AND MATERIALS NEEDED FOR THIS PRACTICE

You will need a pencil and unlined paper.

INSTRUCTIONS

1. Start by sketching construction lines, as shown in Figure 21-22A in the text. These lines will form the boxes for the front, top, and right-side views.

   Darken the lines that make up the object's lines so it is easier to see.

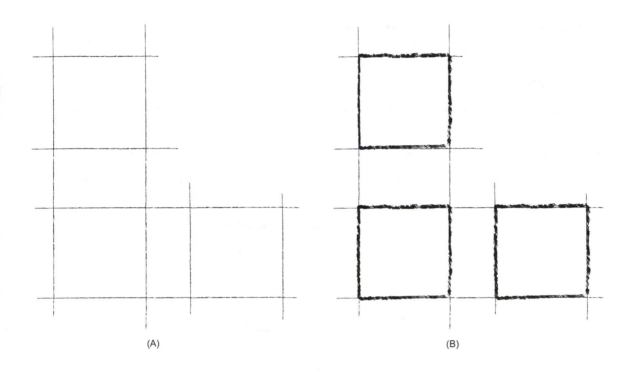

(A) (B)

INSTRUCTOR'S COMMENTS _____
_____
_____

■ PRACTICE 21-4

Name _____   Date _____

Class _____   Instructor _____   Grade _____

OBJECTIVE: After completing this practice, the student will be able to sketch a three-view mechanical drawing of a candlestick holder.

EQUIPMENT AND MATERIALS NEEDED FOR THIS PRACTICE

You will need a pencil and unlined paper.

INSTRUCTIONS

1. Using Figure 21-24, in the text, you will sketch a three-view mechanical drawing of the candlestick holder.

2. To lay out the angle for the candlestick holder, draw a vertical centerline that is 8 in. long.

3. Measure down the correct distance from the top and draw the 4-in.-long bottom line centered on the centerline.

4. Connecting the endpoints of the top and bottom lines will automatically give the angle.

5. Repeat the process using the candlestick holder shown in Figure 21-25 in the text.

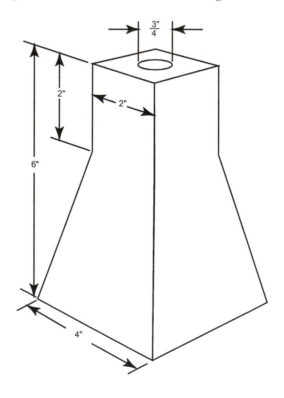

INSTRUCTOR'S COMMENTS _____

_____

_____

■ PRACTICE 21-5

Name _____ Date _____

Class _____ Instructor _____ Grade _____

OBJECTIVE: After completing this practice, the student will be able to sketch irregular shapes and curves using graphing paper.

EQUIPMENT AND TOOLS NEEDED FOR THIS PRACTICE

You will need a pencil and graphing paper.

INSTRUCTIONS

1. Locate a series of points on the graph paper that coincide with points on the curve you are copying. Start with the easy points where the lines on the paper cross at a point on the object.

   For example, one end of the curve starts at the intersection of lines E-3, so put a dot there. Next, the curve is tangent to lines F-2, so put a dot there. Follow the curve around, putting additional dots at the other intersecting points.

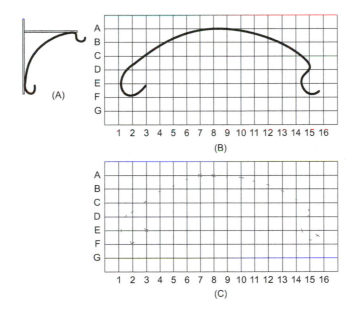

2. Once all of the easy dots are located, make some estimates for the next series of dots. For example, the curve almost touches the 1 line as it crosses the E line. Put a dot there and at similar points where the curve crosses other lines.

3. After you have located all of the points for the curve, sketch a line through all of the points. Refer to the figure below to see how the line should pass through the point. If you are not sure, you can always add additional points that are not on lines to help guide your sketching.

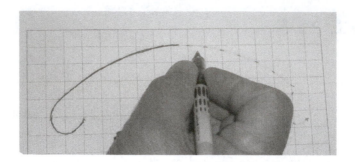

4. Repeat this practice and make a three-view drawing of the candlestick holder shown in Figure 21-31.

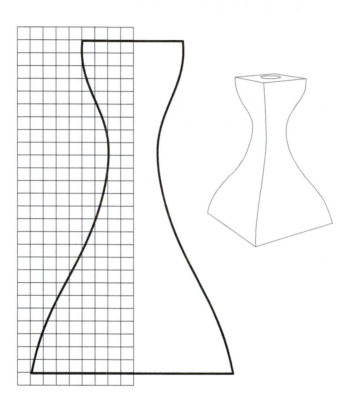

INSTRUCTOR'S COMMENTS _____

_____

_____

CHAPTER 21: QUIZ 1

Name _____ Date _____

Class _____ Instructor _____ Grade _____

INSTRUCTIONS

Carefully read Chapter 21 in the textbook and answer each question.

MATCHING

In the space provided to the left of Column A, write the letter from Column B that best answers or completes the statement in Column A.

	Column A	Column B
_____	1. _____ drawings have been called the universal language; they are produced in a similar format worldwide.	a. object
_____	2. A group of drawings, known as a _____, should contain enough information to enable a welder to produce the weldment.	b. internal
_____	3. Different types of lines are used to represent various parts of the object being illustrated, and the various line types are collectively known as the _____.	c. computers
_____	4. _____ lines show the edge of an object, the intersection of surfaces that form corners or edges, and the extent of a curved surface, such as the sides of a cylinder.	d. cutting plane
_____	5. _____ lines show the same features as object lines except that the corners, edges, and curved surfaces cannot be seen because they are hidden or obscured behind the surface of the object.	e. pictorial and mechanical drawings
_____	6. _____ show the center point of circles, arcs, round, or symmetrical objects.	f. phantom
_____	7. Numbers in the _____ line or next to it give the size or length of an object.	g. scale
_____	8. _____ lines are the lines extending from an object that locate the points being dimensioned.	h. cavalier
_____	9. _____ lines represent an imaginary cut through the object.	i. mechanical
_____	10. Section lines show the surface that has been imaginarily cut away with a cutting plane line to show _____ details.	j. raster
_____	11. _____ lines show that part of an object has been removed when a long uniform object needs to be shortened to fit the drawing page.	k. views

———   12. _____ lines show an alternate position of a moving part or the extent of motion such as the on/off position of a light switch.

l. hidden

———   13. Drawings used for most welding projects can be divided into two categories _____.

m. sketch

———   14. _____ drawings are drawn at a 30° angle so it appears that you are looking at one corner.

n. cutaway view

———   15. On _____ drawings, one surface, usually the front, is drawn flat to the page.

o. set of drawings

———   16. The section view is drawn as if parts of the object were sawn away to reveal internal _____.

p. entire

———   17. The _____ is used to show detail within a part that would be obscured by the part's surface.

q. graph

———   18. Detail views show small details of a part's area and negate the need to draw an enlargement of the _____ part.

r. dimension

———   19. Sometimes it may be necessary to look at other _____ to locate all of the dimensions required to build the object.

s. vector

———   20. When we use a _____, we are saying that the part being drawn is drawn smaller or larger than it really is.

t. centerlines

———   21. A _____ is a quick way of drawing an object.

u. details

———   22. Making a sketch on _____ paper is a way of both making your drawing more accurate and speeding up the sketching process.

v. extension

———   23. _____ have made it much easier to draw plans for projects.

w. break

———   24. Because _____ drawings are seen by the computer as lines, you can zoom in and out, measure, resize, reshape, or rotate the drawing and the lines stay crisp and sharp.

x. isometric

———   25. _____ drawings are commonly known as bitmap drawings because the computer maps the location of every little bit (pixel) of the drawing.

y. alphabet of lines

CHAPTER 21: QUIZ 2

Name _____   Date _____

Class _____   Instructor _____   Grade _____

INSTRUCTIONS

Carefully read Chapter 21 in the textbook and answer each question.

IDENTIFICATION

Identify the numbered items on the drawing by writing the letter next to the identifying term in the space provided.

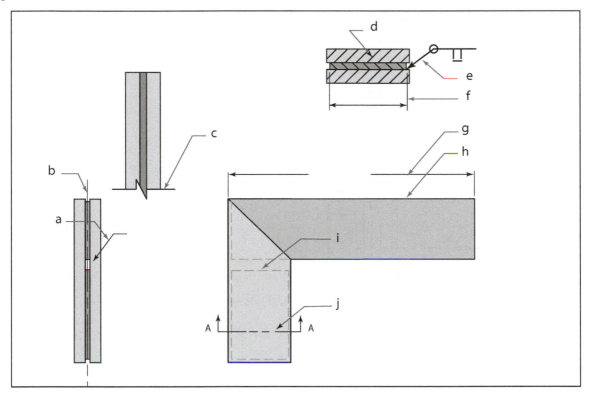

1. _____   HIDDEN LINE

2. _____   ARROW LINE

3. _____   OBJECT LINE

4. _____   CENTER LINE

5. _____   CUTTING PLANE LINE

6. _____   LONG BREAK LINE

7. _____   LEADER

8. _____   EXTENSION LINE

9. _____   DIMENSION LINE

10. _____   SECTION LINE

CHAPTER 21: QUIZ 3

Name _____   Date _____

Class _____   Instructor _____   Grade _____

INSTRUCTIONS

Carefully read Chapter 21 in the textbook and answer each question.

IDENTIFICATION

Write the dimensions in the space provided for the lengths shown by the letters on this drawing.

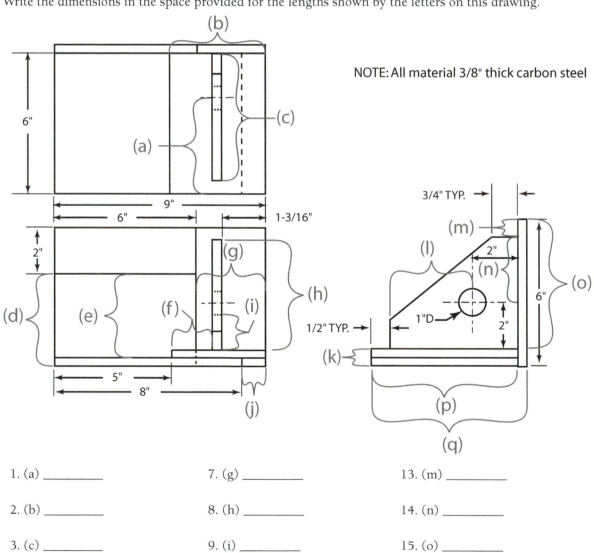

NOTE: All material 3/8" thick carbon steel

1. (a) _____

2. (b) _____

3. (c) _____

4. (d) _____

5. (e) _____

6. (f) _____

7. (g) _____

8. (h) _____

9. (i) _____

10. (j) _____

11. (k) _____

12. (l) _____

13. (m) _____

14. (n) _____

15. (o) _____

16. (p) _____

17. (q) _____

Welding Joint Design and Welding Symbols

■ PRACTICE 22-1

Name _____   Date _____

Class _____   Instructor _____   Grade _____

OBJECTIVE: After completing this practice, you should be able identify the elements that make up a welding symbol.

EQUIPMENT AND MATERIALS NEEDED FOR THIS PRACTICE

Pencil.

INSTRUCTIONS

In the space provided, identify the items shown in the illustration.

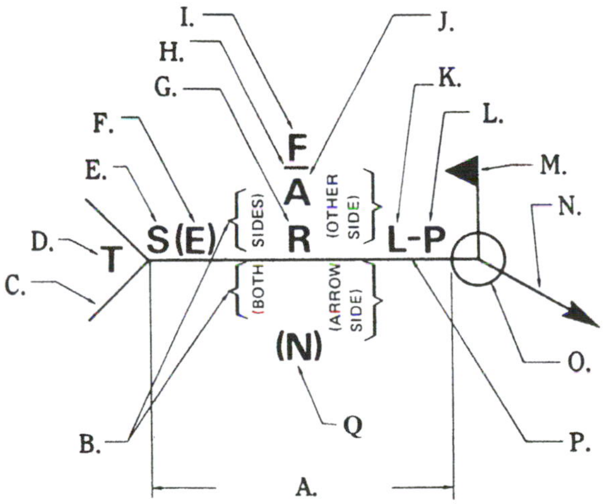

_____ 1. Field weld symbol

_____ 2. Groove angle; included angle of countersink for plug welds

_____ 3. Reference line

_____ 4. Basic weld symbol or detail reference

_____ 5. Pitch (center-to-center spacing) of welds

_____ 6. Specification, process, or other reference of filling

_____ 7. Finish symbol

_____ 8. Tail

_____ 9. Length of weld

_____ 10. Arrow connecting reference line to arrow side member of joint

_____ 11. Weld-all-around symbol

_____ 12. Depth of preparation; size or strength for certain welds

_____ 13. Elements in this area remain as shown when the tail and arrow are reversed

_____ 14. Root opening; depth for plug and slot welds

_____ 15. Contour symbol

_____ 16. Effective throat

INSTRUCTOR'S COMMENTS _____

_____

_____

■ PRACTICE 22-2

Name _____ Date _____

Class _____ Instructor _____ Grade _____

1. From the following weld symbols, draw the welds in their appropriate locations by making a sketch of the joint.

A.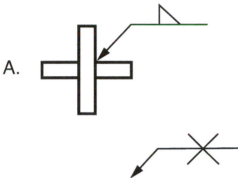

B.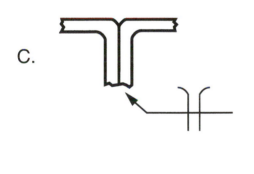

C.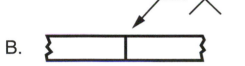

D.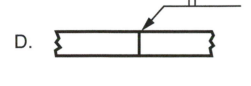

E.

F.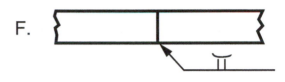

G.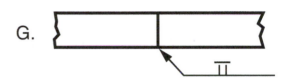

H.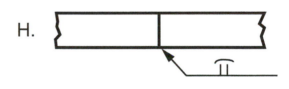

I.

J.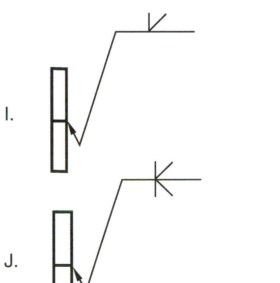

2. Sketch a V-grooved butt joint and label.

3. Sketch a weld on plates in the 1G and 1F positions.

4. Sketch a weld on plates in the 2G and 2F positions.

5. Sketch a weld on plates in the 3G and 3F positions.

6. Sketch a weld on plates in the 4G and 4F positions.

7. Sketch a weld on a pipe in the 1G position.

8. Sketch a weld on a pipe in the 5G position.

9. Sketch a weld on a pipe in the 2G position.

10. Sketch a weld on a pipe in the 6G position.

11. Sketch a weld on a pipe in the 6GR position

CHAPTER 22: QUIZ 1

Name _____   Date _____

Class _____   Instructor _____   Grade _____

INSTRUCTIONS

Carefully read Chapter 22 in the textbook and answer each question.

MATCHING

In the space provided to the left of Column A, write the letter from Column B that best answers or completes the statement in Column A.

Column A	Column B
_____ 1. The term _____ refers to the way pieces of metal are put together or aligned with each other.	a. angle
_____ 2. In a _____ joint, the edges of the metal meet so that the thickness of the joint is approximately equal to the thickness of the metal.	b. edge preparation
_____ 3. In a _____ joint, the edges of the metal overlap so the thickness of the joint is approximately equal to the combined thickness of both pieces of metal.	c. distributed
_____ 4. In a _____ joint, the edge of a piece of metal is placed on the surface of another piece of metal, and usually the parts are placed at a 90° angle with each other.	d. flat
_____ 5. In an outside corner joint, the edges of the metal are brought together at an _____, usually approximately 90° to each other.	e. weld joint design
_____ 6. In an edge joint, the metal surfaces are placed together so that the edges are _____.	f. design
_____ 7. The purpose of a weld joint is to join parts together so that the stresses are _____.	g. control
_____ 8. The welding process to be used has a major effect on the selection of the joint _____.	h. base metal
_____ 9. The area of the metal's surface that is _____ during the welding process is called the faying surface.	i. lap
_____ 10. The faying surface can be shaped before welding to increase the weld's strength; this is called _____.	j. vertical
_____ 11. A weld should be as strong as or stronger than the _____ being joined.	k. butt
_____ 12. With thicker plates or pipe, the edge must be prepared with a _____ on one or both sides.	l. upper

——— 13. _____ is a process of cutting a groove in the back side of a joint that has been welded.

——— 14. Because some metals have specific problems with thermal expansion, crack sensitivity, or distortion, the joint design selected must help _____ these problems.

——— 15. The most ideal welding position for most joints is the _____ position because it allows for larger molten weld pools to be controlled.

——— 16. When welds are made in any position other than the flat position, they are referred to as being done _____.

——— 17. The flat 1G or 1F plate welding position is when welding is performed from the _____ side of the joint and the face of the weld is approximately horizontal.

——— 18. For the horizontal 2G or 2F position, the axis of the weld is approximately _____, but the type of the weld dictates the complete definition.

——— 19. The vertical 3G or 3F plate welding position is when the axis of the weld is approximately _____.

——— 20. The overhead 4G or 4F plate welding position is when welding is performed from the _____ of the joint.

m. underside

n. back gouging

o. melted

p. horizontal

q. out of position

r. tee

s. groove

t. even

CHAPTER 22: QUIZ 2

Name _____   Date _____

Class _____   Instructor _____   Grade _____

INSTRUCTIONS

Carefully read Chapter 22 in the textbook and answer each question.

MATCHING

In the space provided to the left of Column A, write the letter from Column B that best answers or completes the statement in Column A.

Column A

_____ 1. In the horizontal rolled 1G position, the pipe is rolled either continuously or intermittently so that the weld is performed within 0° to 15° of the _____ of the pipe.

_____ 2. In the horizontal fixed 5G position, the pipe is parallel to the horizon, and the weld is made _____ around the pipe.

_____ 3. In the vertical 2G position, the pipe is vertical to the horizon, and the weld is made _____ around the pipe.

_____ 4. In the inclined 6G position, the pipe is _____ in a 45° inclined angle, and the weld is made around the pipe.

_____ 5. In the inclined with a restriction ring 6GR position, the pipe is fixed in a 45° inclined angle, and there is a restricting ring placed around the pipe below the _____.

_____ 6. The type, depth, angle, and location of the groove are usually determined by a _____ that has been qualified for the specific job.

_____ 7. Often the skills or abilities of the welder are a _____ factor in joint design.

_____ 8. _____ must be a consideration for any project to be competitive and cost-effective.

_____ 9. The use of welding symbols enables a designer to indicate clearly to the welder important _____ regarding the weld.

_____ 10. Each type of weld has a specific _____ that is used on drawings to indicate the weld.

_____ 11. The terms arrow side, other side, and both sides are used to indicate the _____ with respect to the joint.

Column B

a. detailed information

b. limiting

c. arrow

d. weld groove

e. parentheses

f. joint design

g. horizontally

h. fixed

i. below

j. backing

k. strengths

——— 12. In the case of fillet and groove welding symbols, the arrow connects the welding symbol reference line to one side of the joint, and the surface of the joint, the arrow point actually touches is considered to be the _____ side of the joint.

l. vertically

——— 13. Dimensions of fillet welds are shown on the _____ side of the reference line as the weld symbol and are shown to the left of the symbol.

m. code or standard

——— 14. Holes in the arrow side member of a joint for plug welding are indicated by placing the weld symbol _____ the reference line.

n. type of test

——— 15. When a definite number of spot welds are desired in a certain joint, the quantity is placed above or below the weld symbol in _____.

o. width

——— 16. The size of seam welds is designated as the _____ of the weld expressed in fractions or decimal hundredths of an inch.

p. combination symbol

——— 17. Joint _____ can be improved by making some type of groove preparation before the joint is welded.

q. top

——— 18. A _____ (strip) is a piece of metal that is placed on the back side of a weld joint.

r. weld location

——— 19. Nondestructive testing (NDT) symbols are used by the designer or engineer to indicate the area to be tested and the _____ to be used.

s. same

——— 20. The welding symbols and nondestructive testing symbols can be combined into one _____, which may help both the welder and inspector to identify welds that need special attention.

t. symbol

CHAPTER 22: QUIZ 3

Name _____ Date _____

Class _____ Instructor _____ Grade _____

INSTRUCTIONS

Carefully read Chapter 22 in the textbook and answer each question.

IDENTIFICATION

Identify the numbered items on the drawing by writing the letter next to the identifying term in the space provided.

1. _____ SINGLE-J

2. _____ SINGLE-U

3. _____ DOUBLE-V

4. _____ SINGLE-BEVEL

5. _____ SQUARE

6. _____ DOUBLE-U

7. _____ DOUBLE-BEVEL

8. _____ SINGLE-V

9. _____ DOUBLE-J

CHAPTER 22: QUIZ 4

Name _____ Date _____

Class _____ Instructor _____ Grade _____

INSTRUCTIONS

Carefully read Chapter 22 in the textbook and answer each question.

IDENTIFICATION

Identify the numbered items on the drawing by writing the letter next to the identifying term in the space provided.

a	b	c	d
e	f	g	h
i	j	k	l

i — RESTRICTION RING, WELD, 45°

j — WELD

k — WELD, 45°, 6G

l — WELD

m — ROLL

Plate Positions	Plate Positions
1. _____ 3F VERTICAL	9. _____ HORIZONTAL FIXED 5G
2. _____ 1G FLAT	10. _____ 45 INCLINED 6G
3. _____ 3G VERTICAL	11. _____ HORIZONTAL ROLLED 1G
4. _____ 4G OVERHEAD	12. _____ VERTICAL 2G
5. _____ 2G HORIZONTAL	13. _____ 45 INCLINED WITH A RESTRICTION RING 6GR
6. _____ 1F FLAT	
7. _____ 2F HORIZONTAL	
8. _____ 4F OVERHEAD	

CHAPTER 22: QUIZ 5

Name _____ Date _____

Class _____ Instructor _____ Grade _____

INSTRUCTIONS
Carefully read Chapter 22 in the textbook and answer each question.

IDENTIFICATION
Identify the numbered items on the drawing by writing the letter next to the identifying term in the space provided.

1. _____ TAIL

2. _____ REFERENCE LINE

3. _____ FIELD WELD SYMBOL

4. _____ CONTOUR SYMBOL

5. _____ ROOT OPENING; DEPTH OF FILLING FOR PLUG AND SLOT WELDS

6. _____ LENGTH OF WELD

7. _____ EFFECTIVE THROAT

8. _____ DEPTH OF PREPARATION; SIZE OR STRENGTH FOR CERTAIN WELDS

9. _____ WELD-ALL-AROUND SYMBOL

10. _____ FINISH SYMBOL

11. _____ GROOVE ANGLE, INCLUDED ANGLE OF COUNTERSINK FOR PLUG WELDS

12. _____ ARROW CONNECTING REFERENCE LINE TO ARROW SIDE MEMBER OF JOINT

13. _____ PITCH (CENTER-TO-CENTER SPACING) OF WELDS

14. _____ ELEMENTS IN THIS AREA REMAIN AS SHOWN WHEN TAIL AND ARROW ARE REVERSED.

15. _____ SPECIFICATION, PROCESS, OR OTHER REFERENCE

16. _____ NUMBER OF SPOT OR PROJECTION WELDS

17. _____ BASIC WELD SYMBOL OR DETAIL REFERENCE

CHAPTER 22: QUIZ 6

Name _____   Date _____

Class _____   Instructor _____   Grade _____

INSTRUCTIONS
Carefully read Chapter 22 in the textbook and answer each question.

IDENTIFICATION
Identify the numbered items on the drawing by writing the letter next to the identifying term in the space provided.

c	a					i	j	k	l		
	b								m	n	o
	d	e	f	g	h						

1. _____   BEVEL

2. _____   PLUG & SLOT

3. _____   V

4. _____   WELD ALL AROUND

5. _____   WELD SURFACE CONTOUR

6. _____   FILLET

7. _____   CONVEX

8. _____   U

9. _____   TYPE OF WELD

10. _____   SQUARE

11. _____   CONCAVE

12. _____   FLUSH

13. _____   J

14. _____   FIELD WELD

15. _____   GROOVE

CHAPTER 22: QUIZ 7

Name _____   Date _____

Class _____   Instructor _____   Grade _____

INSTRUCTIONS

Carefully read Chapter 22 in the textbook and answer each question.

IDENTIFICATION

Identify the numbered items on the drawing by writing the letter next to the identifying term in the space provided.

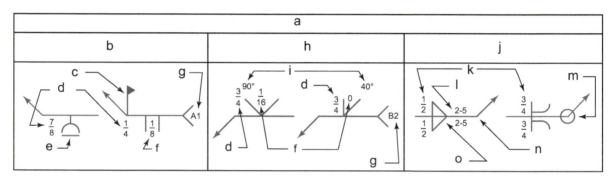

1. _____   WELD ALL AROUND

2. _____   ROOT OPENING

3. _____   OFFSET IF STAGGERED

4. _____   INCREMENT LENGTH

5. _____   INCLUDED ANGLE

6. _____   FIELD WELD

7. _____   SIZE

8. _____   ARROW SIDE OF JOINT

9. _____   BOTH SIDES OF JOINT

10. _____   LOCATION OF WELDS

11. _____   FLUSH

12. _____   SIZE OR GROOVE DEPTH

13. _____   SEE NOTE 5

14. _____   OTHER SIDE OF JOINT

15. _____   PITCH OF INCREMENTS

Fabricating Techniques and Practices

■ PRACTICE 23-1

Name _____   Date _____

Class _____   Instructor _____   Grade _____

OBJECTIVE: After completing this practice, you should be able to lay out square, rectangular, and triangular parts to within ±1/16 in. tolerance in both standard units and S.I. units. Refer to Figure 23-17 in the textbook.

EQUIPMENT AND MATERIALS NEEDED FOR THIS PRACTICE

Piece of metal or paper, soapstone or pencil, tape measure, and square.

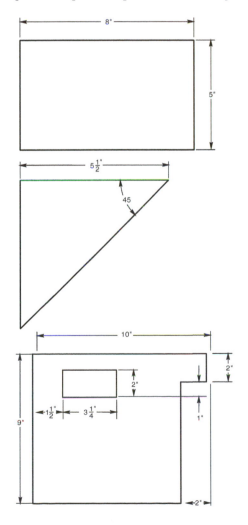

INSTRUCTIONS

1. If you are using a piece of paper, draw two lines to form a 90° (right) angle near one corner of the paper. This will be your baseline.

2. Measure along the baseline or edge of the metal and lay out the lengths of each part.

3. Use the square or squared edge of the metal and lay out the widths of each part.

4. Connect the marks you made for each part by drawing a line between the laid out points.

5. Convert the dimensions into S.I. units and repeat the layout process.

INSTRUCTOR'S COMMENTS _____

_____

_____

■ PRACTICE 23-2

Name _____ Date _____

Class _____ Instructor _____ Grade _____

OBJECTIVE: After completing this practice, you should be able to lay out circles, arcs, and curves to within ±1/16 in. tolerance in both standard units and S.I. units. Refer to Figure 23-20 in the textbook.

EQUIPMENT AND MATERIALS NEEDED FOR THIS PRACTICE

Piece of metal or paper, soapstone or pencil, tape measure, compass or circle template, and square.

INSTRUCTIONS

1. If you are using a piece of paper, draw two lines to form a 90° (right) angle near one corner of the paper. This will be your baseline.

2. Measure along the baseline or edge of the metal and lay out the lengths and centerline for each part.

3. Use the square or squared edge of the metal and lay out the widths and centerline for each part.

4. Connect the marks you made for each part by drawing a line between the laid out points.

5. Use a compass or circle template to lay out the circles and arcs. Be sure that the arcs meet the straight object lines so that they form a smooth tangent.

6. Convert the dimensions into S.I. units and repeat the layout process.

INSTRUCTOR'S COMMENTS _____

_____

_____

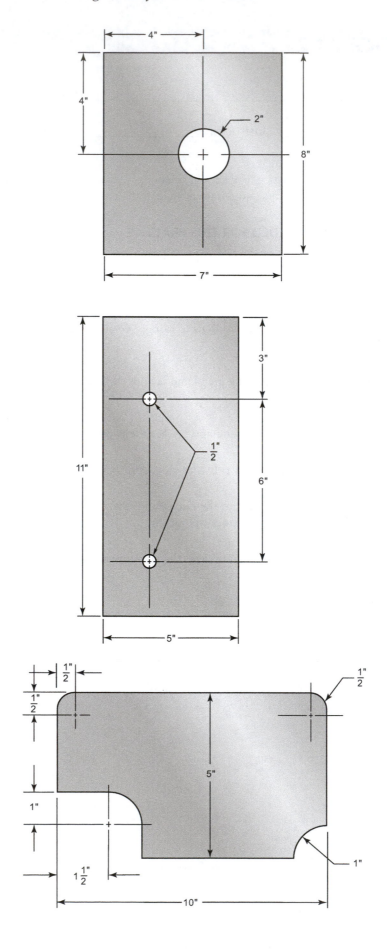

■ PRACTICE 23-3

Name _____ Date _____

Class _____ Instructor _____ Grade _____

OBJECTIVE: After completing this practice, you should be able to lay out parts, nesting them together so that the least amount of material will be wasted. Refer to Figure 23-22 in the textbook.

EQUIPMENT AND MATERIALS NEEDED FOR THIS PRACTICE

Metal or paper that is 8 1/2 in. × 11 in., soapstone or pencil, tape measure, and square.

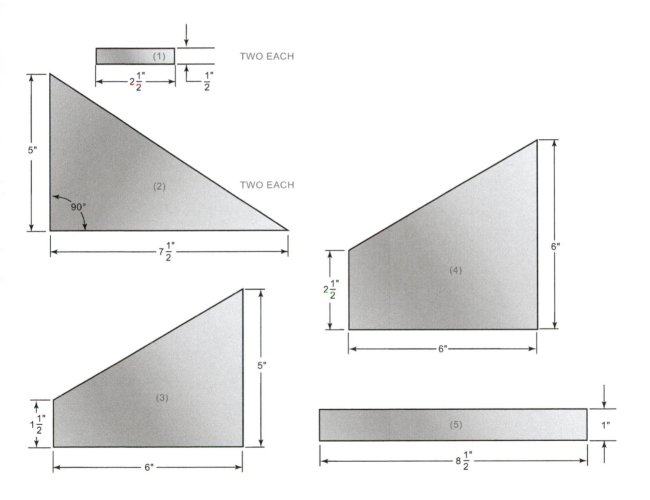

INSTRUCTIONS

1. Sketch the parts using different arrangements until you can arrive at the layout that requires the least space. Assume a 0 in. kerf width.

2. If you are using a piece of paper, draw two lines to form a 90° (right) angle near one corner of the paper. This will be your baseline.

3. Measure along the baseline or edge of the metal and lay out the lengths of each part.

4. Use the square or squared edge of the metal and lay out the widths of each part.

5. Connect the marks you made for each part by drawing a line between the laid out points.

6. Convert the dimensions into S.I. units and repeat the layout process.

INSTRUCTOR'S COMMENTS _____

_____

_____

■ PRACTICE 23-4

Name _____   Date _____

Class _____   Instructor _____   Grade _____

OBJECTIVE: After completing this practice, you should be able to fill out a Bill of Materials. Refer to Table 23-1 in the textbook.

EQUIPMENT AND MATERIALS NEEDED FOR THIS PRACTICE

Paper and pencil.

			BILL OF MATERIALS	
Part	Number Required	Type of Material	Size Standard Units	SI Units

INSTRUCTIONS

1. Assume that all of the parts are made out of 1/2 in. (13 mm) low carbon steel plate and that one set is needed.

2. Write the part number in the first left-hand column.

3. Write the number of units required in the second column.

4. Write the type of material in the center column.

5. Write the overall dimension of the material in the next column.

6. Convert the standard units into S.I. units and write this in the right-hand column.

7. Total the amount of each type of material.

INSTRUCTOR'S COMMENTS _____

_____

_____

■ PRACTICE 23-5

Name _____   Date _____

Class _____   Instructor _____   Grade _____

OBJECTIVE: After completing this practice, you should be able to lay out a shape, allowing space for the material that will be removed in the cut's kerfs.

EQUIPMENT AND MATERIALS NEEDED FOR THIS PRACTICE

Pencil, 8 1/2 in. × 11 in. paper, measuring tape or rule, and square.

INSTRUCTIONS

1. If you are using a piece of paper, draw two lines to form a 90° (right) angle near one corner of the paper. This will be your baseline.

2. Measure along the baseline or edge of the metal and lay out the lengths of each part, adding 3/32 in. to each part for the kerf.

3. Use the square or squared edge of the metal and lay out the widths of each part, adding 3/32 in. to each part for the kerf.

4. Connect the marks you made for each part by drawing a line between the laid out points.

INSTRUCTOR'S COMMENTS _____

_____

_____

■ PRACTICE 23-6

Name _____ Date _____

Class _____ Instructor _____ Grade _____

OBJECTIVE: After completing this practice, you should be able to lay out square, rectangular, and triangular nested parts to within +/−1/16 in. tolerance, assume a 0 in. kerf width in both standard units and S.I. units.

EQUIPMENT AND MATERIALS NEEDED FOR THIS PRACTICE

Paper that is 8 1/2 in. by 11 in., pencil, tape measure or 12 in. rule, and a square.

1.

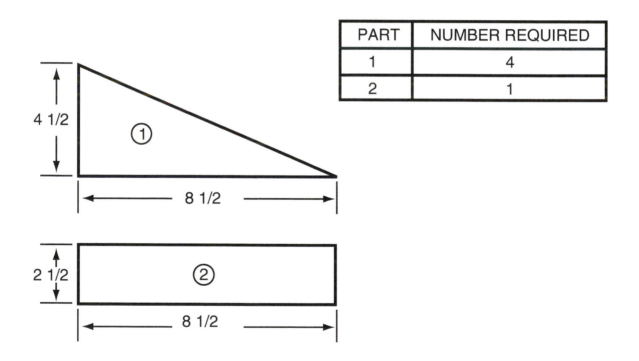

PART	NUMBER REQUIRED
1	4
2	1

2.

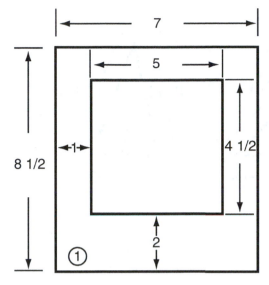

PART	NUMBER REQUIRED
1	1
2	2
3	1
4	1

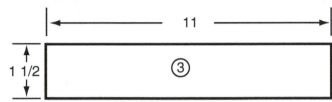

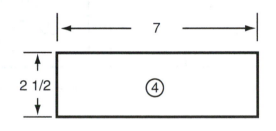

3.

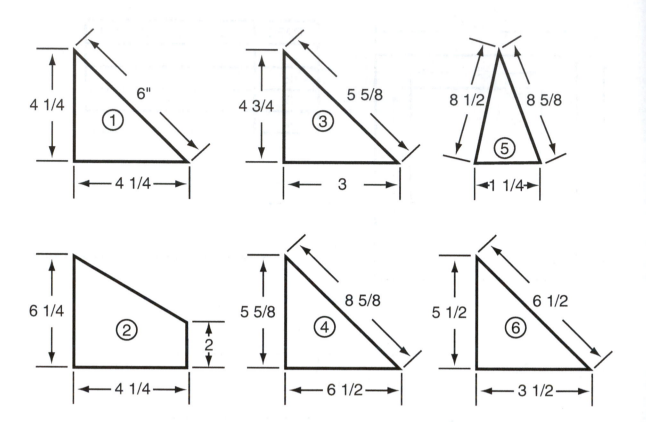

PART	NUMBER REQUIRED
1	2
2	2
3	1
4	1
5	1
6	1

4.

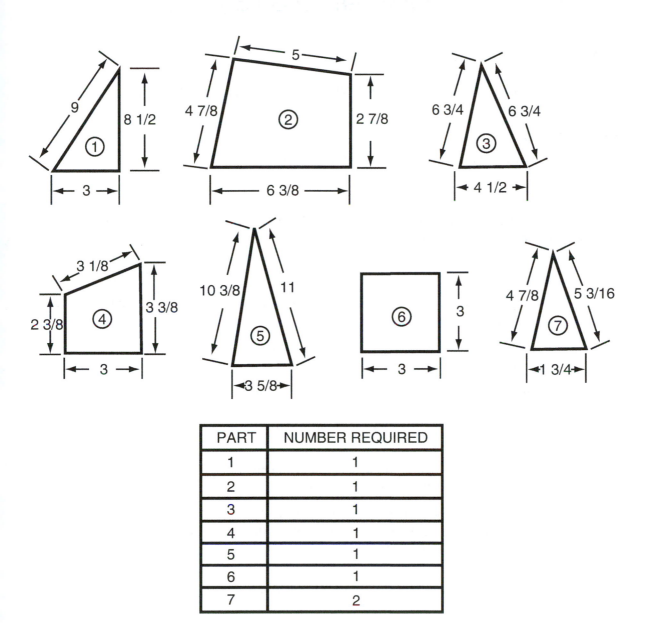

PART	NUMBER REQUIRED
1	1
2	1
3	1
4	1
5	1
6	1
7	2

CHAPTER 23: QUIZ 1

Name _____ Date _____

Class _____ Instructor _____ Grade _____

INSTRUCTIONS

Carefully read Chapter 23 in the textbook and answer each question.

MATCHING

In the space provided to the left of Column A, write the letter from Column B that best answers or completes the statement in Column A.

	Column A	Column B
_____	1. The _____ step in almost every welding operation is the assembly of the parts to be joined by welding.	a. outside
_____	2. _____ is the process of assembling the parts to form a weldment.	b. kerf
_____	3. Fabrication may present some potential safety problems not normally encountered in straight shop welding, and much of the larger fabrication work may need to be performed _____ an enclosed welding booth.	c. custom
_____	4. Welded fabrications can be made from precut and preformed parts, or they can be made from _____ parts.	d. soapstone
_____	5. When a large number of the same items are made in a large-run shop, it may _____ some of the parts to shops that specialize in mass-producing items.	e. tolerance
_____	6. The opposite end of the spectrum from preformed parts is _____ fabrication in which all or most of the assembly is handmade.	f. first
_____	7. Laying out lines and locating points for cutting, bending, drilling, and assembling may be marked with a _____ or a chalk line, scratched with a metal scribe, or punched with a center punch.	g. carbon steel
_____	8. Manual nesting of parts may require several tries at laying out the parts to achieve the lowest possible _____.	h. clamps
_____	9. The _____ is the space created as material is removed during a cut.	i. fitting
_____	10. The most common metal used for weldments is _____, and the most common shapes of metal stock used are plate, sheet, pipe, tubing, and angles.	j. fabrication

———— 11. Plate is usually 3/16 in. (4.8 mm) or thicker and measured in ————.

k. mark

———— 12. Pipe is dimensioned by its diameter and schedule or ————.

l. aligned

———— 13. ———— is the amount that a part can be bigger or smaller than it should be and still be acceptable.

m. face-up

———— 14. The ———— process brings together all the parts of the weldment.

n. outsource

———— 15. The overall dimensions of a weldment could change if the parts are not properly ———— during fit up.

o. strength

———— 16. Identify each part of the assembly and ———— each piece for future reference.

p. tack

———— 17. A variety of ———— can be used to temporarily hold parts in place so that they can be tack welded.

q. assembly

———— 18. ———— is the process of adjusting the parts of a weldment so that they meet the overall tolerance because not all parts fit exactly as they were designed.

r. scrap

———— 19. ———— welds are welds (usually small in size) that are made during the assembly to hold all of the parts of a weldment together so the welding can be finished.

s. hand-cut and hand-formed

———— 20. Placing your helmet ———— will prevent weld spatter, grinding sparks, and other shop dirt from collecting inside.

t. inches and fractions of inches

CHAPTER 23: QUIZ 2

Name _____   Date _____

Class _____   Instructor _____   Grade _____

INSTRUCTIONS

Carefully read Chapter 23 in the textbook and answer each question.

MATCHING

In the space provided to the left of Column A, write the letter from Column B that best answers or completes the statement in Column A.

Column A	Column B
_____ 1. Generally, welding on an assembly should be staggered from one part to another, which will allow both the welding heat and welding stresses to _____.	a. pneumatic
_____ 2. Keep the arc strikes in the welding joint so that they will be _____ as the weld is made.	b. anywhere
_____ 3. _____ work may vary from chipping, cleaning, or grinding the welds to applying paint or other protective surfaces.	c. air compressor or air source
_____ 4. Always read the _____ operating and safety manual for any power tool before using it for the first time.	d. dissipate
_____ 5. The power for hand tools can be provided in one of three ways—corded, cordless, and _____.	e. disadvantage
_____ 6. The major advantage of a corded tool is that it has an endless supply of _____, so it can be used for hours of uninterrupted work.	f. spatter and slag
_____ 7. The major _____ of corded tools is that they can only be used where building or generator power is available.	g. safety
_____ 8. The major advantage of cordless tools is that they can be used _____.	h. finish
_____ 9. The major disadvantage of cordless tools is limited _____ life; they seem to die at the least opportune time.	i. sharpen
_____ 10. The major advantage of pneumatic tools is that they stay _____ even under heavy use.	j. remelted
_____ 11. The major disadvantage of pneumatic tools is that they have to have an external _____.	k. flapper wheels
_____ 12. Pedestal and bench grinders are most often used to grind on smaller parts and to _____ tools and drill bits.	l. turn color

———— 13. Grinding stones range in diameter from 6 inches to 12 inches, are available in various widths, and come in a variety of _____.

m. battery

———— 14. Overheating and loss of temper in the tool or drill bit's edge begins to occur when the part's edge starts to _____.

n. gouge

———— 15. Follow all _____ guidelines from the manufacturers of all of the equipment, tools, machines, grinding stones, grinding discs, and flap wheels.

o. manufacturer's

———— 16. Flap discs are sometimes known as flapper discs or _____.

p. bevels

———— 17. Flap discs can be used to clean up _____ in a groove after the root pass.

q. vibration

———— 18. Cupped grinding stones and fiber discs can be used to grind _____.

r. grits

———— 19. Flap discs are lighter in weight and produce less _____ during use than stones or fiber discs.

s. cool

———— 20. A benefit of flap discs is that they are less likely to _____ a work piece.

t. power

CHAPTER 23: QUIZ 3

Name _____   Date _____

Class _____   Instructor _____   Grade _____

INSTRUCTIONS

Carefully read Chapter 23 in the textbook and answer each question.

MATCHING

In the space provided to the left of Column A, write the letter from Column B that best answers or completes the statement in Column A.

	Column A	Column B
_____	1. Selecting the correct type and style of flap disc impacts _____ because that choice can affect the quality of finish, disc life, and cost/value added to the weldment.	a. grit
_____	2. The disc size must match the grinder's range; the wrong diameter can result in a _____ problem.	b. performance
_____	3. Never use a disc with a lower _____ rating than the grinder's speed because the higher than rated speed can cause the disc to fly apart or explode.	c. pipe and tubing
_____	4. Type 29 flap discs are often used for _____.	d. general-purpose
_____	5. The _____ is the portion of the flap disc that does the work of removing material.	e. larger
_____	6. Ceramic alumina discs are ideal for grinding on stainless steel, aluminum, and other _____ metals.	f. Zirconia alumina
_____	7. _____ discs are ideal for a high rate of stock removal and long life in demanding applications.	g. drill presses or pedestal drills
_____	8. Aluminum oxide discs can be used for _____ metalworking applications.	h. safety
_____	9. The size of the abrasive on a flap disc is expressed as a _____ number.	i. clamp down
_____	10. One type of stationary horizontal power saw for cutting metal is the _____, which has a continuous loop of thin steel.	j. horsepower
_____	11. Power _____ use a reciprocating motion to move the blade back and forth across the stock being cut.	k. stock removal
_____	12. Abrasive blades can be installed on circular saws to cut thin sheet metal and thicker metal sections such as small-diameter _____.	l. reverse

_____ 13. A handheld power shear or a handheld _____ can be used to cut out thin sheet metal sections.

_____ 14. The three major types of drills are portable hand drills, _____, and magnetic base portable drill presses.

_____ 15. A hand drill's capacity is rated in fractions of _____ and/or watts.

_____ 16. Most but not all drills have a forward and _____ function, which makes them ideal for installing and removing screws.

_____ 17. Drill press motors are typically more powerful that any hand drill, so it is important to _____ parts to prevent them from spinning if the drill bit catches.

_____ 18. Magnetic base portable drill presses are ideal for drilling holes _____ than a portable hand drill is capable of or to drill very precisely located holes.

_____ 19. When drilling out of position, all electro-magnet and some permanent-magnet drills must be _____ to prevent them from falling if the magnetic attraction is suddenly lost.

_____ 20. Trusses made for structural steel provide great strength even though they are much _____ than solid steel beams.

m. hard-to-grind

n. safety chained

o. band saw

p. RPM

q. lighter

r. nibbler

s. hacksaws

t. abrasive

CHAPTER 23: QUIZ 4

Name _____ Date _____

Class _____ Instructor _____ Grade _____

INSTRUCTIONS

Carefully read Chapter 23 in the textbook and answer each question.

IDENTIFICATION

Identify the numbered items on the drawing by writing the letter next to the identifying term in the space provided.

1. _____ SHEET LESS THAN $\frac{3"}{16}$

2. _____ SQUARE

3. _____ REINFORCING BAR

4. _____ ZEE BAR

5. _____ WIDE FLANGE

6. _____ RECTANGULAR

7. _____ H-BEAMS

8. _____ PLATE MORE THAN $\frac{3"}{16}$

9. _____ HEXAGON

10. _____ PERFORATED SHEET

11. _____ STANDARD EXPANDED

12. _____ HALF-ROUND

13. _____ HALF-OVAL

14. _____ TEES

15. _____ DIAMOND TREAD

16. _____ FLAT

17. _____ ROUND

18. _____ OCTAGON

19. _____ FLATTENED EXPANDED

20. _____ CHANNELS

SHEET AND PLATE

a   b   c

d   e   f

STRUCTURAL SHAPES

g   h   i

j   k   l

PIPE

m   n   o

TUBING

p   q   r

BAR STOCK

p   q   s

t   u   v

w   x   y

21. _____ STANDARD BEAMS

22. _____ ANGLES

23. _____ STANDARD SCHEDULE 40

24. _____ DOUBLE EXTRA STRONG SCHEDULE 180

25. _____ EXTRA STRONG SCHEDULE 80

CHAPTER 23: QUIZ 5

Name _____   Date _____

Class _____   Instructor _____   Grade _____

INSTRUCTIONS

Carefully read Chapter 23 in the textbook and answer each question.

IDENTIFICATION

Match the letter next to the surface on the PICTORIAL VIEW with the letter on the corresponding surface on the ORTHOGRAPHIC VIEWS.

1. a matches to _____

2. b matches to _____

3. c matches to _____

4. d matches to _____

5. e matches to _____

6. f matches to _____

7. g matches to _____

8. h matches to _____

9. i matches to _____

10. j matches to _____

11. k matches to _____

12. l matches to _____

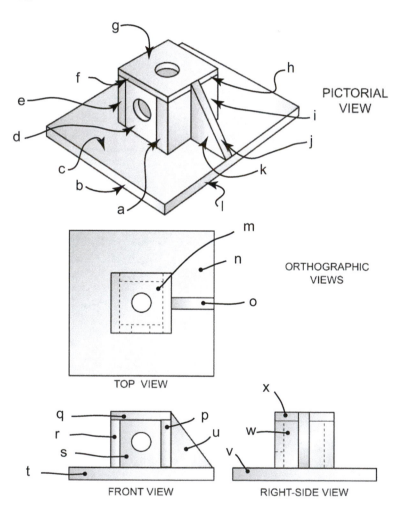

Welding Codes and Standards

■ PRACTICE 24-1

Name _____ Date _____

Class _____ Instructor _____ Grade _____

OBJECTIVE: After completing this practice, you should be able to write a Welding Procedure Specification (WPS).

EQUIPMENT AND MATERIALS NEEDED FOR THIS PRACTICE

You will need a copy of the top half of the WPS form on page 551 in this workbook and the list of instructions in Practice 24-1 on page 656 in the textbook (Welding Principles and Applications 9e) and two or more 8.5 by 11 sheets of paper.

INSTRUCTIONS

Follow the WPS example in Figure 24-3 on pages 652 to 656 in the textbook to create your own unique WPS. You can use the information found in Welding Principles and Applications, welding codes and standards, manufacturers' literature, or material supplied by your instructor to create your WPS. Not all of the blanks will be filled in on the form.

1. The WPS number

2. Date that the WPS was written

3. The welding process to be used

4. The material type and thickness

5. Fillet or groove weld and the joint type

6. Thickness or diameter range qualified

7. Welding position

8. Base metal specification, type, and grade including the P-number, Table 24-2.

9. Backing material, if any

10. AWS filler metal classification number

11. Filler metal specification number listed in Table 24-3

12. F-number listed in Table 24-4

13. A-number listed in Table 24-5

14. Shielding gas(es) and flow rate

Items 15 through 27 are optional and can be completed as a handwritten and sketched document, typed using a computer word and graphic program, or a combination.

INSTRUCTOR'S COMMENTS _____

_____

_____

WELDING PROCEDURE SPECIFICATION (WPS)

Welding Procedures Specifications No: _____(1)_____   Date: ____(2)____

TITLE:

Welding _____(3)_____ of _____(4)_____ to _____(4)_____.

SCOPE:

This procedure is applicable for _____(5)_____

within the range of _____(6)_____ through _____(6)_____.

Welding may be performed in the following positions _____(7)_____.

BASE METAL:

The base metal shall conform to _____(8)_____.

Backing material specification _____(9)_____.

FILLER METAL:

The filler metal shall conform to AWS classification No. _____(10)_____ from

AWS specification _____(11)_____. This filler metal falls into F-number

_____(12)_____ and A-number _____(13)_____.

SHIELDING GAS:

The shielding gas, or gases, shall conform to the following compositions and purity:

_____(14)_____

JOINT DESIGN AND TOLERANCES: (15)

PREPARATION OF BASE METAL: (16)

ELECTRICAL CHARACTERISTICS: (17)

The current shall be _____(18)_____.

The base metal shall be on the _____(19)_____ side of the line.

PREHEAT: (20)

BACKING GAS: (21)

WELDING TECHNIQUE: (22)

INTERPASS TEMPERATURE: (23)

CLEANING: (24)

INSPECTION: (25)

REPAIR: (26)

SKETCHES: (27)

WELDING PROCEDURE SPECIFICATION (WPS)

Welding Procedures Specifications No: (1)_____   Date: (2)_____

TITLE:
Welding (3)_____ of (4)_____ to (4)_____ .

SCOPE:
This procedure is applicable for (5)_____
within the range of (6)_____   through (6)_____ .

Welding may be performed in the following positions (7)_____ .

BASE METAL:
The base metal shall conform to (8)_____ .

Backing material specification (9)_____ .

FILLER METAL:
The filler metal shall conform to AWS classification No. (10)_____ from
AWS specification (11)_____ . This filler metal falls into F-number
(12)_____ and A-number (13)_____ .

SHIELDING GAS:
The shielding gas, or gases, shall conform to the following compositions and purity:
(14)_____

NOTE: The items below (15 through 27) are optional.

JOINT DESIGN AND TOLERANCES: (15)

PREPARATION OF BASE METAL: (16)

ELECTRICAL CHARACTERISTICS: (17)
The current shall be (18)_____ .

The base metal shall be on the (19)_____ side of the line.

PREHEAT: (20)

BACKING GAS: (21)

WELDING TECHNIQUE: (22)

INTERPASS TEMPERATURE: (23)

CLEANING: (24)

INSPECTION: (25)

REPAIR: (26)

SKETCHES: (27)

■ PRACTICE 24-2

Name _____   Date _____

Class _____   Instructor _____   Grade _____

OBJECTIVE: After completing this practice, you should be able to write a Procedure Qualification Record (PQR).

EQUIPMENT AND MATERIALS NEEDED FOR THIS PRACTICE

You will need a copy of the PQR form on page 553 in this workbook and the list of the instructions in Practice 24-2 on pages 656 to 661 in the textbook (Welding Principles and Applications 9e).

INSTRUCTIONS

Fill in the blanks on the PQR form on page 553 with the welding machine settings, welding position, filler metal specifications, base metal specifications and dimensions, test results, and any other relevant information required on the PQR form. You can use some of the information used in Practice 24-1. Not all of the blanks will be filled in on the forms.

INSTRUCTOR'S COMMENTS _____

_____

_____

PROCEDURE QUALIFICATION RECORD (PQR)

Welding Qualification Record No: ___(1)___ WPS No: ___(2)___ Date: ___(3)___
Material specification ___(4)___ to _____
P-No. ___(5)___ to P-No. _____ Thickness and O.D. ___(6)___
Welding process: Manual ___(7)___ Automatic ___(8)___
Thickness Range ___(9)___

Filler Metal

Specification No. ___(10)___ Classification ___(11)___ F-number ___(12)___
A-number ___(13)___ Filler Metal Size ___(14)___ Trade Name ___(15)___
Describe filler metal (if not covered by AWS specification) ___(16)___

Flux or Atmosphere

Shielding Gas ___(17)___ Flow Rate ___(18)___ Purge ___(19)___
Flux Classification ___(20)___ Trade Name ___(21)___

Welding Variables

Joint Type ___(22)___ Position ___(29)___
Backing ___(23)___ Preheat ___(30)___
Passes and Size ___(24)___ Bead Type ___(31)___
No. of Arcs ___(25)___ Current ___(32)___
Ampere ___(26)___ Volts ___(33)___
Travel Speed ___(27)___ Oscillation ___(34)___
Interpass Temperature Range ___(28)___

Weld Results

Appearance ___(35)___ Weld Size ___(36)___

Guided-Bend Test

Type	Result	Type	Result
(37)	(38)	(37)	(38)

Tensile Test

Specimen No.	Dimensions Width | Thickness	Area	Ultimate Total Load, lb.	Ultimate Unit Stress, psi	Character of Failure and Location
(39)	(40)	(41)	(42)	(43)	(44)

Welder's Name ___(45)___ Identification No. ___(46)___ Laboratory Test No. _____
By virtue of these test welder meets performance requirements.
Test Conducted by ___(47)___ Address _____
per ___(48)___ Date ___(49)___
We certify that the statements in this record are correct and that the test
welds performed and tested are in accordance with the WPS.
Manufacture ___(50)___
Signed by _____
Date _____

PROCEDURE QUALIFICATION RECORD (PQR)

Welding Qualification Record No: (1)_____ WPS No: (2)_____ Date: (3)_____

Material specification   (4)_____ to (4)_____

P-No. (5)_____ to P-No. (5)_____ Thickness and O.D. (6)_____

Welding process: Manual (7)_____ Automatic (8)_____

Thickness Range (9)_____

Filler Metal

Specification No. (10)_____ Classification (11)_____ F-number (12)_____

A-number (13)_____ Filler Metal Size (14)_____ Trade Name (15)_____

Describe filler metal (if not covered by AWS specification) (16)_____

_____

Flux or Atmosphere

Shielding Gas (17)_____ Flow Rate (18)_____ Purge (19)_____

Flux Classification (20)_____ Trade Name (21)_____

Welding Variables

Joint Type (22)_____	Position (29)_____
Backing (23)_____	Preheat (30)_____
Passes and Size (24)_____	Bead Type (31)_____
No. of Arcs (25)_____	Current (32)_____
Ampere (26)_____	Volts (33)_____
Travel Speed (27)_____	Oscillation (34)_____
Interpass Temperature Range (28)_____	

Weld Results

Appearance (35)_____ Weld Size (36)_____

Guided-Bend Test

Type	Result	Type	Result
(37)	(38)	(37)	(38)

Tensile Test

Specimen No.	Dimensions Width | Thickness	Area	Ultimate Total Load, lb.	Ultimate Unit Stress, psi	Character of Failure and Location
(39)	(40)	(41)	(42)	(43)	(44)

Welder's Name   (45)_____ Identification No. (46)_____ Laboratory Test No. _____

By virtue of these test welder meets performance requirements.

Test Conducted by   (47)_____ Address _____

                        per (48)_____ Date (49)_____

We certify that the statements in this record are correct and that the test
welds performed and tested are in accordance with the WPS.

Manufacture   (50)_____

Signed by _____

Date _____

CHAPTER 24: QUIZ 1

Name _____   Date _____

Class _____   Instructor _____   Grade _____

INSTRUCTIONS

Carefully read Chapter 24 in the textbook and answer each question.

MATCHING

In the space provided to the left of Column A, write the letter from Column B that best answers or completes the statement in Column A.

	Column A	Column B
_____	1. A welding _____ is a detailed listing of the rules or principles that are to be applied to a specific classification or type of product.	a. safest
_____	2. A welding _____ is a detailed statement of the legal requirements for a specific classification or type of weld to be made on a specific product.	b. governing
_____	3. Products manufactured to code or specification requirements commonly must be _____ to ensure compliance.	c. AWS D1.1
_____	4. Many _____ agencies require a specific code or standard to be followed.	d. witnessed
_____	5. A bonding or insuring company must feel that the product is the _____ that can be produced.	e. inspected and tested
_____	6. The American Petroleum Institute's API Standard 1104 is a commonly used code for _____.	f. differences
_____	7. The American Welding Society's _____ is a commonly used code for bridges, buildings, and other structural steel.	g. qualified
_____	8. _____ is a program that can be used by welding students to demonstrate their skills on a standardized weld test to demonstrate their qualifications.	h. specification
_____	9. A _____ is a set of written instructions by which a sound weld is made.	i. test samples
_____	10. _____ is the standard federal government, military, or aerospace terminology denoting a WPS.	j. tentative WPS
_____	11. The WPS must be _____ to prove or verify that the list of variables—amperage, voltage, filler, and so on—will provide a sound weld.	k. code or standard
_____	12. A record of all the parameters used to produce the test welds should be recorded on a form called the _____.	l. Welding Schedule

_____ 13. Test samples are welded in accordance with the _____, and the welding parameters are recorded on the PQR.

_____ 14. Test samples must be _____ by an authorized person from an independent testing lab, the customer, an insurance company, or other individual(s) as specified by the code or listing agency.

_____ 15. If the _____ pass the applicable test, then the procedure has completed qualification.

_____ 16. If the test samples do not pass the applicable test, then the tentative WPS value parameters are changed as deemed feasible, and the test samples are then rewelded and _____ to determine if they do or do not meet applicable requirements.

_____ 17. A qualified WPS is usable for an _____ length of time, usually until it is replaced by a process considered more efficient for the product.

_____ 18. Ideally, the WPS should include all of the information required to make the _____.

_____ 19. There are large _____ among various codes.

_____ 20. Some codes may require a written procedure for each _____.

m. position

n. Procedure Qualification Record (PQR)

o. indefinite

p. welding procedure specification

q. retested

r. pipelines

s. AWS SENSE

t. weld

CHAPTER 24: QUIZ 2

Name _____   Date _____

Class _____   Instructor _____   Grade _____

INSTRUCTIONS

Carefully read Chapter 24 in the textbook and answer each question.

IDENTIFICATION

Identify the numbered items on the drawing by writing the letter next to the identifying term in the space provided.

	Type of Material
P-1	a
P-3	b
P-4	b
P-5	c
P-6	d
P-7	e
P-8	f
P-9	g
P-10	h
P-21	i
P-31	j
P-41	k

	Filler Metal Type
l	Aluminum and Aluminum-Alloy Wire Electrodes and Rods
m	Aluminum and Aluminum-Alloy SMAW Electrodes
n	Brazing and Braze Welding Filler Metal
o	Carbon Steel SMAW Electrodes
p	Carbon Steel FCAW Electrodes
q	Carbon Steel SAW Wires and Fluxes
r	Carbon Steel GMAW Electrodes and Rods
s	Carbon and Low-Alloy Steel Rods for OFW
t	Alloy Steel for SMAW
u	Low-Alloy Steel Electrodes and Fluxes for SAW
v	Low-Alloy Steel Electrodes and Rods for GMAW
w	Low-Alloy Steel for FCAW

Group Designation	Metal Types	AWS Electrode Classification
F1	Carbon steel	x
F2	Carbon steel	y
F3	Carbon steel	z
F4	Carbon steel	aa
F5	Stainless steel	bb
F6	Stainless steel	cc
F22	Aluminum	dd

1. _____ A5.10

2. _____ EXX15, EXX16, EXX18

3. _____ Nickel alloy steel

4. _____ Aluminum and aluminum-base alloys

5. _____ A5.29

6. _____ EXX12, EXX13, EXX14

7. _____ A5.1

8. _____ Alloy steel

9. _____ A5.18

10. _____ ERXXX

11. _____ Nickel

12. _____ High-alloy steel—predominantly martensitic

13. _____ A5.2

14. _____ EXX20, EXX24, EXX27, EXX28

15. _____ A5.20

16. _____ Carbon steel

17. _____ A5.28

18. _____ EXX10, EXX11

19. _____ A5.3

20. _____ Low-alloy steel

21. _____ ERXXXX

22. _____ A5.5

23. _____ Specialty high-alloy steels

24. _____ A5.23

25. _____ High-alloy steel—austenitic

26. _____ Copper and copper alloy

27. _____ A5.8

28. _____ EXXX15, EXXX16

29. _____ A5.17

30. _____ High-alloy steel—predominantly ferritic

Testing and Inspection

Etching Solutions for Microscopic Examination of Metals	
Metal to Be Etched	**Etchant Mixture**
Aluminum	45 cc hydrochloric acid 15 cc nitric acid 15 cc hydrofluoric acid 25 cc water
Brass and copper alloys	25 cc ammonium hydroxide 25 cc hydrogen peroxide 25 cc water
Iron and steel	5 cc nitric acid 100 cc ethyl alcohol
Nickel and its alloys	50 cc of 70% nitric acid 50 cc of 50% acetic acid
Stainless steels	30 cc hydrochloric acid 10 cc nitric acid 80 cc glycerol

Mixing chemicals may be hazardous. Do not attempt to mix these solutions without proper supervision and instructions by someone trained to handle these materials, such as a chemist.

SAFETY PRECAUTIONS

When working with etchants, use extreme caution. Protective clothing is required (safety glasses, face shield, gloves, and an apron). Any mixing, handling, or use of an etchant is to be done in a well-ventilated area.

CAUTION **As an added safety precaution always start the mixture by pouring the water into the mixing container first. THEN pour the acid slowly and carefully while stirring the mixture.**

CHAPTER 25: QUIZ 1

Name _____ Date _____

Class _____ Instructor _____ Grade _____

INSTRUCTIONS
Carefully read Chapter 25 in the textbook and answer each question.

MATCHING
In the space provided to the left of Column A, write the letter from Column B that best answers or completes the statement in Column A.

Column A

Column B

_____ 1. The extent to which a welder and product are subjected to testing and inspection depends on the intended _____ of the product.

a. porosity

_____ 2. Mechanical testing (DT) methods, except for hydrostatic testing, result in the product being _____.

b. nondestructive

_____ 3. _____ testing (NDT) does not destroy the part being tested.

c. internal

_____ 4. A _____, according to AWS, is "a discontinuity or discontinuities that by nature or accumulated effect render a part or product unable to meet minimum applicable acceptance standards or specifications.

d. removed

_____ 5. _____ results when gas that was dissolved in the molten weld pool forms bubbles that are trapped as the metal cools to become solid.

e. coalescence

_____ 6. _____ are nonmetallic materials, such as slag and oxides, that are trapped in the weld metal, between weld beads, or between the weld and the base metal.

f. service

_____ 7. Inadequate joint penetration occurs when the depth that the weld penetrates the joint is _____ than that needed to fuse through the plate or into the preceding weld.

g. contaminants

_____ 8. Incomplete fusion is the lack of _____ between the molten filler metal and previously deposited filler metal and/or the base metal.

h. underfill

_____ 9. Even though arc strikes can be ground smooth, they cannot be _____.

i. inference

_____ 10. Overlap, also called _____, occurs in fusion welds when weld deposits are larger than the joint is conditioned to accept.

j. inclusions

_____ 11. Undercut is a groove melted into the base metal adjacent to the weld toe or weld root and left _____ by weld metal.

k. pulled apart

——— 12. _____ are the tiny cracks that develop in the weld craters as the weld pool shrinks and solidifies.

l. parallel to

——— 13. _____ on a groove weld is when the weld metal deposited is inadequate to bring the weld's face or root surfaces to a level equal to that of the original plane or plate surface.

m. destroyed

——— 14. Some welding problems result from _____ plate defects that the welder cannot control.

n. fatigue

——— 15. Laminations differ from lamellar tearing because they are more extensive and involve thicker layers of nonmetallic _____.

o. crater cracks

——— 16. When laminations intersect a joint being welded, the heat and stresses of the weld may cause some laminations to become _____.

p. defect

——— 17. Lamellar tears appear as cracks _____ and under the steel surface.

q. delaminated

——— 18. Because most destructive testing results in some degree of damage or total destruction of the part being tested, only an _____ can be made about whether the other weldments are fit for service because they were not actually tested.

r. unfilled

——— 19. When tensile tests are performed, the specimen is placed in the tensile testing machine and _____.

s. less

——— 20. _____ testing is used to determine how well a weld can resist repeated fluctuating stresses or cyclic loading.

t. cold lap

CHAPTER 25: QUIZ 2

Name _____ Date _____

Class _____ Instructor _____ Grade _____

INSTRUCTIONS
Carefully read Chapter 25 in the textbook and answer each question.

MATCHING
In the space provided to the left of Column A, write the letter from Column B that best answers or completes the statement in Column A.

Column A	Column B
_____ 1. The three methods of testing welded butt joints are: (1) the nick-break test; (2) the guided-bend test; and (3) the _____ test.	a. convex
_____ 2. With the _____ test, a force is then applied, and the specimen is ruptured by one or more blows of a hammer.	b. location
_____ 3. Once a guided-bend test is completed and the specimen is removed from the jig, the _____ surface is examined for cracks or other discontinuities and judged acceptable or unacceptable according to specified criteria.	c. surface
_____ 4. _____ lines are drawn on the weld face of a free-bend specimen.	d. fluorescent
_____ 5. The percent of elongation of an alternate bend specimen is obtained by dividing the elongation by the initial gauge length and _____ by 100.	e. nick-break
_____ 6. In a fillet weld break test, a force is applied to the specimen until the specimen _____, and then the break surface is examined for soundness.	f. liquid
_____ 7. Specimens to be tested by etching are etched for two purposes: (1) to determine the soundness of a weld or (2) to determine the _____ of a weld.	g. very near
_____ 8. When diluting an acid, always pour the acid _____ into the water while continuously stirring the water.	h. impact
_____ 9. One common _____ test is the Izod test, in which a notched specimen is struck by an anvil mounted on a pendulum.	i. discontinuities
_____ 10. Nondestructive testing of welds is a method used to test materials for _____ defects such as cracks, arc strikes, undercuts, and lack of penetration.	j. multiplying

———— 11. _____ inspection is the most frequently used nondestructive testing method and is the first step in almost every other inspection process.

k. inside

———— 12. Two types of penetrants are now in use to locate minute surface cracks and porosity, the color-contrast version and the _____ version.

l. slowly

———— 13. Magnetic particle inspection uses finely divided ferromagnetic particles (powder) to indicate defects open to the surface or just below the surface on _____ materials.

m. gauge

———— 14. Radiographic inspection (RT) is a method for detecting flaws _____ weldments.

n. Brinell

———— 15. Radiography gives a picture of all _____ that are parallel (vertical) or nearly parallel to the source.

o. strength

———— 16. _____ uses electronically produced high-frequency sound waves, which penetrate metals and many other materials at speeds of several thousand feet (meters) per second.

p. ultrasonic inspection

———— 17. Leak checking can be performed by filling the welded container with either a gas or a _____.

q. free-bend

———— 18. Eddy current inspection is effective in testing nonferrous and ferrous materials for internal and external cracks, slag inclusions, porosity, and lack of fusion that are on or _____ the surface.

r. magnetic

———— 19. Hardness is the resistance of metal to penetration and is an index of the wear resistance and _____ of the metal.

s. visual

———— 20. The two types of hardness testing machines in common use are the Rockwell and the _____ testers.

t. ruptures

CHAPTER 25: QUIZ 3

Name _____ Date _____

Class _____ Instructor _____ Grade _____

INSTRUCTIONS

Carefully read Chapter 25 in the textbook and answer each question.

IDENTIFICATION

Identify the numbered items on the drawing by writing the letter next to the identifying term in the space provided.

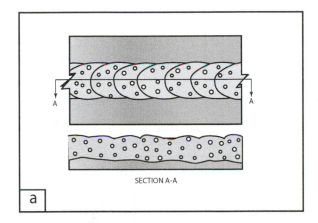

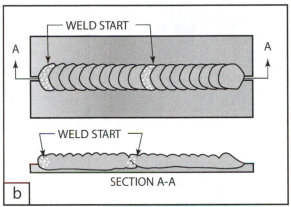

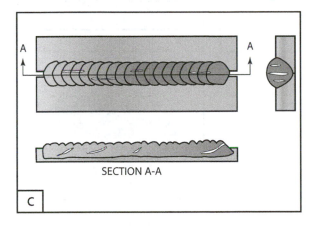

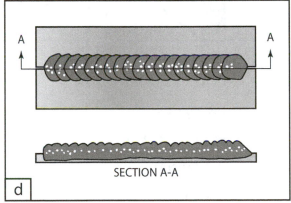

1. _____ Piping or Wormhole Porosity.

2. _____ Uniformly Scattered Porosities

3. _____ Linear Porosity

4. _____ Clustered Porosity

CHAPTER 25: QUIZ 4

Name _____   Date _____

Class _____   Instructor _____   Grade _____

INSTRUCTIONS

Carefully read Chapter 25 in the textbook and answer each question.

IDENTIFICATION

Identify the numbered items on the drawing by writing the letter next to the identifying term in the space provided.

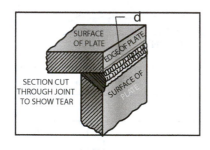

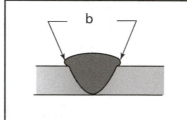

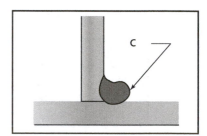

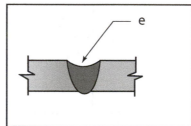

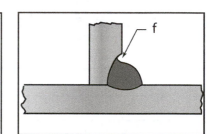

1. _____ CRATER OR STAR CRACK

2. _____ UNDERCUT

3. _____ LAMELLAR TEAR

4. _____ DELAMINATION

5. _____ ARC STRIKE

6. _____ UNDERFILL

7. _____ OVERLAP

8. _____ LAMINATION

9. _____ ROLLOVER

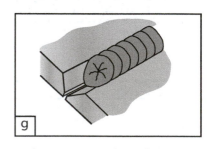

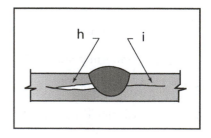

CHAPTER 25: QUIZ 5

Name _____ Date _____

Class _____ Instructor _____ Grade _____

INSTRUCTIONS

Carefully read Chapter 25 in the textbook and answer each question.

IDENTIFICATION

Identify the numbered items on the drawing by writing the letter next to the identifying term in the space provided.

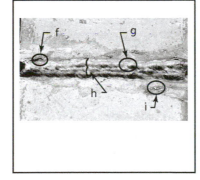
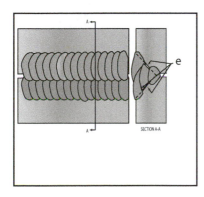
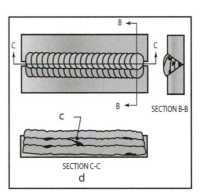
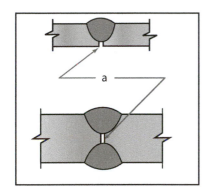
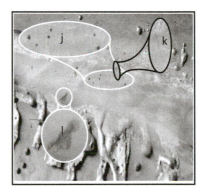
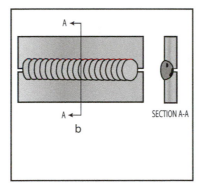

1. _____ MULTIPLE PASS

2. _____ TRANSVERS CRACK

3. _____ UNDERCUT

4. _____ SINGLE WELD BEAD

5. _____ UNIFORM SCATTERED POROSITY

6. _____ UNEVEN BEAD SHAPE

7. _____ NO INTERPASS FUSION

8. _____ POOR RESTART

9. _____ SLAG INCLUSION

10. _____ ARC STRIKE

11. _____ INCOMPLETE ROOT PENETRATION

12. _____ STOP AND RESTART

Welding Metallurgy

MELTING POINTS OF METALS AND ALLOYS		
	Melting Point	
Metal or alloy	°F	°C
Aluminum, cast (8% copper)	1175	635
Aluminum, pure	1220	660
Aluminum (5% silicon)	1118	603
Brass, naval	1625	885
Brass, yellow	1660	904
Bronze, aluminum	1905	1041
Bronze, manganese	1600	871
Bronze, phosphor	1830 to 1922	999 to 1050
Bronze, tobin	1625	885
Chromium	2740	1504
Copper	1981	1083
Iron, cast	2300	1260
Iron, malleable	2300	1260
Iron, pure	2786	1530
Iron, wrought	2750	1510
Lead	620	327
Magnesium	1200	649
Manganese	2246	1230
Molybdenum	4532	2500
Monel metal	2480	1360
Nickel	2686	1452
Nickel silver (18% nickel)	2030	1110
Silver, pure	1762	961
Silver solders (50% silver)	1160 to 1275	627 to 690
Solder (50–50)	420	216
Stainless steel (18–8)	2550	1399
Stainless steel, low carbon (18–8)	2640	1449
Steel, high carbon (0.55–0.83% carbon)	2500 to 2550	1371 to 1399
Steel, low carbon (maximum 0.30% carbon)	2600 to 2750	1427 to 1510
Steel, medium carbon (0.30–0.55% carbon)	2550 to 2600	1399 to 1427
Steel, cast	2600 to 2750	1427 to 1510
Steel, manganese	2450	1343
Steel, nickel (3.5% nickel)	2600	1427
Tantalum	5160	2849
Tin	420	232
Titanium	3270	1799
Tungsten	6152	3400
Vanadium	3182	1750
White metal	725	385
Zinc	786	419

HEAT COLORS AT GIVEN TEMPERATURES OF FERROUS MATERIALS		
Temper Color	**Temperature**	
	°F	**°C**
Faint straw	400	204
Straw	440	227
Dark straw	460	238
Very deep straw	480	249
Brown yellow	500	260
Bronze or brown purple	520	271
Peacock or full purple	540	282
Bluish purple	550	288
Blue	570	299
Full blue	590	310
Very dark blue	600	316
Light blue	640	338
Faint red (visible in dark)	750	399
Faint red	900	482
Blood red	1050	565
Dark cherry	1175	635
Medium cherry	1250	677
Cherry or full red	1375	746
Bright cherry	1450	788
Bright red	1550	843
Salmon	1650	899
Orange	1725	940
Lemon	1825	996
Light yellow	1975	1079
White	2200	1204

SURFACE COLOR OF SOME COMMON METALS			
Metals	**Color of Unfinished, Unbroken Surface**	**Color and Structure of Newly Fractured Surface**	**Color of Freshly Filed Surface**
White cast iron	Dull gray	Silvery white; crystalline	Silvery white
Gray cast iron	Dull gray	Dark silvery; crystalline	Light silvery gray
Malleable iron	Dull gray	Dark gray; finely crystalline	Light silvery gray
Wrought iron	Light gray	Bright gray	Light silvery gray
Low carbon and cast steel	Dark gray	Bright gray	Bright silvery gray
High carbon steel	Dark gray	Light gray	Bright silvery gray
Stainless steel	Dark gray	Medium gray	Bright silvery gray
Copper	Reddish-brown to green	Bright red	Bright copper color
Brass and bronze	Reddish-yellow, yellow-green, or brown	Red to yellow	Reddish-yellow to yellowish-white
Aluminum	Light gray	White; finely crystalline	White
Monel metal	Dark gray	Light gray	Light gray
Nickel	Dark gray	Off white	Bright silvery white
Lead	White to gray	Light gray; crystalline	White
Copper-nickel (70–30)	Gray	Light gray	Bright silvery white

IDENTIFICATION OF METALS BY CHIP TEST	
Metals	**Chip Characteristics**
White cast iron	Chips are small, brittle fragments. Chipped surfaces not smooth.
Gray cast iron	Chips are about 1/8 in. in length. Metal not easily chipped, so chips break off and prevent smooth cut.
Malleable iron	Chips are 1/4 to 3/8 in. in length (larger than chips from cast iron). Metal is tough and hard to chip.
Wrought iron	Chips have smooth edges. Metal is easily cut or chipped; chip can be made as a continuous strip.
Low carbon and cast steel	Chips have smooth edges. Metal is easily cut or chipped; chip can be made as a continuous strip.
High carbon steel	Chips show a fine-grain structure. Edges of chips are lighter in color than chips of low carbon steel. Metal is hard, but can be chipped in a continuous strip.
Copper	Chips are smooth, with sawtooth edges where cut. Metal is easily cut; chip can be cut as a continuous strip.
Brass and bronze	Chips are smooth, with sawtooth edges. These metals are easily cut, but chips are more brittle than chips of copper. Continuous strip is not easily cut.
Aluminum and aluminum alloys	Chips are smooth, with sawtooth edges. Chip can be cut as a continuous strip.
Monel	Chips have smooth edges. Continuous strip can be cut. Metal chips easily.
Nickel	Chips have smooth edges. Continuous strip can be cut. Metal chips easily.
Lead	Chips of any shape may be obtained because the metal is so soft that it can be cut with a knife.

IDENTIFICATION OF METALS BY OXYACETYLENE TORCH TEST	
Metals	**Reactions When Heated by Oxyacetylene Torch**
White cast iron	Metal becomes dull red before melting. Melts at moderate rate. A medium tough film of slag develops. Molten metal is watery, reddish-white in color, and does not show sparks. When flame is removed, depression in surface of metal under flame disappears.
Gray cast iron	Pool of molten metal is quiet, rather watery, but with heavy, tough film forming on surface. When torch flame is raised, depression in surface of metal disappears instantly. Molten pool takes time to solidify, and gives off no sparks.
Malleable iron	Metal becomes red before melting; melts at moderate rate. A medium tough film of slag develops, but can be broken up. Molten pool is straw-colored, watery, and leaves blow holes when it boils. Center of pool does not give off sparks, but the bright outside portion does.
Wrought iron	Metal becomes bright red before it melts. Melting occurs quietly and rapidly, without sparking. There is a characteristic slag coating, greasy or oily in appearance, with white lines. The straw-colored molten pool is not viscous, is usually quiet but may have a tendency to spark; is easily broken up.
Low carbon and cast steel	Melts quickly under the torch, becoming bright red before it melts. Molten pool is liquid, straw-colored, gives off sparks when melted, and solidifies almost instantly. Slag is similar to the molten metal and is quiet.
High carbon steel	Metal becomes bright red before melting, melts rapidly. Melting surface has cellular appearance, and is brighter than molten metal of low carbon steel; sparks more freely, and sparks are whiter. Slag is similar to the molten metal and is quiet.
Stainless steels	Reactions vary depending upon the composition.
Copper	Metal has high heat conductivity; therefore, larger flame is required to produce fusion than would be required for the same size piece of steel. Copper color may become intense before metal melts; metal metals slowly, and may turn black and then red. There is little slag. Molten pool shows mirror-like surface directly under flame, and tends to bubble. Copper that contains small amounts of other metals melts more easily, solidifies more slowly than pure copper.
Brass and bronze	These metals melt very rapidly, becoming noticeably red before melting. True brass gives off white fumes when melting. Bronze flows very freely when melting, and may fume slightly.
Aluminum and aluminum alloys	Melting is very rapid, with no apparent change in color of metal. Molten pool is same color as unheated metal and is fluid; stiff black scum forms on surface, tends to mix with the metal, and is difficult to remove.
Monel	Melts more slowly than steel, becoming red before melting. Slag is gray scum, quiet, and hard to break up. Under the scum, molten pool is fluid and quiet.
Nickel	Melts slowly (about like Monel), becoming red before melting. Slag is gray scum, quiet, and hard to break up. Under the scum, molten pool is fluid and quiet.
Lead	Melts at very low temperature, with no apparent change in color. Molten metal is white and fluid under a thin coat of dull gray slag. At higher temperature, pool boils and gives off poisonous fumes.

CHAPTER 26: QUIZ 1

Name _____   Date _____

Class _____   Instructor _____   Grade _____

INSTRUCTIONS

Carefully read Chapter 26 in the textbook and answer each question.

MATCHING

In the space provided to the left of Column A, write the letter from Column B that best answers or completes the statement in Column A.

	Column A	Column B
——	1. Metals gain their desirable mechanical and chemical properties as a result of how they are shaped or formed, their alloying elements, and how they were _____.	a. grain
——	2. Heat is the quantity of thermal energy and _____ is the level of thermal activity.	b. metallurgist
——	3. _____ heat is the heat required to change matter from one state to another, and it does not result in a temperature change.	c. heat-treated
——	4. As matter becomes warmer, its atoms vibrate at a _____ frequency.	d. mechanical
——	5. If the _____ properties of a metal are known, then a product can be constructed that will meet specific engineering specifications, and a safe and sound structure can be constructed.	e. continuous
——	6. All thermal processes including welding affect the original _____ structure of metals.	f. brittleness
——	7. Steel making begins with the _____ of the ore.	g. latent
——	8. During ingot casting, the molten metal is poured into large molds where it is allowed to cool and solidify into block shapes called _____.	h. process
——	9. _____ casting is the most efficient way that molten steel is initially formed into a shape that can be formed into a finished product.	i. fracturing
——	10. Forming _____ are used to convert the ingots into usable products like plates, sheets, pipe, rods, and wire.	j. pull
——	11. The size and shape of every type of metal's grain structure are the result of the _____ that was used to form its shape— casting, rolling, drawing, extruding, or forging.	k. temperature

—— 12. It is the responsibility of the _____ or engineer to select a metal that has the best group of properties for any specific job.

—— 13. Hardness may be defined as resistance to _____.

—— 14. _____ is the ease with which a metal will crack or break apart without noticeable deformation.

—— 15. _____ is the ability of a metal to be permanently twisted, drawn out, bent, or changed in shape without cracking or breaking.

—— 16. Toughness is the property that allows a metal to withstand forces, sudden shock, or bends without _____.

—— 17. Tensile strength refers to the property of a material that resists forces applied to _____ metal apart.

—— 18. Compressive strength is the property of a material to resist being _____.

—— 19. Shear strength of a material is a measure of how well a part can withstand forces acting to cut or _____ it apart.

—— 20. Torsional strength is the property of a material to withstand a _____ force.

l. twisting

m. ingots

n. ductility

o. slice

p. higher

q. mills

r. crushed

s. penetration

t. mining

CHAPTER 26: QUIZ 2

Name _____ Date _____

Class _____ Instructor _____ Grade _____

INSTRUCTIONS

Carefully read Chapter 26 in the textbook and answer each question.

MATCHING

In the space provided to the left of Column A, write the letter from Column B that best answers or completes the statement in Column A.

Column A	Column B
—— 1. Elasticity is the ability of a material to return to its _____ form after removal of the load.	a. crystal lattices
—— 2. Solids that are crystalline in form have an _____ arrangement of their atoms.	b. carbon
—— 3. The fundamental building blocks of all metals are atoms arranged in very precise three-dimensional patterns called _____.	c. temperature
—— 4. An _____ is a metal with one or more elements added to it, resulting in a significant change in the metal's properties.	d. eutectic
—— 5. A _____ composition is the lowest possible melting temperature of an alloy.	e. strength
—— 6. Iron is a pure metal element containing no measurable carbon and is relatively soft, but when alloyed with as little as 0.80% _____ can become tool steel.	f. orderly
—— 7. Welders must understand numerous methods used to strengthen metals because improper welding techniques can significantly weaken a metal, resulting in a weld that will _____.	g. metallurgical
—— 8. It is possible to replace some of the atoms in the crystal lattice with atoms of another metal in a process called _____.	h. grain
—— 9. Precipitation hardening or _____ is the heat treatment used to strengthen many aluminum alloys.	i. hardest
—— 10. Two phases or constituents may exist in equilibrium, depending on the alloy's _____ and composition.	j. solid-solution hardening
—— 11. Quenching is the process of rapidly _____ a metal by one of several methods.	k. cracking
—— 12. _____ is the process of reheating a part that has been hardened through heating and quenched.	l. reduce

—— 13. Martensite is the _____ of the transformation products of austenite and is formed only on cooling below a certain temperature known as the M_3 temperature (approximately 400°F to 600°F [200°C to 315°C] for carbon steels).

m. original

—— 14. Martensite formation can be minimized by _____ the steel to slow the cooling rates.

n. austenitic

—— 15. Sheets, bars, and tubes are intentionally cold-worked to increase their _____ because cold working will strengthen almost all metals and their alloys.

o. preheating

—— 16. When metals are heated, the rate of _____ growth increases with temperature and the length of time at that temperature.

p. alloy

—— 17. Preheat is used to _____ the rate at which welds cool.

q. tempering

—— 18. Generally, preheating provides two beneficial effects—lower residual stresses and reduced _____.

r. age hardening

—— 19. The most commonly used temperature range for stress relief steel is between 1100°F and 1150°F (593°C and 620°C), which is high enough to drop the yield residual stresses by 80% and low enough to prevent any harmful _____ changes in most steels.

s. cooling

—— 20. Annealing, frequently referred to as full annealing, involves heating the structure of a metal to a high enough temperature, slightly above Ac_3, to turn it completely _____.

t. fail

CHAPTER 26: QUIZ 3

Name _____ Date _____

Class _____ Instructor _____ Grade _____

INSTRUCTIONS

Carefully read Chapter 26 in the textbook and answer each question.

MATCHING

In the space provided to the left of Column A, write the letter from Column B that best answers or completes the statement in Column A.

Column A	Column B
____ 1. When welding, the length of time a metal is at an elevated temperature and the rate at which it is cooled have a significant effect on the _____ and surrounding metal's properties.	a. austenite
____ 2. The high temperatures along the sides of a weld in steel transform into _____ grain structure.	b. ductility
____ 3. If the austenite is cooled too quickly, it will transform into _____, which is a very large, hard, and brittle grain structure with almost no ductility.	c. larger
____ 4. At temperatures between 480°F and 900°F (250°C and 500°C), bainite grain structure is formed, which is hard and strong but has some _____.	d. heat-affected zone
____ 5. At temperatures below the bainite range for an alloy, _____ grain structures are formed, which is a layered combination of cementite (Fe_3C) and ferrite (Fe).	e. oxyacetylene welding (OFW)
____ 6. Grain transformation is a function of both time and _____.	f. increases
____ 7. When given enough time, crystals tend to grow _____, just like the increased time allowed the coarse pearlite grain structures to develop.	g. plasma arc welding (PAW)
____ 8. The lowest temperature along the sides of the weld that has any changes sets the outer extremity of the zone of change called the _____, HAZ.	h. weld
____ 9. Some metals are easily affected even by _____ temperature changes, whereas others are more resistant.	i. dry
____ 10. Some high-intensity heat sources, such as _____, are very concentrated and can have an HAZ area that is only a few thousandths of an inch wide.	j. air

—— 11. _____ is a much less intense heating source, and the resulting HAZ will be very large.

k. martensite

—— 12. The larger the piece of metal being welded, the greater its ability to absorb _____ without a significant change in temperature.

l. carbon

—— 13. As the temperature of the base metal _____—whether from preheating or postheating or the welding process itself—the larger the HAZ.

m. arcs

—— 14. Many welding problems and defects result from undesirable gases that can _____ in the weld metal.

n. solidified

—— 15. Nitrogen comes from _____ drawn into the arc stream.

o. pearlite

—— 16. The common source of oxygen contamination, air, reaches the weld because of poor shielding or excessively long _____.

p. hydrogen

—— 17. Hydrogen problems are avoidable by keeping organic materials away from weld joints, keeping the welding consumables _____, and preheating the components to be welded.

q. heat

—— 18. Cold cracking is the result of _____ dissolving in the weld metal and then diffusing into the heat-affected zone.

r. dissolve

—— 19. Hot cracks are caused by tearing the metal along partially fused grain boundaries of welds that have not completely _____.

s. small

—— 20. Problems are minimized by using very low carbon steel called extra low carbon (ELC) steels because without _____, the chromium carbides cannot form.

t. temperature

CHAPTER 26: QUIZ 4

Name _____   Date _____

Class _____   Instructor _____   Grade _____

INSTRUCTIONS
Carefully read Chapter 26 in the textbook and answer each question.

IDENTIFICATION
Identify the numbered items on the drawing by writing the letter next to the identifying term in the space provided.

1. _____ BCC NON MAGNETIC
2. _____ FCC IRON
3. _____ HEAT PURE IRON
4. _____ BEAKERS
5. _____ INVISIBLE LIGHT
6. _____ 1670°F
7. _____ THERMOMETER
8. _____ 212°F
9. _____ HOT PLATES
10. _____ STEAM
11. _____ CHANGE IN SENSIBLE HEAT
12. _____ VISIBLE LIGHT
13. _____ 32°F
14. _____ BCC MAGNETIC
15. _____ 1418°F
16. _____ HEAT INPUT
17. _____ NO CHANGE IN SENSIBLE HEAT
18. _____ BCC IRON
19. _____ ICE
20. _____ HEAT WATER
21. _____ 2552°F
22. _____ AREA OF SENSIBLE HEAT CHANGE
23. _____ AREA OF LATENT HEAT CHANGE
24. _____ WATER
25. _____ 2795°F

CHAPTER 26: QUIZ 5

Name _____   Date _____

Class _____   Instructor _____   Grade _____

INSTRUCTIONS

Carefully read Chapter 26 in the textbook and answer each question.

IDENTIFICATION

Identify the numbered items on the drawing by writing the letter next to the identifying term in the space provided.

1. _____ MOLTEN INSIDE

2. _____ MOLD

3. _____ CAST GRAIN STRUCTURE

4. _____ TUNDISH

5. _____ MOLTEN STEEL

6. _____ SLAB

7. _____ COOLING AREA

8. _____ SOLID OUTSIDE

9. _____ FORGED GRAIN STRUCTURE

10. _____ DELAMINATION

11. _____ LADLE

12. _____ PLATE

13. _____ FLAME CUT

14. _____ SLAG

15. _____ ROLLED GRAIN STRUCTURE

16. _____ POROSITY AND SLAG INCLUSIONS

17. _____ LAMINATION

18. _____ ROLLING MILL

CHAPTER 26: QUIZ 6

Name _____   Date _____

Class _____   Instructor _____   Grade _____

INSTRUCTIONS

Carefully read Chapter 26 in the textbook and answer each question.

IDENTIFICATION

Identify the numbered items on the drawing by writing the letter next to the identifying term in the space provided.

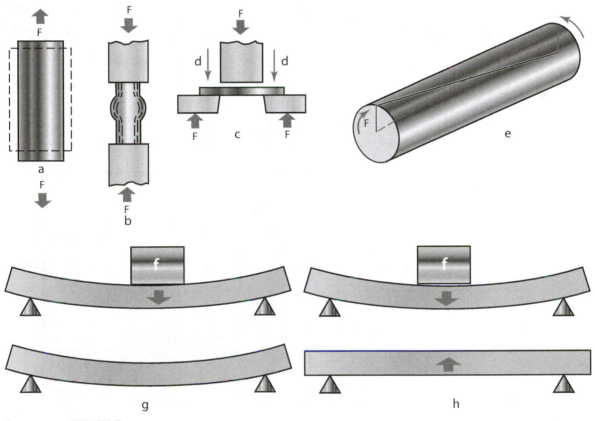

1. _____   TENSILE

2. _____   PERMANENT DEFORMATION

3. _____   COMPRESSIVE

4. _____   BEAM RETURNS TO ORIGINAL FORM

5. _____   TORSIONAL

6. _____   SHEAR

7. _____   LOAD FORCE

8. _____   CLAMP

CHAPTER 26: QUIZ 7

Name _____   Date _____

Class _____   Instructor _____   Grade _____

INSTRUCTIONS

Carefully read Chapter 26 in the textbook and answer each question.

IDENTIFICATION

Identify the numbered items on the drawing by writing the letter next to the identifying term in the space provided.

1. _____   α & LIQUID

2. _____   SOLID β

3. _____   100% TIN (SN)

4. _____   °C

5. _____   α

6. _____   °F

7. _____   β & LIQUID

8. _____   TEMPERATURE

9. _____   SOLID$_α$

10. _____   FACE-CENTERED CUBIC UNIT CELL

11. _____   ATOMS OF THE BASE METAL

12. _____   LEAD (Pb) 100%

13. _____   ATOMS OF THE HARDENING ALLOY

14. _____   BODY-CENTERED CUBIC UNIT CELL

15. _____   SOLID $_{α\&β}$

16. _____   HEXAGONAL CLOSE-PACKED CUBIC UNIT CELL

17. _____   LIQUID

18. _____   EUTECTIC COMPOSITION

CHAPTER 26: QUIZ 8

Name _____ Date _____

Class _____ Instructor _____ Grade _____

INSTRUCTIONS
Carefully read Chapter 26 in the textbook and answer each question.

IDENTIFICATION
Identify the numbered items on the drawing by writing the letter next to the identifying term in the space provided.

1. _____ UPPER TRANSFORMATION TEMPERATURE A_3

2. _____ STEAM POCKET

3. _____ STRESS-RELIEVING RANGE

4. _____ BREAK THIS DIRECTION

5. _____ CARBURIZING RANGE

6. _____ ELONGATION v(%), TENSILE STRENGTH (Kpsi)

7. _____ TRANSFORMATION RANGE

8. _____ TENSILE YIELD STRENGTH

9. _____ ANNEALING AND NORMALIZING

10. _____ PERCENT CARBON

11. _____ NITRIDING RANGE

12. _____ ELONGATION

13. _____ HOT METAL

14. _____ TENSILE STRENGTH

15. _____ A_2 MAGNETIC POINT A_2

16. _____ WATER

17. _____ YIELD AND ULTIMATE STRENGTH (Kpsi)

18. _____ BRINE

19. _____ TEMPERING TEMPERATURE

20. _____ ELONGATION IN 2" (50 mm)

21. _____ SPHEROIDIZING RANGE

22. _____ ELONGATION AND REDUCTION IN AREA (%)

23. _____ A_1 LOWER TRANSFORMATION TEMPERATURE A_1

24. _____ ULTIMATE TENSILE STRENGTH

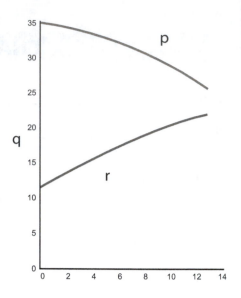

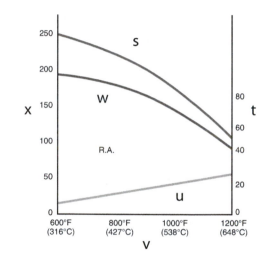

CHAPTER 26: QUIZ 9

Name _____   Date _____

Class _____   Instructor _____   Grade _____

INSTRUCTIONS

Carefully read Chapter 26 in the textbook and answer each question.

IDENTIFICATION

Identify the numbered items on the drawing by writing the letter next to the identifying term in the space provided.

1. _____ Tempered Martensite

2. _____ Minutes

3. _____ Bainite

4. _____ Cementite

5. _____ Very Slow Cooling

6. _____ Time

7. _____ Rapid Quenching

8. _____ Days

9. _____ Transformation Occurs

10. _____ Ferrite (a)

11. _____ Martensite

12. _____ Moderate Cooling

13. _____ Fractions of a Second

14. _____ Line M_F

15. _____ Hours

16. _____ Austenite

17. _____ Line M_S

18. _____ Pearlite

19. _____ Seconds

20. _____ Austenite Is Stable Above 1340°F

21. _____ Coarse Pearlite

22. _____ Entectoid Temperature

CHAPTER 26: QUIZ 10

Name _____   Date _____

Class _____   Instructor _____   Grade _____

INSTRUCTIONS

Carefully read Chapter 26 in the textbook and answer each question.

IDENTIFICATION

Identify the numbered items on the drawing by writing the letter next to the identifying term in the space provided.

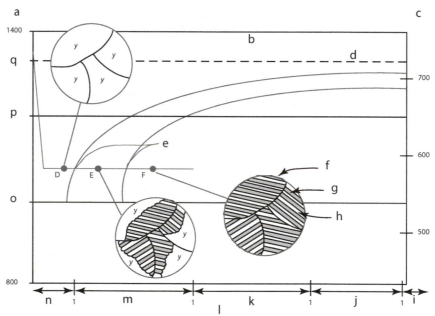

1. _____ Fine Pearlite	10. _____ Seconds
2. _____ °F	11. _____ Eutectoid Temperature
3. _____ Time	12. _____ Fractions of a Second
4. _____ Days	13. _____ Minutes
5. _____ 1000	14. _____ Austenite Is Stable Above 1340°F
6. _____ Cementite	15. _____ Ferrite (a)
7. _____ Hours	16. _____ 1340
8. _____ Transformation Occurs	17. _____ °C
9. _____ 1200	

CHAPTER 26: QUIZ 11

Name _____   Date _____

Class _____   Instructor _____   Grade _____

INSTRUCTIONS
Carefully read Chapter 26 in the textbook and answer each question.

IDENTIFICATION
Identify the numbered items on the drawing by writing the letter next to the identifying term in the space provided.

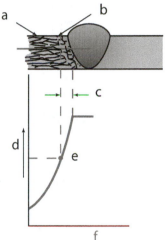

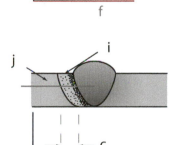

1. _____   TEMPERATURE

2. _____   HEAT-AFFECTED ZONE

3. _____   POSITION

4. _____   WELD METAL

5. _____   RECRYSTALLIZED METAL

6. _____   OVERAGED METAL

7. _____   ZONES

8. _____   COLD-WORKED METAL

9. _____   RECRYSTALLIZATION TEMPERATURE

10. _____   AUSTENITIC; GRAIN GROWTH AT HIGH TEMPERATURE, FINE GRAIN AT LOW TEMPERATURE

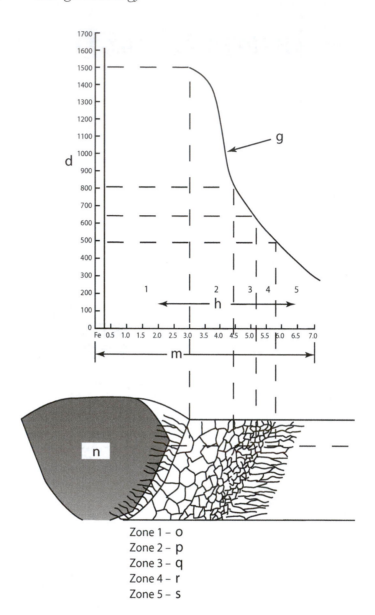

Zone 1 – o
Zone 2 – p
Zone 3 – q
Zone 4 – r
Zone 5 – s

11. _____ COLD-WORKED STEEL 0.2% CARBON

12. _____ OVERAGING TEMPERATURE

13. _____ RECRYSTALLIZATION

14. _____ WEIGHT PERCENTAGE CARBON

15. _____ AUSTENITE + FERRITE; GRAIN REFINED AND GRAIN GROWTH

16. _____ AGE-HARDENED METAL

17. _____ MAXIMUM TEMPERATURE IN PLATE

18. _____ LIQUID METAL AND THE BEGINNING OF GRAIN GROWTH

CHAPTER 26: QUIZ 12

Name _____ Date _____

Class _____ Instructor _____ Grade _____

INSTRUCTIONS

Carefully read Chapter 26 in the textbook and answer each question.

IDENTIFICATION

Identify the numbered items on the drawing by writing the letter next to the identifying term in the space provided.

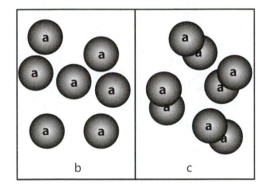

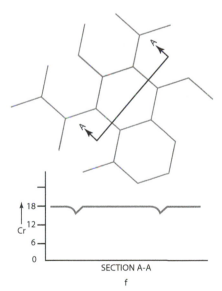

1. _____ HOT CRACK (LONGITUDINAL)   5. _____ MOLECULAR HYDROGEN

2. _____ BEFORE SENSITIZATION   6. _____ AFTER SENSITIZATION

3. _____ WELD CRATER   7. _____ ATOMIC HYDROGEN

4. _____ H

Weldability of Metals

STANDARD STEEL AND STEEL ALLOY NUMBERING SYSTEM	
Class	Number
Plain carbon steels	10XX
Free-cutting carbon steels	11XX
Manganese steels	13XX
Nickel steels	20XX
Nickel-chromium steels	30XX
Molybdenum steels	40XX
Chrome-molybdenum steels	41XX
Nickel-chrome-molybdenum steels	43XX, 47XX
Nickel-molybdenum steels	46XX, 48XX
Chromium steels	50XX
Chromium-vanadium steels	60XX
Heat-resisting casting alloys	70XX
Nickel-chrome-molybdenum steels	80XX, 93XX, 98XX
Silicon-manganese steels	90XX

STANDARD ALUMINUM AND ALUMINUM ALLOY NUMBER DESIGNATIONS	
Major Alloying Element	Number
Aluminum (99% minimum)	1XXX
Copper	2XXX
Manganese	3XXX
Silicon	4XXX
Magnesium	5XXX
Magnesium-silicon	6XXX
Zinc	7XXX
Other element	8XXX
Unused class	9XXX

STANDARD WROUGHT ALUMINUM AND MAGNESIUM COMPOSITIONS TEMPER DESIGNATION

Designation	Temper
F	As fabricated
O	Annealed
H	Strain hardened
H1	Strain hardened only
H2	Strain hardened, then partially annealed
H3	Strain hardened, then stabilized
W	Solution heat treated but with unstable temper
T	Thermally treated to produce stable tempers other than F, O, or H
T2	Annealed (for castings only)
T3	Solution heat treated, then cold worked
T4	Solution heat treated, then naturally aged
T5	Artificially aged
T6	Solution heat treated, then artificially aged
T7	Solution heat treated, then stabilized
T8	Solution heat treated, then cold worked, then artificially aged
T9	Solution heat treated, then artificially aged, then cold worked
T10	Artificially aged, then cold worked

STANDARD COPPER AND COPPER ALLOY SERIES DESIGNATIONS

Series	Alloying Element
100	None or very slight amount
200, 300, 400, and 665 to 699	Zinc
500	Tin
600 to 640	Aluminum
700 to 735	Nickel
735 to 799	Nickel and zinc

LETTERS USED TO IDENTIFY ALLOYING ELEMENTS IN MAGNESIUM ALLOYS

Letter	Alloying Element
A	Aluminum
B	Bismuth
C	Copper
D	Cadmium
E	Rare earth
F	Iron
H	Thorium
K	Zirconium
L	Beryllium
M	Manganese
N	Nickel
P	Lead
Q	Silver
R	Chromium
S	Silicon
T	Tin
Z	Zinc

STANDARD STAINLESS STEEL NUMBERING SYSTEM

Austenitic	Common Application and Characteristics
301	A general utility stainless steel, easily worked
302	Readily fabricated, for decorative or corrosion resistance
304, 304LC	A general utility stainless steel, easily worked
308	Used where corrosion resistance better than 1800 is needed
309	High-scaling resistance and good strength at high temperatures
310	More chromium and nickel for greater resistance to scaling in high heat
316, 316LC	Excellent resistance to chemical corrosion
317	Higher alloy than 316 for better corrosion resistance
321	Titanium stabilized to prevent carbide precipitation
347	Columbian stabilized to prevent carbide precipitation

Martensitic	
403	Used for forged turbine blades
410	General purpose, low priced, heat treatable
414	Nickel added; for knife blades, springs
416	Free machining
420	Higher carbon for cutlery and surgical instruments
431	High mechanical properties
440A	For instruments, cutlery, valves
440B	Higher carbon than 440A
440C	Higher carbon than 440A or B for high hardness
501	Less resistance to corrosion than chromium nickel types
502	Less resistance to corrosion than chromium nickel types

Ferritic	
405	Nonhardening when air cooled from high temperatures
406	For electrical resistances
430	Easily formed alloy, for automobile trim
430F	Free machining variety of 430 grade
446	High resistance to corrosion and scaling up to 215°F

CHAPTER 27: QUIZ 1

Name _____ Date _____

Class _____ Instructor _____ Grade _____

INSTRUCTIONS

Carefully read Chapter 27 in the textbook and answer each question.

MATCHING

In the space provided to the left of Column A, write the letter from Column B that best answers or completes the statement in Column A.

	Column A	Column B
_____	1. _____ weldability means that almost any process can be used to produce acceptable welds and that little effort is needed to control the procedures.	a. fit for service
_____	2. _____ weldability means that the processes used are limited and that the preparation of the joint and the procedure used to fabricate it must be controlled very carefully or the weldment will not function as intended.	b. cracking
_____	3. _____ is defined by the American Welding Society (AWS) as "the capacity of a metal to be welded under the fabrication conditions imposed into a specific, suitably designed structure and to perform satisfactorily in the intended service."	c. plain carbon steels
_____	4. Most welding processes produce a thermal cycle in which the metals are heated over a range of _____.	d. good
_____	5. A wide range of welding conditions can exist for welding methods when joining metals with good weldability; however, if weldability is a problem, then _____ usually will be necessary.	e. unified numbering system
_____	6. If the wrong filler metal is selected, then the weld can have major defects and will not be _____.	f. poor
_____	7. It is important that the correct shielding gas be used in sufficient quantity and with the correct application to protect the _____.	g. adjustments
_____	8. _____ is a common problem when welding on brittle metals such as cast iron or some high-strength alloys.	h. AISI
_____	9. The selection of the welding process to be used for any metal, thickness, and application often requires some _____.	i. low
_____	10. The two primary numbering systems classify the types of steel according to their basic _____ composition.	j. temperatures

—— 11. The _____ steel classification system was developed by the Society of Automotive Engineers.

        k. high

—— 12. The _____ steel classification system is sponsored by the American Iron and Steel Institute.

        l. high-manganese

—— 13. The _____ (UNS) eventually will replace the AISI and other systems.

        m. medium

—— 14. Steels alloyed with carbon and only a low concentration of silicon and manganese are known as _____.

        n. weldability

—— 15. _____-carbon (mild) steels have carbon content of less than 0.30% and can be welded easily by all welding processes.

        o. difficult

—— 16. The welding of _____-carbon steels, with 0.30% to 0.50% carbon content, is best accomplished by the various fusion processes, depending on the carbon content of the base metal.

        p. SAE

—— 17. _____-carbon steels usually have a carbon content of 0.50% to 0.90% and are much more difficult to weld than either the low- or medium-carbon steels.

        q. weight

—— 18. Because tool steel has a carbon content from 0.8% to 1.50%, it is very _____ to weld.

        r. molten weld metal

—— 19. _____ steel contains 12% or more manganese and a carbon content ranging from 1% to 1.4% and is used for wear resistance in applications involving impact.

        s. chemical

—— 20. Low-alloy steels are used increasingly because of requirements for high strength with less _____.

        t. research

CHAPTER 27: QUIZ 2

Name _____ Date _____

Class _____ Instructor _____ Grade _____

INSTRUCTIONS
Carefully read Chapter 27 in the textbook and answer each question.

MATCHING
In the space provided to the left of Column A, write the letter from Column B that best answers or completes the statement in Column A.

Column A

_____ 1. Stainless steels consist of four groups of _____: austenitic, ferritic, martensitic, and precipitation hardening.

_____ 2. The most widely used stainless steels are the _____ types.

_____ 3. Because ferritic stainless steels contain almost no nickel, they are _____ than austenitic steels.

_____ 4. _____ stainless steels are also low in nickel but contain more carbon than the ferritic and are used in applications requiring both wear resistance and corrosion resistance.

_____ 5. Chromium-molybdenum steel is used for high-temperature service and for _____.

_____ 6. All five types of cast iron have high _____ contents, usually ranging from 1.7% to 4%.

_____ 7. The major purpose of preheating and postheating of cast iron is to control the _____.

_____ 8. Because cast iron is _____, it does not bend before it breaks; therefore, broken parts can usually be fitted back together like the pieces of a puzzle.

_____ 9. A number of small welds are better than one or two large welds because the small welds do not have as much _____ and are less likely to cause postweld cracking.

_____ 10. Cracks in parts that cannot be preheated to the desired level can still be welded, but the welds will be very hard and are more likely to _____.

_____ 11. Even though the part cannot be preheated to the desired level, it cannot be welded _____; it must be heated to at least 75°F (24°C) or higher before starting to weld.

_____ 12. _____ welds are short welds that start ahead of the ending point of the first weld and go back to the end of the first weld.

Column B

a. type

b. stress

c. alloys

d. surface contamination

e. carbon

f. corrosion

g. back-stepping

h. chromium-nickel austenitic

i. strength

j. cold

k. higher

l. aircraft parts

———— 13. Metals that are not primarily composed of iron are known as _____ metals.

m. martensitic

———— 14. When you are welding copper, the welding current should be considerably _____ than when welding steel.

n. oxygen

———— 15. One of the characteristics of aluminum and its alloys is that it has a great affinity for _____.

o. cheaper

———— 16. Two of the most important properties of titanium are its extremely high strength-to-weight ratio (in alloy form) and its generally excellent _____ resistance.

p. difficult

———— 17. Magnesium must be alloyed with other elements to provide the necessary _____ for most applications.

q. nonferrous

———— 18. Repair, or maintenance, welding is one of the most _____ types of welding.

r. brittle

———— 19. All _____ must be removed before welding to prevent the possibility of injuring your health from exposure to materials released during welding.

s. recrack

———— 20. Before the joint can be prepared for welding, you must try to identify the _____ of metal.

t. rate of temperature change

CHAPTER 27: QUIZ 3

Name _____ Date _____

Class _____ Instructor _____ Grade _____

INSTRUCTIONS

Carefully read Chapter 27 in the textbook and answer each question.

IDENTIFICATION

Identify the numbered items on the drawing by writing the letter next to the identifying term in the space provided.

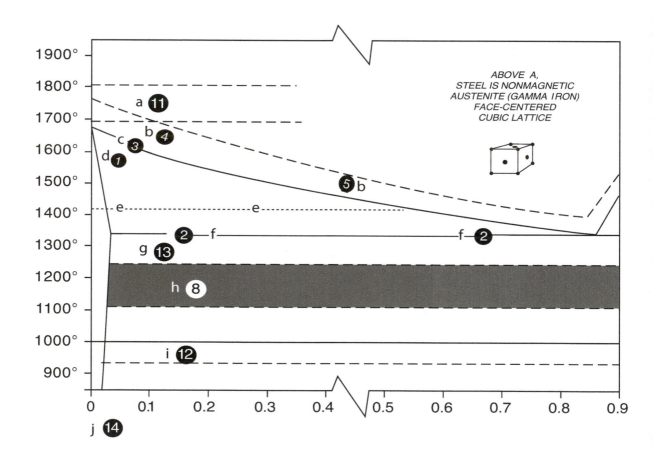

1. _____ SPHEROIDIZING RANGE

2. _____ CARBURIZING RANGE

3. _____ ANNEALING AND
NORMALIZING

4. _____ TRANSFORMATION RANGE

5. _____ A_2 MAGNETIC POINT A_2

6. _____ STRESS-RELIEVING RANGE

7. _____ NITRIDING RANGE

8. _____ PERCENT CARBON

9. _____ LOWER TRANSFORMATION
TEMPERATURE A_1

10. _____ PREHEATING RANGE FOR WELDING

11. _____ UPPER TRANSFORMATION TEMPERATURE A_3

CHAPTER 27: QUIZ 4

Name _____   Date _____

Class _____   Instructor _____   Grade _____

INSTRUCTIONS

Carefully read Chapter 27 in the textbook and answer each question.

IDENTIFICATION

Identify the numbered items on the drawing by writing the letter next to the identifying term in the space provided.

1. _____ LOW-CARBON STEEL

2. _____ GRAY CAST IRON

3. _____ HIGH-CARBON STEEL

4. _____ MALLEABLE IRON

5. _____ WHITE CAST IRON

6. _____ RESHAPE A CHISEL BY GRINDING OFF THE MUSHROOMED HEAD

7. _____ CRACKS IN THE HARDENED HEAD WILL ALLOW CHIPS TO BREAK OFF THE MUSHROOMED HEAD IF THEY ARE NOT GROUND OFF.

CHAPTER 27: QUIZ 5

Name _____   Date _____

Class _____   Instructor _____   Grade _____

INSTRUCTIONS

Carefully read Chapter 27 in the textbook and answer each question.

IDENTIFICATION

The metal type and thickness affect the type of groove preparation that should be used for welding. In the space provided below put an X beside the type of groove shape for each type and thickness of metal.

| Metal | Thickness in inches | | | |
	1/8–1/4	1/4–1/2	1/2–1	Over 1
Mild Steel	1	2	3	4
Aluminum	5	6	7	8
Magnesium	9	10	11	12
Stainless Steel	13	14	15	16
Cast iron	17	18	19	20

1. _____ Square groove, _____ V-groove, _____ U-groove

2. _____ Square groove, _____ V-groove, _____ U-groove

3. _____ Square groove, _____ V-groove, _____ U-groove

4. _____ Square groove, _____ V-groove, _____ U-groove

5. _____ Square groove, _____ V-groove, _____ U-groove

6. _____ Square groove, _____ V-groove, _____ U-groove

7. _____ Square groove, _____ V-groove, _____ U-groove

8. _____ Square groove, _____ V-groove, _____ U-groove

9. _____ Square groove, _____ V-groove, _____ U-groove

10. _____ Square groove, _____ V-groove, _____ U-groove

11. _____ Square groove, _____ V-groove, _____ U-groove

12. _____ Square groove, _____ V-groove, _____ U-groove

13. _____ Square groove, _____ V-groove, _____ U-groove

14. _____ Square groove, _____ V-groove, _____ U-groove

15. _____ Square groove, _____ V-groove, _____ U-groove

16. _____ Square groove, _____ V-groove, _____ U-groove

17. _____ Square groove, _____ V-groove, _____ U-groove

18. _____ Square groove, _____ V-groove, _____ U-groove

19. _____ Square groove, _____ V-groove, _____ U-groove

20. _____ Square groove, _____ V-groove, _____ U-groove

Filler Metal Selection

American Welding Society and American Society for Testing and Materials Specifications		
AWS Designation	**ASTM Designation**	**Title**
A5.1	A 233	Mild Steel Covered Arc Welding Electrodes
A5.2	A 251	Iron and Steel Gas Welding Rods
A5.3	B 184	Aluminum and Aluminum-Alloy Arc Welding Electrodes
A5.4	A 298	Corrosion-Resisting Chromium and Chromium-Nickel Steel Covered Welding Electrodes
A5.5	A 316	Low-Alloy Steel Covered Arc Welding Electrodes
A5.6	B 255	Copper and Copper-Alloy Arc Welding Electrodes
A5.7	B 259	Copper and Copper-Alloy Welding Rods
A5.8	B 260	Brazing Filler Metal
A5.9	A 371	Corrosion-Resisting Chromium and Chromium-Nickel Steel Welding Rods and Bare Electrodes
A5.10	B 285	Aluminum and Aluminum-Alloy Welding Rods and Bare Electrodes
A5.11	B 295	Nickel and Nickel-Alloy Covered Welding Electrodes
A5.12	B 297	Tungsten Arc Welding Electrodes
A5.13	A 399	Surfacing Welding Rods and Electrode
A5.14	B 304	Nickel and Nickel-Alloy Bare Welding Rods and Electrodes
A5.15	A 398	Welding Rods and Covered Electrodes for Welding Cast Iron
A5.16	B 232	Titanium and Titanium-Alloy Bare Welding Rods and Electrodes
A5.17	A 558	Bare Mild Steel Electrodes and Fluxes for Submerged Arc Welding
A5.18	A 559	Mild Steel Electrodes for Gas Metal Arc Welding

CHAPTER 28: QUIZ 1

Name _____ Date _____

Class _____ Instructor _____ Grade _____

INSTRUCTIONS

Carefully read Chapter 28 in the textbook and answer each question.

MATCHING

In the space provided to the left of Column A, write the letter from Column B that best answers or completes the statement in Column A.

Column A	Column B
_____ 1. The most widely used numbering and lettering system for filler metal identification is the one developed by the _____.	a. reduced
_____ 2. Examples of _____ information that may be given by manufacturers may include some or all of the following: welding electrode manipulation techniques, joint design, prewelding preparation, postwelding procedures, types of equipment that can be welded, welding currents, and welding positions.	b. estimating
_____ 3. An example of _____ information provided by the electrode manufacturers might be the welding amperage range setting for each size of welding electrode.	c. flux covering
_____ 4. The information supplied by the manufacturer can be used for a variety of purposes, including _____ the pounds of electrodes needed for a job.	d. alloys
_____ 5. A standard weld metal test for a good weld is the minimum _____ strength, psi (N/mm^2)—the load in pounds (megapascal [MPa]) that would be required to break a section of sound weld that has a cross-sectional area of 1 sq in.	e. general
_____ 6. As the percentage of carbon increases, the tensile strength increases, the hardness increases, and ductility is _____.	f. technical
_____ 7. To determine the weldability of a specific alloy, formulas have been developed to calculate the _____ equivalence.	g. preheating
_____ 8. The carbon equivalent (CE) of an alloy is an indication of how the weld will affect the surrounding metal, called the _____ (HAZ).	h. shielding gases
_____ 9. Shielded metal arc welding (SMAW) electrodes have two parts—the inner core wire and a _____.	i. data sheets
_____ 10. A _____ is the primary metal source for a weld.	j. thickness

——— 11. Heat generated by the arc causes some constituents in the flux covering to decompose and others to vaporize, forming _____.

k. tensile

——— 12. Elements in the flux are mixed with the filler metal, some of which stay in the weld metal as _____.

l. modified

——— 13. Welding fluxes can affect the penetration and _____ of the weld bead.

m. amperage

——— 14. Selecting the best filler metal for a job is seldom delegated to the _____ in large shops.

n. carbon

——— 15. When choosing a welding electrode, a welder should consider the power range—what is the _____ range on the welder and its duty cycle?

o. heat-affected zone

——— 16. When choosing a welding electrode, a welder must consider the _____ of metal—the penetration characteristics of each welding electrode may differ.

p. softness

——— 17. On low-carbon steel plate 1 in. (25 mm) thick or more, _____ is required with most welding electrodes.

q. welder

——— 18. Mechanical properties such as tensile strength, yield strength, hardness, toughness, ductility, and impact strength can be _____ by the selection of specific welding electrodes.

r. core wire

——— 19. The hardness or _____ of the weld greatly affects any grinding, drilling, or machining.

s. contour

——— 20. The characteristics of each manufacturer's filler metals can be compared to one another by using _____ supplied by the manufacturer.

t. American Welding Society (AWS)

CHAPTER 28: QUIZ 2

Name _____   Date _____

Class _____   Instructor _____   Grade _____

INSTRUCTIONS

Carefully read Chapter 28 in the textbook and answer each question.

MATCHING

In the space provided to the left of Column A, write the letter from Column B that best answers or completes the statement in Column A.

	Column A	Column B
_____	1. The AWS classification system for filler metal uses a series of letters and numbers in a code that gives the important information about the _____.	a. tensile
_____	2. E—Indicates an arc welding _____ (E).	b. rutile
_____	3. R—Indicates a _____ (R) that is heated by some source other than electric current flowing directly through it.	c. iron powder
_____	4. ER—Indicates a filler metal that is supplied for use in either an _____ (ER) form.	d. polarity
_____	5. The AWS numbering system for A5.1 and A5.5 carbon and low-alloy steel-covered electrodes uses the letter E followed by a series of numbers to indicate the minimum _____ strength of a good weld, the position(s) in which the electrode can be used, the type of flux coating, and the type(s) of welding current.	e. Lo-Hi
_____	6. The E6010 electrodes are designed to be used with DCEP polarity and have an _____-based flux (cellulose, $C_6H_{10}O_5$).	f. filler metal
_____	7. The E6011 electrodes are designed to be used with _____ or DCEP reverse polarity and have an organic-based flux.	g. moisture
_____	8. The E6012 electrodes are designed to be used with AC or DCEN polarity and to have _____-based flux (titanium dioxide TiO_2).	h. PAW
_____	9. The E6013 electrodes are designed to be used with AC or DC, either _____.	i. electrode
_____	10. E7014 electrodes are commonly used for welding on heavy sheet metal, ornamental iron, machinery, frames, and general _____ work.	j. self

———— 11. The E7024 electrodes are designed to be used with AC or DC, either polarity, and they have a rutile-based flux with _____ added.

k. hydrogen

———— 12. The E7016 electrodes are designed to be used with AC or DCEP polarity and have a low-_____-based (mineral) flux.

l. DCEP

———— 13. The E7018 electrodes are designed to be used with AC or _____ polarity and have a low-hydrogen-based flux with iron powder added.

m. deoxidizer

———— 14. E7018 electrodes are sometimes referred to as _____ rods because they allow very little hydrogen into the weld pool.

n. FCAW

———— 15. The AWS specification for carbon steel filler metals for gas shielded welding wire is A5.18; filler metal classified within these specifications can be used for GMAW, GTAW, and _____ processes.

o. rod

———— 16. The AWS specification for carbon steel filler metals for flux cored arc welding wire is A5.36/A5.36M, and flux cored filler metals classified within this specification can be used for the _____ process.

p. single

———— 17. Heated storage lockers are available to keep all wire electrodes dry to prevent surface oxidation and keep _____ out of the flux core of FCAW electrodes.

q. organic

———— 18. E70T-1 and E71T-1 have a high-level _____ in the flux core.

r. repair

———— 19. E70T-2 and E71T-2 are highly deoxidized flux cored filler metal that can be used for _____ pass welds only.

s. electrode or rod

———— 20. E70T-4 and E71T-4 are _____-shielding, flux cored filler metal.

t. AC

CHAPTER 28: QUIZ 3

Name _____ Date _____

Class _____ Instructor _____ Grade _____

INSTRUCTIONS

Carefully read Chapter 28 in the textbook and answer each question.

MATCHING

In the space provided to the left of Column A, write the letter from Column B that best answers or completes the statement in Column A.

	Column A	Column B
_____	1. Metal cored electrodes are similar to FCA welding electrodes in that they are _____ in design and are filled with a mixture.	a. lowest
_____	2. The AWS specification for _____–covered arc electrodes is A5.4, and for stainless steel bare, cored, and stranded electrodes and welding rods is A5.9.	b. A5.3
_____	3. The AWS identification system for covered nonferrous electrodes is based on the atomic symbol or symbols of the major alloy(s) or the metal's _____ number.	c. identification
_____	4. The AWS specifications for aluminum and aluminum alloy filler metals are _____ for covered arc welding electrodes and A5.10 for bare welding rods and electrodes.	d. contaminated
_____	5. The aluminum-covered arc welding electrodes do not use the _____ before the electrode number.	e. strength
_____	6. ER1100 aluminum has the _____ percentage of alloy agents of all of the aluminum alloys, and it melts at 1215°F (657°C).	f. cast iron
_____	7. ER4043 is a _____-purpose welding filler metal.	g. tubular
_____	8. ER5356 has 4.5% to 5.5% _____ added to improve the tensile strength.	h. fire
_____	9. ER5556 has 4.7% to 5.5% magnesium and 0.5% to 1.0% manganese added to produce a weld with high _____.	i. stainless steel
_____	10. The ENi nickel arc welding electrodes are designed to be used with AC or DCEP polarity and are used for _____ repair.	j. hydrogen
_____	11. ECuAl—The aluminum bronze welding electrodes are designed to be used with DCEP polarity and have _____ as their major alloy.	k. letter E

—— 12. Hardfacing or wear-resistant electrodes are the most popular _____ electrodes; however, there are also cutting and brazing electrodes.

—— 13. The joining of magnesium alloys by torch welding or brazing is possible without a _____ hazard because the melting point of magnesium is 1202°F (651°C) to 858°F (459°C) below its boiling point, where magnesium may start to burn.

—— 14. To help control postweld hydrogen cracking-related problems use welding filler metal and fluxes classified as low _____.

—— 15. To help control postweld hydrogen cracking-related problems follow proper storage and handling procedures for filler metal and fluxes to prevent them from becoming _____ with sources of hydrogen such as moisture, oil, grease, or other hydrocarbons.

l. general

m. copper

n. magnesium

o. special-purpose

CHAPTER 28: QUIZ 4

Name _____ Date _____

Class _____ Instructor _____ Grade _____

INSTRUCTIONS

Carefully read Chapter 28 in the textbook and answer each question.

IDENTIFICATION

Identify the numbered items on the drawing by writing the letter next to the identifying term in the space provided.

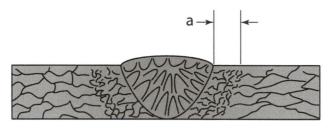

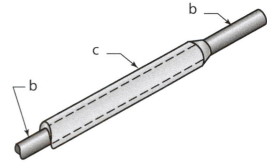

1. _____ SMALL DROPLETS

2. _____ CORE WIRE

3. _____ HEAT-AFFECTED ZONE

4. _____ SLAG

5. _____ FLUX COVERING

6. _____ MOLTEN POOL

7. _____ MnO

8. _____ GASEOUS SHIELD

9. _____ WELD

10. _____ LARGE GLOBS

11. _____ SiO_2

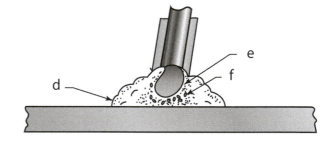

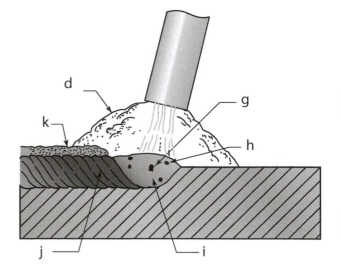

CHAPTER 28: QUIZ 5

Name _____ Date _____

Class _____ Instructor _____ Grade _____

INSTRUCTIONS

Carefully read Chapter 28 in the textbook and answer each question.

IDENTIFICATION

Identify the numbered items on the drawing by writing the letter next to the identifying term in the space provided.

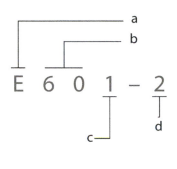

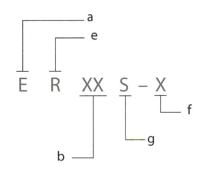

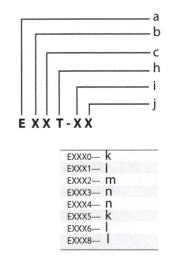

1. _____ DCRP only

2. _____ DESIGNATES A CUT LENGTH "ROD" TYPE ELECTRODE (THE "R" MAY BE OMITTED ON SOME ELECTRODES)

3. _____ SOLID WIRE

4. _____ ELECTRODE

5. _____ AC and DCSP

6. _____ WELDING CURRENT

7. _____ TENSILE STRENGTH

8. _____ AC and DC

9. _____ USABILITY DESIGNATOR

10. _____ WELDING POSITION

11. _____ THIS DESIGNATOR IDENTIFIES THE ELECTRODE AS A FLUX CORED ELECTRODE

12. _____ AC and DCRP

13. _____ INDICATES THE FILLER METAL COMPOSITION AND WELD'S MECHANICAL PROPERTIES

14. _____ SHIELDING GAS DESIGNATOR

Welding Automation and Robotics

CHAPTER 29: QUIZ 1

Name _____ Date _____

Class _____ Instructor _____ Grade _____

INSTRUCTIONS

Carefully read Chapter 29 in the textbook and answer each question.

MATCHING

In the space provided to the left of Column A, write the letter from Column B that best answers or completes the statement in Column A.

Column A

_____ 1. More and more modern businesses are using _____ (CAD) to improve products.

_____ 2. _____ (CAM) technology can aid in selecting assembly methods, planning the product flow through the manufacturing steps, and scheduling the various operations to reduce the actual production costs.

_____ 3. A _____ process is one that is completely performed by hand.

_____ 4. A _____ process is one in which the filler metal is fed into the weld automatically.

_____ 5. A _____ process is one in which the joining is performed by equipment requiring the welding operator to observe the progress of the weld and make adjustments as required.

_____ 6. An _____ process is a dedicated process (designed to do only one type of welding on a specific part) that does not require adjustments to be made by the operator during the actual welding cycle.

_____ 7. _____ processes are similar to automatic joining except that they are flexible and more easily adjusted or changed; unlike automatic joining, there is no dedicated machine for each product.

Column B

a. machine joining

b. industrial robot

c. remote

d. computer-aided design

e. Welding Procedure Specification

f. market analysis

g. prototype

_____ 8. An _____ is a reprogrammable, multifunctional manipulator designed to move material, parts, tools, or specialized devices through variable programmed motions for the performance of a variety of tasks.

_____ 9. Early robot programming took a great deal of time and involved a lot of _____.

_____ 10. Another method of programming the robot is to teach it to move its arm along the desired path by physically directing its movements by hand or by using a _____ control panel.

_____ 11. Another method of programming a robot is to use computer _____ to produce the program.

_____ 12. Evaluating the need for the use of a robot and selecting a robot to meet that need is a _____ process that involves many factors.

_____ 13. Premature investment in automation can be more costly than it can be productive, so a comprehensive _____ of present and future needs must be the first step to ensure that your business is not overextended.

_____ 14. The design of the weldment must be _____ with a robotic system.

_____ 15. The robot is the prime component in an automated _____ (a manufacturing unit consisting of two or more workstations).

_____ 16. 3-D metal printing machines are great for rapidly producing a _____ of an item or part.

_____ 17. All personnel should be instructed in the safe _____ of the robot.

_____ 18. All personnel should be instructed in the _____ of an emergency power shutoff.

_____ 19. _____ should be mounted around the floor and work area to stop all movement when unauthorized personnel are detected in the work area during the operation.

_____ 20. A _____ (WPS) contains the minimum and maximum setting for all of the welding parameters required to make an acceptable weld.

h. computer-aided manufacturing

i. sensors

j. work cell

k. semiautomatic joining

l. location

m. complex

n. automatic joining

o. operation

p. manual joining

q. compatible

r. software

s. Automated joining

t. trial and error

CHAPTER 29: QUIZ 2

Name _____   Date _____

Class _____   Instructor _____   Grade _____

INSTRUCTIONS
Carefully read Chapter 29 in the textbook and answer each question.

IDENTIFICATION
Identify the numbered items on the drawing by writing the letter next to the identifying term in the space provided.

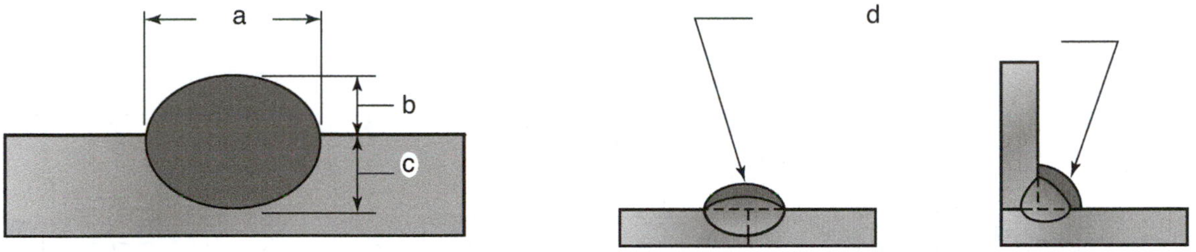

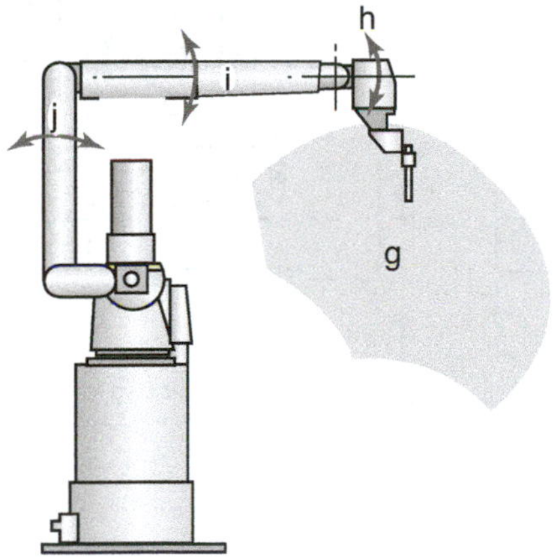

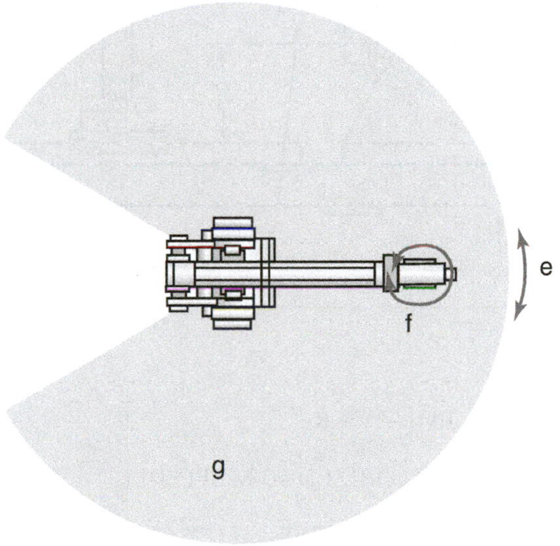

1. _____   R-AXIS

2. _____   GYRO-AXIS

3. _____   WIDTH

4. _____   WELD REINFORCEMENT FROM ADDITIONAL FILLER METAL

5. _____   WORKING RANGE

6. _____   U-AXIS

7. _____   PENETRATION

8. _____   θ-AXIS

9. _____   L-AXIS

10. _____   REINFORCEMENT

CHAPTER 29: QUIZ 3

Name _____ Date _____

Class _____ Instructor _____ Grade _____

INSTRUCTIONS

Carefully read Chapter 29 in the textbook and answer each question.

IDENTIFICATION

Identify the numbered items on the drawing by writing the letter next to the identifying term in the space provided.

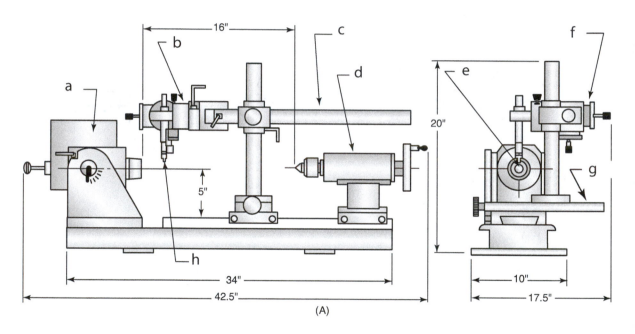

(A)

1. _____ TAILSTOCK

2. _____ MOTORIZED 50-M SPINDLE

3. _____ WELDING TORCH (BY CUSTOMER)

4. _____ THETA TORCH SLIDE

5. _____ 1" THRU HOLE

6. _____ CROSSLIDE

7. _____ X-Y-Z TORCH SLIDES

8. _____ "S" TORCH STAND

CHAPTER 29: QUIZ 4

Name _____ Date _____

Class _____ Instructor _____ Grade _____

INSTRUCTIONS

Carefully read Chapter 29 in the textbook and answer each question.

IDENTIFICATION

Identify the numbered items on the drawing by writing the letter next to the identifying term in the space provided.

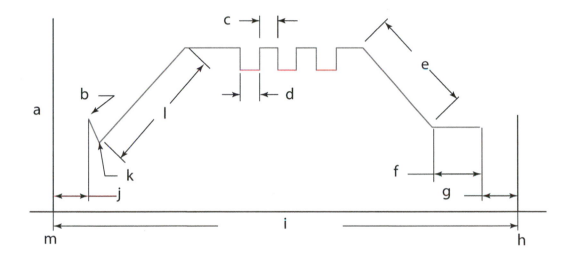

1. _____ HIGH PULSE TIME

2. _____ INITIAL CURRENT

3. _____ HOT START

4. _____ FINAL CURRENT

5. _____ CURRENT SETTING

6. _____ PREPURGE TIME

7. _____ POSTPURGE TIME

8. _____ DOWNSLOPE

9. _____ WELDING CYCLE TIME

10. _____ LOW PULSE TIME

11. _____ START

12. _____ STOP

13. _____ UPSLOPE

CHAPTER 30

Other Welding Processes

CHAPTER 30: QUIZ 1

Name _____ Date _____

Class _____ Instructor _____ Grade _____

INSTRUCTIONS

Carefully read Chapter 30 in the textbook and answer each question.

MATCHING

In the space provided to the left of Column A, write the letter from Column B that best answers or completes the statement in Column A.

Column A	Column B
_____ 1. Some of welding processes require less skill or knowledge to set up and operate, such as resistance spot welding (RSW), while others demand a great deal of technical information and training, such as _____ (EBW).	a. continuously
_____ 2. Submerged arc welding (SAW), electroslag welding (ESW), and electrogas welding (EGW) are constant potential (CP) welding processes that are used to fabricate large, _____ sections.	b. electron beam welding
_____ 3. Submerged arc welding (SAW) is a fusion welding process in which the heat is produced from an arc between the work and a _____ fed filler metal electrode.	c. joint preparation
_____ 4. Fluxes are classified according to the _____ properties of the weld metal deposited.	d. horizontal
_____ 5. An advantage of SAW is that the absence of the _____ means that many welders can work close to each other without the problem of arc flash.	e. rocker arm
_____ 6. A disadvantage of SAW is that it is restricted to flat position and _____ fillets.	f. lap seam
_____ 7. One reason handheld SA welding is increasing in usage is that there are little if any _____ to exhaust.	g. electrons
_____ 8. The heat for electroslag welding (ESW) welding is produced as a result of the electrical resistance of the _____.	h. thick
_____ 9. The electroslag welding process has the advantage of minimum _____.	i. production rate
_____ 10. The electroslag welding process has the disadvantage that lengthy _____ times are needed.	j. molten flux

_____ 11. The major difference between electrogas welding (EGW) and
electroslag welding is that there is an arc between the _____ and
the molten weld pool with EGW.

k. seam weld

_____ 12. Resistance welding (RW) is a process in which _____ between
the parts' surfaces occurs as a result of the heat produced by the
electric resistance between the parts' contacting surfaces.

l. Resistance spot
welding (RSW)

_____ 13. _____ is the most common of the various resistance welding
processes.

m. mechanical

_____ 14. The three types of spot welding machines commonly used are
_____, press type, and portable type.

n. worktable

_____ 15. Multiple-spot welders are used when a high _____ is a
requirement.

o. electrode

_____ 16. Seam welding is similar in some ways to spot welding except that
the spots are spaced so closely together that they actually overlap
one another to make a continuous _____.

p. gun

_____ 17. The types of seam welds used in most seam welding processes are
the _____, the flanged seam, and the mashed seam.

q. arc light

_____ 18. Electron beam welding utilizes the energy from a fast-moving
beam of _____ focused on the base material.

r. setup

_____ 19. The electron beam welding _____ consists of a filament, a
cathode, an anode, and a focusing coil.

s. fusion

_____ 20. Electron beam equipment is designed and manufactured so that
the _____ can be rotated beneath the gun for circular welds or
driven along a path that corresponds to, or is parallel to, the
centerline of the chamber for linear welds.

t. fumes or smoke

CHAPTER 30: QUIZ 2

Name _____ Date _____

Class _____ Instructor _____ Grade _____

INSTRUCTIONS

Carefully read Chapter 30 in the textbook and answer each question.

MATCHING

In the space provided to the left of Column A, write the letter from Column B that best answers or completes the statement in Column A.

Column A	Column B
_____ 1. Ultrasonic welding (USW) is a process for joining similar and dissimilar metals by introducing high-frequency _____ into the overlapping metals in the area to be joined.	a. narrow
_____ 2. Ultrasonic welding has many applications in the assembly of _____ products.	b. atoms
_____ 3. Inertia welding is a form of _____ welding where one workpiece is fixed in a stationary holding device and the other is clamped in a spindle chuck.	c. hardened
_____ 4. With inertia welding, the weld zone is very _____ and has a fine-grained structure with no melt product or grain growth.	d. vibrations
_____ 5. Some of the advantages of the inertia welding process are its superior welds, fast production welds, and its _____ operation.	e. molten metal
_____ 6. In laser beam welding, fusion is obtained by directing a highly concentrated beam of coherent _____ to a very small spot.	f. temperature
_____ 7. Because the laser beam is a light beam, it can operate in air or any transparent material, and the source of the beam need not even be _____ to the work	g. ceramics
_____ 8. The laser is based on the principle that _____ in certain crystals and gases can be made to release a coherent, monochromatic (single-wavelength) energy when they are excited.	h. electrical
_____ 9. The laser beam can be used to heat the surface of the materials so they can be _____.	i. thermite
_____ 10. In plasma arc welding (PAW), a plasma jet is produced by forcing _____ to flow along an arc restricted electromagnetically as it passes through a nozzle.	j. light

——— 11. With stud welding, when the end of the stud and the underlying spot on the surface of the work have been properly _____, they are brought together under pressure.

k. force

——— 12. Thermal spraying is the process of spraying _____ onto a surface to form a coating.

l. friction

——— 13. A thermal spraying installation requires, at a minimum, the following equipment: air _____, air control unit, air flowmeter, oxyfuel gas or arc equipment, and exhaust equipment.

m. water-cooled jacket

——— 14. The development of equipment, materials, and methods has greatly broadened the scope of powder spraying, so now a wide range of alloys and _____ can be applied at speeds and costs that are economically feasible.

n. sprayed

——— 15. Some thermospray guns require no _____, just two lightweight hoses to supply oxygen and fuel gas.

o. heated

——— 16. Metal that is _____ is usually applied in layers less than 0.010 in. (0.25 mm) thick.

p. air

——— 17. The plasma spraying process (PSP), makes use of a spray gun that uses an electric arc contained within a _____.

q. clean

——— 18. With cold welding (CW), the coalescence of the metal surfaces occurs as a result of the _____ applied.

r. compressor

——— 19. Thermite welding (TW) is a process that uses an exothermic reaction to develop a high _____.

s. gas

——— 20. _____ welding of rails is used throughout the world to join lengths of rails into continuous track.

t. close

CHAPTER 30: QUIZ 3

Name _____   Date _____

Class _____   Instructor _____   Grade _____

INSTRUCTIONS

Carefully read Chapter 30 in the textbook and answer each question.

MATCHING

In the space provided to the left of Column A, write the letter from Column B that best answers or completes the statement in Column A.

	Column A	Column B
_____	1. Hardfacing is defined as the process of obtaining desired properties or _____ by applying, using oxyfuel or arc welding, an integral layer of metal of one composition onto a surface, an edge, or the point of a base metal of another composition.	a. sweating or tinning
_____	2. Steel or special hardfacing alloys should be used where the surface must resist hard or _____ wear.	b. flat
_____	3. In hardfacing operations, oxyfuel welding permits the surfacing layer to be deposited by flowing molten filler metal into the underlying surface; this method of surfacing is called _____.	c. quality
_____	4. Hardfacing by the arc welding method has many advantages, including high rates of _____, flexibility of operation, and ease of mechanization.	d. base metal
_____	5. The type of service to which a part is to be exposed governs the degree of _____ required of the surfacing deposit.	e. hardfacing
_____	6. _____ electrodes may be classified into the following three general groups: resistance to severe abrasion; resistance to both impact and moderate abrasion; and resistance to severe impact and moderately severe abrasion.	f. abrasive
_____	7. Because most hardfacing electrodes are too fluid for out-of-position welding with SMAW, the work should be arranged in the _____ position.	g. filler metal
_____	8. Care must be exercised when using the GMA, FCA, and GTA welding processes for hardfacing to avoid _____ of the weld.	h. laser beam
_____	9. With friction stir welding (FSW), a weld is formed between the joining surfaces as the result of the mechanical stirring of the metal without the metal _____.	i. dimensions
_____	10. Magnetic pulse welding is a _____ form of welding; neither surface that is being joined melts.	j. surface

——— 11. The hybrid laser welding process combines a _____ with GTAW, GMAW, or PAW processes.

k. deposition

——— 12. Laser beam welding melts the _____ to form the weld.

l. solid-state

——— 13. _____ cladding can be performed by combining the laser beam with the addition of GMA welding filler metal.

m. dilution

——— 14. The combination of a laser beam with heat of the arc from a GTA welding torch can be used with or without _____ being added.

n. melting

——— 15. When a filler wire is automatically fed into the GTA welding arc, it is referred to as _____.

o. Cold Wire Gas Tungsten Arc Welding

CHAPTER 30: QUIZ 4

Name _____ Date _____

Class _____ Instructor _____ Grade _____

INSTRUCTIONS

Carefully read Chapter 30 in the textbook and answer each question.

IDENTIFICATION

Identify the numbered process by writing the letter next to the identifying term in the space provided.

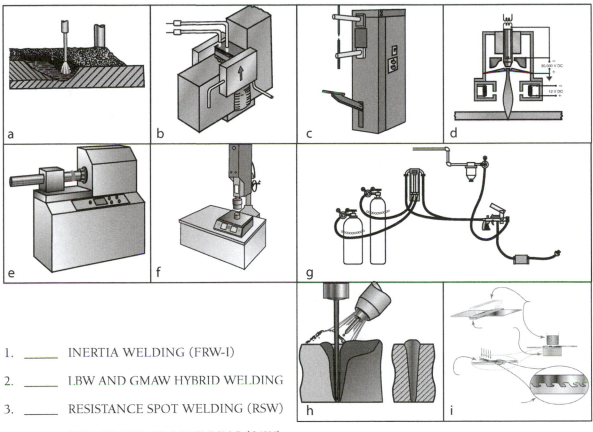

1. _____ INERTIA WELDING (FRW-I)

2. _____ LBW AND GMAW HYBRID WELDING

3. _____ RESISTANCE SPOT WELDING (RSW)

4. _____ SUBMERGED ARC WELDING (SAW)

5. _____ ELECTROSLAG WELDING (ESW)

6. _____ MAGNETIC PULSE WELDING (MPW)

7. _____ ELECTRONE BEAM WELDING (EBW)

8. _____ THERMAL SPRAYING (THSP)

9. _____ ULTRASONICC WELDING (USW)

Oxyacetylene Welding

■ PRACTICE 31-1

Name _____ Date _____

Class _____ Instructor _____ Grade _____

OBJECTIVE: After completing this practice, you will be able to safely set up an oxyfuel torch set.

EQUIPMENT AND MATERIALS NEEDED

Disassembled oxyfuel torch set, two regulators, two reverse flow valves, one set of hoses, a torch body, a welding tip, two cylinders, a portable cart or supporting wall with safety chain, and a wrench.

INSTRUCTIONS

1. Safety chain the cylinders in the cart or to a wall. Then remove the valve protection caps.

2. Crack the cylinder valve on each cylinder for a second to blow away dirt that may be in the valve.

   CAUTION Always stand to one side. Point the valve away from anyone in the area and be sure there are no sources of ignition when cracking the valve.

3. Attach the regulators to the cylinder valves. The regulator nuts should be started by hand and then tightened with a wrench.

4. Attach a reverse flow valve or flashback arrestor, if the torch does not have them built in, to the hose connection on the regulator or to the hose connection on the torch body, depending on the type of reverse flow valve in the set. Occasionally, test each reverse flow valve by blowing through it to make sure it works properly.

5. Connect the hoses. The red hose has a left-hand nut and attaches to the fuel gas regulator. The green hose has a right-hand nut and attaches to the oxygen regulator.

6. Attach the torch to the hoses. Connect both hose nuts finger tight before using a wrench to tighten either one.

7. Check the tip seals for nicks or O rings, if used, for damage. Check the owner's manual, or a supplier, to determine if the torch tip should be tightened by hand only or should be tightened with a wrench.

CAUTION   **Tightening a tip the incorrect way may be dangerous and might damage the equipment. Check all connections to be sure they are tight. The oxyfuel equipment is now assembled and ready for use.**

CAUTION   **Connections should not be overtightened. If they do not seal properly, repair or replace them.**

CAUTION   **Leaking cylinder valve stems should not be repaired. Turn off the valve, disconnect the cylinder, mark the cylinder, and notify the supplier to come and pick up the bad cylinder.**

The assembled oxyfuel welding equipment is now tested and ready to be ignited and adjusted.

INSTRUCTOR'S COMMENTS _____

_____

_____

■ PRACTICE 31-2

Name _____   Date _____

Class _____   Instructor _____   Grade _____

OBJECTIVE: After completing this practice, you should be able to safely turn on and test oxyfuel equipment for gas leaks.

EQUIPMENT AND MATERIALS NEEDED FOR THIS PRACTICE

1. Oxyfuel equipment that is properly assembled.

2. A nonadjustable tank wrench.

3. A leak-detecting solution.

INSTRUCTIONS

1. Back out the regulator pressure adjusting screws until they are loose.

2. Standing to one side of the regulator, open the cylinder valve SLOWLY so that the pressure rises on the gauge slowly.

    CAUTION If the valve is opened quickly, the regulator or gauge may be damaged, or the gauge may explode.

3. Open the oxygen valve all the way until it stops turning.

4. Open the acetylene or other fuel gas valve 1/4 turn or just enough to get gas pressure. If the cylinder valve does not have a handwheel, use a nonadjustable wrench and leave it in place on the valve stem while the gas is on.

    CAUTION The acetylene valve should never be opened more than 1-1/2 turns, so that in an emergency, it can be turned off quickly.

5. Open one torch valve and point the tip away from any source of ignition. Slowly turn in the pressure adjusting screw until gas can be heard escaping from the torch. The gas should flow long enough to allow the hose to be completely purged (emptied) of air and replaced by the gas before the torch valve is closed. Repeat this process with the other gas.

6. After purging is completed, and with both torch valves off, adjust both regulators to read 5 psig (35 kPag).

7. Spray a leak-detecting solution on each hose and regulator connection and on each valve stem on the torch and cylinders. Watch for bubbles, which indicate a leak. Turn off the cylinder valve before tightening any leaking connections.

CAUTION Connections should not be overtightened. If they do not seal properly, repair or replace them.

CAUTION Leaking cylinder valve stems should not be repaired. Turn off the valve, disconnect the cylinder, mark the cylinder, and notify the supplier to come and pick up the bad cylinder.

The assembled oxyfuel welding equipment is now tested and ready to be ignited and adjusted.

INSTRUCTOR'S COMMENTS _____

_____

_____

■ PRACTICE 31-3

Name _____   Date _____

Class _____   Instructor _____   Grade _____

OBJECTIVE: After completing this practice, you should be able to safely light and adjust an oxyacetylene flame.

EQUIPMENT AND MATERIALS NEEDED FOR THIS PRACTICE

1. Oxyfuel welding equipment that is properly assembled and tested.

2. A spark lighter.

3. Gas welding goggles.

4. Gloves and proper protective clothing.

INSTRUCTIONS

1. Wearing proper clothing, gloves, and gas welding goggles, turn both regulator adjusting screws in until the working pressure gauges read 5 psig (35 kPag). If you mistakenly turn on more than 5 psig (35 kPag), open the torch valve to allow the pressure to drop as the adjusting screw is turned outward.

2. Turn on the torch fuel-gas valve just enough so that some gas escapes.

    CAUTION Be sure the torch is pointed away from any sources of ignition or any object or person that might be damaged or harmed by the flame when it is lit.

3. Using a spark lighter, light the torch. Hold the lighter near the end of the tip but not covering the end.

    CAUTION A spark lighter is the only safe device to use when lighting any torch.

4. With the torch lit, increase the flow of acetylene until the flame almost stops smoking.

5. Slowly turn on the oxygen and adjust the torch to a neutral flame. Look for white acetylene cones, NOT BLUE.

    This flame setting uses the minimum gas flow rate for this specific tip. The fuel flow should never be adjusted to a rate below the point where the smoke stops. This is the minimum flow rate at which the cool gases will pull the flame heat out of the tip. If excessive heat is allowed to build up in a tip, it can cause a backfire or flashback.

The maximum gas flow rate gives a flame which, when adjusted to the neutral setting, does not settle back on the tip.

INSTRUCTOR'S COMMENTS _____

_____

_____

■ PRACTICE 31-4

Name _____   Date _____

Base Metal Thickness _____   Filler Metal Diameter_____

Class _____   Instructor _____   Grade _____

Welding Principles and Applications

MATERIAL:
16-GAUGE MILD STEEL SHEET

PROCESS:
OXYFUEL WELDING FLAT

NUMBER:	DRAWN BY:
PRACTICE 31-4	MIKE MORAN

OBJECTIVE: After completing this practice, you should be able to push a weld pool on 16-gauge carbon steel sheet.

EQUIPMENT AND MATERIALS NEEDED FOR THIS PRACTICE

1. Properly setup oxyacetylene welding equipment.

2. Proper safety protection (welding goggles, safety glasses, spark lighter, and pliers). Refer to Chapter 2 in the text for more specific safety information.

3. Two or more pieces of 16-gauge carbon steel sheet, approximately 3 in. (76 mm) wide by 6 in. (152 mm) long.

INSTRUCTIONS

1. Start at one end and hold the torch at a 45° angle in the direction of the weld.

2. When the metal starts to melt, move the torch in a circular pattern down the sheet toward the other end. If the size of the molten weld pool changes, speed up or slow down to keep it the same size all the way down the sheet.

3. Repeat this practice until you can keep the width of the molten weld pool uniform and the direction of travel in a straight line.

4. Turn off the gas cylinders and regulators and clean up your work area when you are finished welding.

INSTRUCTOR'S COMMENTS _____

_____

_____

■ PRACTICE 31-5

Name _____ Date _____

Base Metal Thickness _____ Filler Metal Diameter_____

Class _____ Instructor _____ Grade _____

Welding Principles and Applications

MATERIAL:
16-GAUGE MILD STEEL SHEET

PROCESS:
OXYFUEL WELDING FLAT

| NUMBER: | DRAWN BY: |
| PRACTICE 31-5 | GAVIN DUBOIS |

OBJECTIVE: After completing this experiment, you should be able to demonstrate the controlling of the weld pool by changing the torch angle and torch height while pushing a pool on 16-gauge carbon steel sheet.

EQUIPMENT AND MATERIALS NEEDED FOR THIS EXPERIMENT

1. Properly setup oxyacetylene welding equipment.

2. Proper safety protection (welding goggles, safety glasses, spark lighter, and pliers). Refer to Chapter 2 in the text for more specific safety information.

3. One or more pieces of 16-gauge carbon steel sheet, approximately 6 in. (152 mm) long.

INSTRUCTIONS

CHANGING TORCH ANGLE

1. Start at one end and hold the torch at a 45° angle in the direction of the weld.

2. When the metal starts to melt, move the torch in a circular pattern down the sheet toward the other end. As the weld is moved toward the other end, gradually change the torch angle from the starting 45° to an angle of 90° then to an ending angle of 30°.

3. Repeat this part of the experiment until you can control the width of the molten weld pool by changing the torch angle.

INSTRUCTOR'S COMMENTS _____

_____

_____

CHANGING TORCH HEIGHT

1. Start at one end and hold the torch with the inner cone about 1/8 in. (3 mm) above the metal surface at a 45° angle in the direction of the weld.

2. When the metal starts to melt, move the torch in a circular pattern down the sheet toward the other end. As the weld is moved toward the other end, gradually change the torch height from the starting 1/8 in. (3 mm) to an ending height of 1/2 in. (13 mm).

3. Repeat this part of the experiment until you can control the width of the molten weld pool by changing the torch height.

4. Turn off the gas cylinders and regulators and clean up your work area when you are finished welding.

INSTRUCTOR'S COMMENTS _____

_____

_____

■ PRACTICE 31-6

Name _____ Date _____

Base Metal Thickness _____ Filler Metal Diameter _____

Class _____ Instructor _____ Grade _____

OBJECTIVE: After completing this practice, you should be able to make stringer beads on sheet steel in the flat position.

EQUIPMENT AND MATERIALS NEEDED FOR THIS PRACTICE

1. Properly setup oxyacetylene welding equipment.

2. Proper safety protection (welding goggles, safety glasses, spark lighter, and pliers). Refer to Chapter 2 in the text for more specific safety information.

3. One or more pieces of 16-gauge carbon steel sheet, approximately 3 in. (76 mm) wide by 6 in. (152 mm) long.

4. 1/16 in. (2 mm), 3/32 in. (2.4 mm), and 1/8 in. (3 mm) diameter RG45 filler wire. Determine by trial and error the size of filler rod with which you are most comfortable.

INSTRUCTIONS

1. Use a clean piece of 16-gauge mild steel and a torch adjusted to a neutral flame.

2. Hold the torch at a 45° angle to the metal with the inner cone about 1/8 in. (3 mm) above the metal surface.

3. The end of the filler rod should always be kept inside the protective envelope of the flame. Experiment with the three different filler rods for your preference of filler wire.

4. The end of the filler rod should be dipped into the leading edge of the molten weld pool. If the filler rod touches the hot metal around the molten weld pool, it will stick. When this happens, move the flame directly to the end of the filler rod to melt and free it.

5. Move the torch in a circular pattern down the sheet toward the other end while adding filler rod.

6. The width and buildup of the welding bead should be uniform.

7. Repeat this practice until you can keep the width of the bead uniform and the direction of travel in a straight line.

8. Turn off the gas cylinders and regulators and clean up your work area when you are finished welding.

INSTRUCTOR'S COMMENTS _____

_____

_____

■ PRACTICE 31-7

Name _____ Date _____

Base Metal Thickness _____ Filler Metal Diameter_____

Class _____ Instructor _____ Grade _____

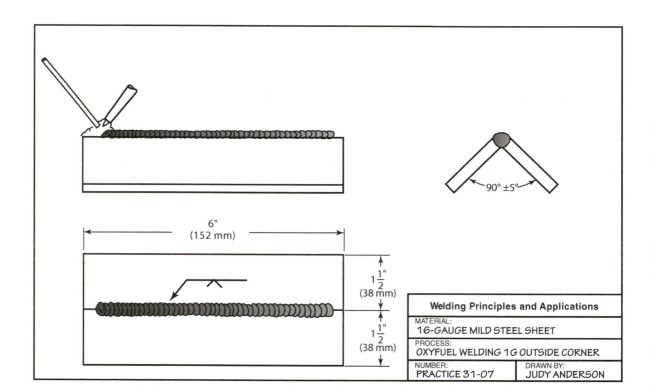

Welding Principles and Applications

MATERIAL:	16-GAUGE MILD STEEL SHEET
PROCESS:	OXYFUEL WELDING 1G OUTSIDE CORNER
NUMBER: PRACTICE 31-07	DRAWN BY: JUDY ANDERSON

OBJECTIVE: After completing this practice, you should be able to weld outside corner joints in the flat position.

EQUIPMENT AND MATERIALS NEEDED FOR THIS PRACTICE

1. Properly setup oxyacetylene welding equipment.

2. Proper safety protection (welding goggles, safety glasses, spark lighter, and pliers). Refer to Chapter 2 in the text for more specific safety information.

3. Two or more pieces of 16-gauge carbon steel sheet, approximately 1-1/2 in. (38 mm) wide by 6 in. (152 mm) long.

4. 1/16 in. (2 mm), 3/32 in. (2.4 mm), and 1/8 in. (3 mm) diameter RG45 filler wire. Determine by trial and error the size of filler rod with which you are most comfortable.

INSTRUCTIONS

1. Place one of the pieces of metal in a jig or on a firebrick and hold or brace the other piece of metal horizontally on it.

2. Tack the ends of the two sheets together.

3. Then set it upright and put two or three more tacks on the joint.

4. Hold the torch at a 45° angle to the metal with the inner cone about 1/8 in. (3 mm) above the metal surface and make a uniform weld along the joint.

5. Repeat this weld until the weld can be made without defects.

6. Turn off the gas cylinders and regulators and clean up your work area when you are finished welding.

INSTRUCTOR'S COMMENTS _____

_____

_____

■ PRACTICE 31-8

Name _____   Date _____

Base Metal Thickness _____   Filler Metal Diameter_____

Class _____   Instructor _____   Grade _____

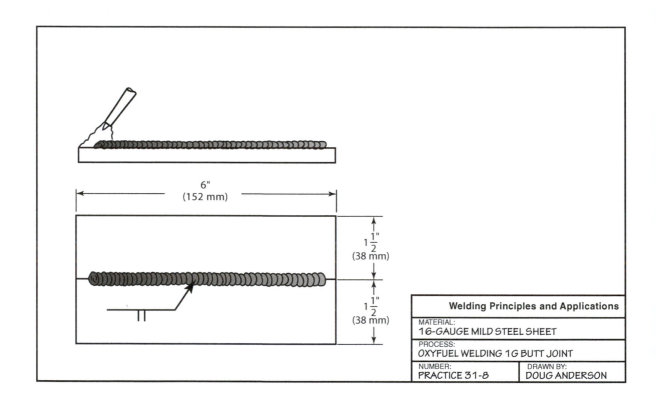

6"
(152 mm)

$1\frac{1}{2}$"
(38 mm)

$1\frac{1}{2}$"
(38 mm)

Welding Principles and Applications

MATERIAL:
16-GAUGE MILD STEEL SHEET

PROCESS:
OXYFUEL WELDING 1G BUTT JOINT

NUMBER:
PRACTICE 31-8

DRAWN BY:
DOUG ANDERSON

OBJECTIVE: After completing this practice, you should be able to weld butt joints in the flat position.

EQUIPMENT AND MATERIALS NEEDED FOR THIS PRACTICE

1. Properly setup oxyacetylene welding equipment.

2. Proper safety protection (welding goggles, safety glasses, spark lighter, and pliers). Refer to Chapter 2 in the text for more specific safety information.

3. Two or more pieces of 16-gauge carbon steel sheet, approximately 1-1/2 in. (38 mm) wide by 6 in. (152 mm) long.

4. 1/16 in. (2 mm), 3/32 in. (2.4 mm), and 1/8 in. (3 mm) diameter RG45 filler wire. Determine by trial and error the size of filler rod with which you are most comfortable.

INSTRUCTIONS

1. Place the two pieces of metal in a jig or on a firebrick and tack weld both ends together.

2. The tack on the ends can be made by simply heating the ends and allowing them to fuse together or by placing a small drop of filler metal on the sheet and heating the filler metal until it fuses to the sheet. The latter way is especially convenient if you have to use one hand to hold the sheets together and the other to hold the torch.

3. After both ends are tacked together, place one or two small tacks along the joint to prevent warping during welding.

4. With the sheets tacked together, start welding from one end to the other.

5. Repeat this weld until you can make a welded butt joint that is uniform in width and reinforcement and has no visual defects. The penetration of this practice weld may vary.

6. Turn off the gas cylinders and regulators and clean up your work area when you are finished welding.

INSTRUCTOR'S COMMENTS _____

_____

_____

■ PRACTICE 31-9

Name _____ Date _____

Base Metal Thickness _____ Filler Metal Diameter_____

Class _____ Instructor _____ Grade _____

OBJECTIVE: After completing this practice, you should be able to make a butt joint in the flat position with 100% penetration.

EQUIPMENT AND MATERIALS NEEDED FOR THIS PRACTICE

1. Properly setup oxyacetylene welding equipment.

2. Proper safety protection (welding goggles, safety glasses, spark lighter, and pliers). Refer to Chapter 2 in the text for more specific safety information.

3. Two or more pieces of 16-gauge carbon steel sheet, approximately 1-1/2 in. (38 mm) wide by 6 in. (152 mm) long.

4. 1/16 in. (2 mm), 3/32 in. (2.4 mm), and 1/8 in. (3 mm) diameter RG45 filler rods. Determine by trial and error the size of filler rod with which you are most comfortable.

INSTRUCTIONS

1. Place the two pieces of metal in a jig or on a firebrick and tack weld both ends together in a butt joint configuration.

2. After both ends are tacked together, place one or two additional tacks along the joint to prevent warping.

3. Apply heat for a long enough period of time so that a keyhole appears before adding filler metal. This will ensure 100% penetration of the joint.

4. Continue making keyholes all the way along the joint and filling them with filler metal.

5. Inspect the bottom side of the weld for 100% penetration and visual defects. Repeat this practice until you can consistently make welds with 100% penetration and that are visually defect free.

6. Turn off the gas cylinders and regulators and clean up your work area when you are finished welding.

INSTRUCTOR'S COMMENTS _____

_____

_____

■ PRACTICE 31-10

Name _____ Date _____

Base Metal Thickness _____ Filler Metal Diameter _____

Class _____ Instructor _____ Grade _____

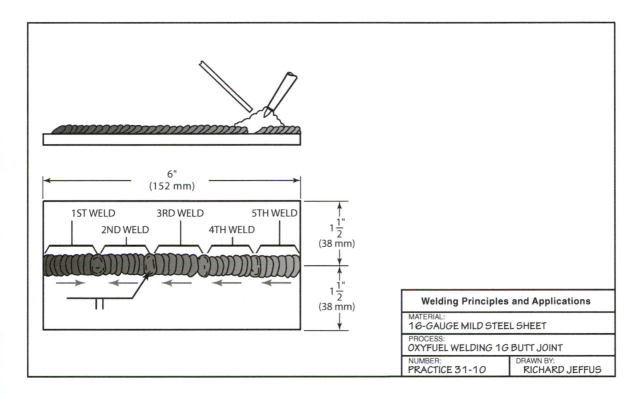

6"
(152 mm)

1ST WELD	3RD WELD	5TH WELD
2ND WELD	4TH WELD	

$1\frac{1}{2}$"
(38 mm)

$1\frac{1}{2}$"
(38 mm)

Welding Principles and Applications

MATERIAL:
16-GAUGE MILD STEEL SHEET

PROCESS:
OXYFUEL WELDING 1G BUTT JOINT

| NUMBER: PRACTICE 31-10 | DRAWN BY: RICHARD JEFFUS |

OBJECTIVE: After completing this practice, you should be able to weld butt joints on gauge material with minimum distortion.

EQUIPMENT AND MATERIALS NEEDED FOR THIS PRACTICE

1. Properly setup oxyacetylene welding equipment.

2. Proper safety protection (welding goggles, safety glasses, spark lighter, and pliers). Refer to Chapter 2 in the text for more specific safety information.

3. Two or more pieces of 16-gauge carbon steel sheet, approximately 1-1/2 in. (38 mm) wide by 6 in. (152 mm) long.

4. 1/16 in. (2 mm), 3/32 in. (2.4 mm), and 1/8 in. (3 mm) diameter RG45 filler wire. Determine by trial and error the size of filler rod with which you are most comfortable.

INSTRUCTIONS

1. Tack weld the two plates together.

2. Use a back-stepping technique. Distortion can be controlled by back stepping, proper tacking, and clamping.

3. Practice this weld until you can pass a visual inspection for distortion.

4. Turn off the gas cylinders and regulators and clean up your work area when you are finished welding.

INSTRUCTOR'S COMMENTS _____

_____

_____

■ PRACTICE 31-11

Name _____   Date _____

Base Metal Thickness _____   Filler Metal Diameter _____

Class _____   Instructor _____   Grade _____

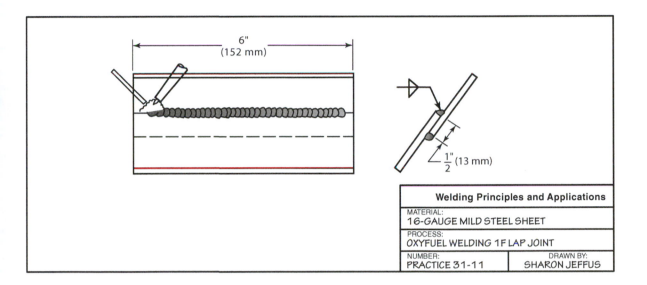

OBJECTIVE: After completing this practice, you should be able to weld lap joints on gauge material in the flat position.

EQUIPMENT AND MATERIALS NEEDED FOR THIS PRACTICE

1. Properly setup oxyacetylene welding equipment.

2. Proper safety protection (welding goggles, safety glasses, spark lighter, and pliers). Refer to Chapter 2 in the text for more specific safety information.

3. Two or more pieces of 16-gauge carbon steel sheet, approximately 1-1/2 in. (38 mm) wide by 6 in. (152 mm) long.

4. 1/16 in. (2 mm), 3/32 in. (2.4 mm), and 1/8 in. (3 mm) diameter RG45 filler wire. Determine by trial and error the size of filler rod with which you are most comfortable.

INSTRUCTIONS

1. Place the two pieces of metal on a firebrick and tack both ends.

2. Starting at one end, make a uniform weld along the joint. Both sides of the joint can be welded.

3. Repeat this weld until the weld can be made without defects.

4. Turn off the gas cylinders and regulators and clean up your work area when you are finished welding.

INSTRUCTOR'S COMMENTS _____

_____

_____

■ PRACTICE 31-12

Name _____   Date _____

Base Metal Thickness _____   Filler Metal Diameter_____

Class _____   Instructor _____   Grade _____

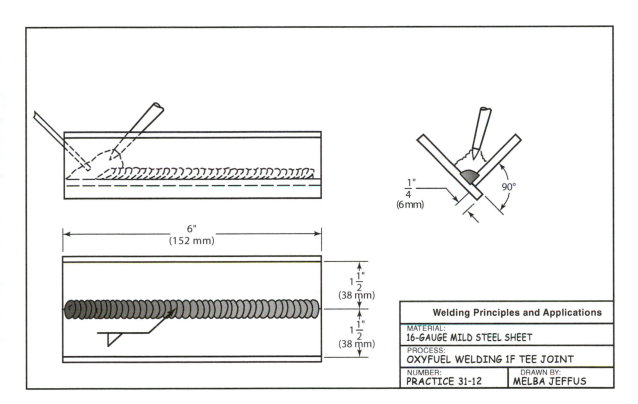

Welding Principles and Applications

MATERIAL:
16-GAUGE MILD STEEL SHEET

PROCESS:
OXYFUEL WELDING 1F TEE JOINT

NUMBER:	DRAWN BY:
PRACTICE 31-12	MELBA JEFFUS

OBJECTIVE: After completing this practice, you should be able to weld tee joints on gauge materials in the flat position.

EQUIPMENT AND MATERIALS NEEDED FOR THIS PRACTICE

1. Properly setup oxyacetylene welding equipment.

2. Proper safety protection (welding goggles, safety glasses, spark lighter, and pliers). Refer to Chapter 2 in the text for more specific safety information.

3. Two or more pieces of 16-gauge carbon steel sheet, approximately 1-1/2 in. (38 mm) wide by 6 in. (152 mm) long.

4. 1/16 in. (2 mm), 3/32 in. (2.4 mm), and 1/8 in. (3 mm) diameter RG45 filler wire. Determine by trial and error the size of filler rod with which you are most comfortable.

INSTRUCTIONS

1. Place the first piece of metal flat on a firebrick and hold or brace the second piece vertically on the first piece. The vertical piece should be within 5° of square to the bottom sheet.

2. Tack the two sheets at the ends.

3. Starting at one end, make a uniform weld along the joint.

4. Repeat this weld until the weld can be made without defects.

5. Turn off the gas cylinders and regulators and clean up your work area when you are finished welding.

INSTRUCTOR'S COMMENTS _____

_____

_____

■ PRACTICE 31-13

Name _____   Date _____

Base Metal Thickness _____   Filler Metal Diameter_____

Class _____   Instructor _____   Grade _____

OBJECTIVE: After completing this practice, you should be able to make a stringer bead at a 45° angle.

EQUIPMENT AND MATERIALS NEEDED FOR THIS PRACTICE

1. Properly setup oxyacetylene welding equipment.

2. Proper safety protection (welding goggles, safety glasses, spark lighter, and pliers). Refer to Chapter 2 in the text for more specific safety information.

3. One or more pieces of 16-gauge carbon steel sheet, approximately 3 in. (76 mm) wide by 6 in. (152 mm) long.

4. 1/16 in. (2 mm), 3/32 in. (2.4 mm), and 1/8 in. (3 mm) diameter RG45 filler rods. Determine by trial and error the size of filler rod with which you are most comfortable.

INSTRUCTIONS

1. Position the workpiece at a 45° angle to the surface of the table.

2. Use the torch and add filler metal as you did in Practice 31-6.

3. It may be necessary to flash the torch away from the metal to avoid overheating. Move the hotter primary flame away from the puddle while still keeping the puddle within the secondary flame.

4. Establish a rhythm of moving the torch and adding filler metal so as to keep the bead uniform.

5. Inspect the completed bead for uniformity and freedom from defects.

6. Repeat this practice until you can consistently make welds that are visually defect free.

7. Turn off the gas cylinders and regulators and clean up your work area when you are finished welding.

INSTRUCTOR'S COMMENTS _____

_____

_____

■ PRACTICE 31-14

Name _____   Date _____

Base Metal Thickness _____   Filler Metal Diameter_____

Class _____   Instructor _____   Grade _____

OBJECTIVE: After completing this practice, you should be able to make a stringer bead in the vertical position.

EQUIPMENT AND MATERIALS NEEDED FOR THIS PRACTICE

1. Properly setup oxyacetylene welding equipment.

2. Proper safety protection (welding goggles, safety glasses, spark lighter, and pliers). Refer to Chapter 2 in the text for more specific safety information.

3. One or more pieces of 16-gauge carbon steel sheet, approximately 3 in. (76 mm) wide by 6 in. (152 mm) long.

4. 1/16 in. (2 mm), 3/32 in. (2.4 mm), and 1/8 in. (3 mm) diameter RG45 filler rods. Determine by trial and error the size of filler rod with which you are most comfortable.

INSTRUCTIONS

1. Position the workpiece in a vertical position.

2. Use the torch and add filler metal as you did in Practice 31-6. Try making beads from the bottom up and from the top down.

3. It may be necessary to flash the torch away from the metal to avoid overheating. Move the hotter primary flame away from the puddle while still keeping the puddle within the secondary flame.

4. Establish a rhythm of moving the torch and adding filler metal so as to keep the bead uniform.

5. Inspect the completed bead for uniformity and freedom from defects.

6. Repeat this practice until you can consistently make welds that are visually defect free.

7. Turn off the gas cylinders and regulators and clean up your work area when you are finished welding.

INSTRUCTOR'S COMMENTS _____

_____

_____

■ PRACTICE 31-15

Name _____ Date _____

Base Metal Thickness _____ Filler Metal Diameter_____

Class _____ Instructor _____ Grade _____

OBJECTIVE: After completing this practice, you should be able to make a butt joint at a 45° angle.

EQUIPMENT AND MATERIALS NEEDED FOR THIS PRACTICE

1. Properly setup oxyacetylene welding equipment.

2. Proper safety protection (welding goggles, safety glasses, spark lighter, and pliers). Refer to Chapter 2 in the text for more specific safety information.

3. Two or more pieces of 16-gauge carbon steel sheet, approximately 1-1/2 in. (38 mm) wide by 6 in. (152 mm) long.

4. 1/16 in. (2 mm), 3/32 in. (2.4 mm), and 1/8 in. (3 mm) diameter RG45 filler rods. Determine by trial and error the size of filler rod with which you are most comfortable.

INSTRUCTIONS

1. Place the two pieces of metal in a jig or on a firebrick and tack weld both ends together in a butt joint configuration.

2. After both ends are tacked together, place one or two additional tacks along the joint to prevent warping.

3. Position the tacked workpiece at a 45° angle to the surface of the table.

4. Use the torch and add filler metal as you did in Practice 31-8.

5. Establish a rhythm of moving the torch and adding filler metal so as to keep the bead uniform.

6. Inspect the completed bead for uniformity and freedom from defects.

7. Repeat this practice until you can consistently make welds that are visually defect free.

8. Turn off the gas cylinders and regulators and clean up your work area when you are finished welding.

INSTRUCTOR'S COMMENTS _____

_____

_____

■ PRACTICE 31-16

Name _____ Date _____

Base Metal Thickness _____ Filler Metal Diameter_____

Class _____ Instructor _____ Grade _____

OBJECTIVE: After completing this practice, you should be able to make a butt joint in a vertical position.

EQUIPMENT AND MATERIALS NEEDED FOR THIS PRACTICE

1. Properly setup oxyacetylene welding equipment.

2. Proper safety protection (welding goggles, safety glasses, spark lighter, and pliers). Refer to Chapter 2 in the text for more specific safety information.

3. Two or more pieces of 16-gauge carbon steel sheet, approximately 1-1/2 in. (38 mm) wide by 6 in. (152 mm) long.

4. 1/16 in. (2 mm), 3/32 in. (2.4 mm), and 1/8 in. (3 mm) diameter RG45 filler rods. Determine by trial and error the size of filler rod with which you are most comfortable.

INSTRUCTIONS

1. Place the two pieces of metal in a jig or on a firebrick and tack weld both ends together in a butt joint configuration. After both ends are tacked together, place one or two additional tacks along the joint to prevent warping.

2. Position the tacked workpiece in a vertical position.

3. Use the torch and add filler metal as you did in Practice 31-8. Try making the bead from the bottom up and from the top down.

4. Establish a rhythm of moving the torch and adding filler rod so as to keep the bead uniform.

5. Inspect the completed bead for uniformity and freedom from visual defects.

6. Repeat this practice until you can consistently make welds with 100% penetration and that are visually defect free.

7. Turn off the gas cylinders and regulators and clean up your work area when you are finished welding.

INSTRUCTOR'S COMMENTS _____

_____

_____

■ PRACTICE 31-17

Name _____ Date _____

Base Metal Thickness _____ Filler Metal Diameter_____

Class _____ Instructor _____ Grade _____

OBJECTIVE: After completing this practice, you should be able to make a butt joint in a vertical position with 100% penetration.

EQUIPMENT AND MATERIALS NEEDED FOR THIS PRACTICE

1. Properly setup oxyacetylene welding equipment.

2. Proper safety protection (welding goggles, safety glasses, spark lighter, and pliers). Refer to Chapter 2 in the text for more specific safety information.

3. Two or more pieces of 16-gauge carbon steel sheet, approximately 1-1/2 in. (38 mm) wide by 6 in. (152 mm) long.

4. 1/16 in. (2 mm), 3/32 in. (2.4 mm), and 1/8 in. (3 mm) diameter RG45 filler rods. Determine by trial and error the size of filler rod with which you are most comfortable.

INSTRUCTIONS

1. Place the two pieces of metal in a jig or on a firebrick and tack weld both ends together in a butt joint configuration. After both ends are tacked together, place one or two additional tacks along the joint to prevent warping.

2. Position the tacked workpiece in a vertical position.

3. Use the torch and add filler metal as you did in Practice 31-9. Try making the bead from the bottom up and from the top down.

4. Apply heat for a long enough period of time so that a keyhole appears before adding filler metal. This will ensure 100% penetration of the joint.

5. Inspect the bottom side of the weld for 100% penetration and visual defects.

6. Repeat this practice until you can consistently make welds with 100% penetration and that are visually defect free.

7. Turn off the gas cylinders and regulators and clean up your work area when you are finished welding.

INSTRUCTOR'S COMMENTS _____

_____

_____

■ PRACTICE 31-18

Name _____ Date _____

Base Metal Thickness _____ Filler Metal Diameter_____

Class _____ Instructor _____ Grade _____

OBJECTIVE: After completing this practice, you should be able to make a lap joint at a 45° angle.

EQUIPMENT AND MATERIALS NEEDED FOR THIS PRACTICE

1. Properly setup oxyacetylene welding equipment.

2. Proper safety protection (welding goggles, safety glasses, spark lighter, and pliers). Refer to Chapter 2 in the text for more specific safety information.

3. Two or more pieces of 16-gauge carbon steel sheet, approximately 1-1/2 in. (38 mm) wide by 6 in. (152 mm) long.

4. 1/16 in. (2 mm), 3/32 in. (2.4 mm), and 1/8 in. (3 mm) diameter RG45 filler rods. Determine by trial and error the size of filler rod with which you are most comfortable.

INSTRUCTIONS

1. Place the two pieces of metal in a jig or on a firebrick and tack weld both ends together in a lap joint configuration. After both ends are tacked together, place one or two additional tacks along the joint to prevent warping.

2. Position the tacked workpiece at a 45° to the surface of the table.

3. Use the torch and add filler metal as you did in Practice 31-11.

4. It may be necessary to flash off the torch to control the heat of the molten weld pool.

5. Inspect the bead for uniformity and for visual defects.

6. Repeat this practice until you can consistently make welds that are visually defect free.

7. Turn off the gas cylinders and regulators and clean up your work area when you are finished welding.

INSTRUCTOR'S COMMENTS _____

_____

_____

■ PRACTICE 31-19

Name _____   Date _____

Base Metal Thickness _____   Filler Metal Diameter_____

Class _____   Instructor _____   Grade _____

OBJECTIVE: After completing this practice, you should be able to make a lap joint in the vertical position.

EQUIPMENT AND MATERIALS NEEDED FOR THIS PRACTICE

1. Properly setup oxyacetylene welding equipment.

2. Proper safety protection (welding goggles, safety glasses, spark lighter, and pliers). Refer to Chapter 2 in the text for more specific safety information.

3. Two or more pieces of 16-gauge carbon steel sheet, approximately 1-1/2 in. (38 mm) wide by 6 in. (152 mm) long.

4. 1/16 in. (2 mm), 3/32 in. (2.4 mm), and 1/8 in. (3 mm) diameter RG45 filler rods. Determine by trial and error the size of filler rod with which you are most comfortable.

INSTRUCTIONS

1. Place the two pieces of metal in a jig or on a firebrick and tack weld both ends together in a lap joint configuration. After both ends are tacked together, place one or two additional tacks along the joint to prevent warping.

2. Position the tacked workpiece in a vertical position. Try making the bead from the bottom up and from the top down.

3. Use the torch and add filler metal as you did in Practice 31-11.

4. It may be necessary to flash off the torch to control the heat of the molten weld pool.

5. Inspect the bead for uniformity and for visual defects.

6. Repeat this practice until you can consistently make welds that are visually defect free.

7. Turn off the gas cylinders and regulators and clean up your work area when you are finished welding.

INSTRUCTOR'S COMMENTS _____

_____

_____

■ PRACTICE 31-20

Name _____ Date _____

Base Metal Thickness _____ Filler Metal Diameter_____

Class _____ Instructor _____ Grade _____

OBJECTIVE: After completing this practice, you should be able to make a tee joint at a 45° angle.

EQUIPMENT AND MATERIALS NEEDED FOR THIS PRACTICE

1. Properly setup oxyacetylene welding equipment.

2. Proper safety protection (welding goggles, safety glasses, spark lighter, and pliers). Refer to Chapter 2 in the text for more specific safety information.

3. Two or more pieces of 16-gauge carbon steel sheet, approximately 1-1/2 in. (38 mm) wide by 6 in. (152 mm) long.

4. 1/16 in. (2 mm), 3/32 in. (2.4 mm), and 1/8 in. (3 mm) diameter RG45 filler rods. Determine by trial and error the size of filler rod with which you are most comfortable.

INSTRUCTIONS

1. Place the two pieces of metal in a jig or on a firebrick and tack weld them together in a tee joint configuration.

2. Position the tacked workpiece at a 45° angle to the surface of the table.

3. Use the torch and add filler metal as you did in Practice 31-12.

4. As you make the bead, be sure that it has uniform width and reinforcement.

5. Inspect the bead for uniformity and for visual defects.

6. Repeat this practice until you can consistently make welds that are visually defect free.

7. Turn off the gas cylinders and regulators and clean up your work area when you are finished welding.

INSTRUCTOR'S COMMENTS _____

_____

_____

■ PRACTICE 31-21

Name _____ Date _____

Base Metal Thickness _____ Filler Metal Diameter_____

Class _____ Instructor _____ Grade _____

OBJECTIVE: After completing this practice, you should be able to make a tee joint in a vertical position.

EQUIPMENT AND MATERIALS NEEDED FOR THIS PRACTICE

1. Properly setup oxyacetylene welding equipment.

2. Proper safety protection (welding goggles, safety glasses, spark lighter, and pliers). Refer to Chapter 2 in the text for more specific safety information.

3. Two or more pieces of 16-gauge carbon steel sheet, approximately 1-1/2 in. (38 mm) wide by 6 in. (152 mm) long.

4. 1/16 in. (2 mm), 3/32 in. (2.4 mm), and 1/8 in. (3 mm) diameter RG45 filler rods. Determine by trial and error the size of filler rod with which you are most comfortable.

INSTRUCTIONS

1. Place the two pieces of metal in a jig or on a firebrick and tack weld them together in a tee joint configuration.

2. Position the tacked workpiece in a vertical position. Try making the bead from the bottom up and from the top down.

3. Use the torch and add filler metal as you did in Practice 31-12.

4. As you make the bead, be sure that it has uniform width and reinforcement.

5. Inspect the bead for uniformity and for visual defects.

6. Repeat this practice until you can consistently make welds that are visually defect free.

7. Turn off the gas cylinders and regulators and clean up your work area when you are finished welding.

INSTRUCTOR'S COMMENTS _____
_____
_____

■ PRACTICE 31-22

Name _____   Date _____

Base Metal Thickness _____   Filler Metal Diameter_____

Class _____   Instructor _____   Grade _____

OBJECTIVE: After completing this practice, you should be able to make a horizontal stringer bead at a 45° reclining angle.

EQUIPMENT AND MATERIALS NEEDED FOR THIS PRACTICE

1. Properly setup oxyacetylene welding equipment.

2. Proper safety protection (welding goggles, safety glasses, spark lighter, and pliers). Refer to Chapter 2 in the text for more specific safety information.

3. One or more pieces of 16-gauge carbon steel sheet, approximately 3 in. (76 mm) wide by 6 in. (152 mm) long.

4. 1/16 in. (2 mm), 3/32 in. (2.4 mm), and 1/8 in. (3 mm) diameter RG45 filler rods. Determine by trial and error the size of filler rod with which you are most comfortable.

INSTRUCTIONS

1. Position the workpiece in a 45° horizontal reclining position.

2. Use the torch and add filler metal as you did in Practice 31-6.

3. Add the filler metal along the top leading edge of the molten weld pool.

4. You may have to flash the torch off the molten weld pool occasionally to control the heat and avoid sagging.

5. Inspect the bead for uniformity and for visual defects.

6. Repeat this practice until you can consistently make welds that are visually defect free.

7. Turn off the gas cylinders and regulators and clean up your work area when you are finished welding.

INSTRUCTOR'S COMMENTS _____

_____

_____

■ PRACTICE 31-23

Name _____ Date _____

Base Metal Thickness _____ Filler Metal Diameter _____

Class _____ Instructor _____ Grade _____

OBJECTIVE: After completing this practice, you should be able to make a horizontal stringer bead in the 2G position.

EQUIPMENT AND MATERIALS NEEDED FOR THIS PRACTICE

1. Properly setup oxyacetylene welding equipment.

2. Proper safety protection (welding goggles, safety glasses, spark lighter, and pliers). Refer to Chapter 2 in the text for more specific safety information.

3. One or more pieces of 16-gauge carbon steel sheet, approximately 3 in. (76 mm) wide by 6 in. (152 mm) long.

4. 1/16 in. (2 mm), 3/32 in. (2.4 mm), and 1/8 in. (3 mm) diameter RG45 filler rods. Determine by trial and error the size of filler rod with which you are most comfortable.

INSTRUCTIONS

1. Position the workpiece in a horizontal position.

2. Use the torch and add filler metal as you did in Practice 31-6.

3. Add the filler metal along the top leading edge of the molten weld pool.

4. You may have to flash the torch off the molten weld pool occasionally to control the heat and avoid sagging.

5. Inspect the bead for uniformity and for visual defects.

6. Repeat this practice until you can consistently make welds that are visually defect free.

7. Turn off the gas cylinders and regulators and clean up your work area when you are finished welding.

INSTRUCTOR'S COMMENTS _____

_____

_____

■ PRACTICE 31-24

Name _____   Date _____

Base Metal Thickness _____   Filler Metal Diameter_____

Class _____   Instructor _____   Grade _____

OBJECTIVE: After completing this practice, you should be able to make a butt joint in the horizontal position.

EQUIPMENT AND MATERIALS NEEDED FOR THIS PRACTICE

1. Properly setup oxyacetylene welding equipment.

2. Proper safety protection (welding goggles, safety glasses, spark lighter, and pliers). Refer to Chapter 2 in the text for more specific safety information.

3. Two or more pieces of 16-gauge carbon steel sheet, approximately 1-1/2 in. (38 mm) wide by 6 in. (152 mm) long.

4. 1/16 in. (2 mm), 3/32 in. (2.4 mm), and 1/8 in. (3 mm) diameter RG45 filler rods. Determine by trial and error the size of filler rod with which you are most comfortable.

INSTRUCTIONS

1. Tack the two pieces together in a butt joint configuration and position the workpiece in a horizontal position.

2. Use the torch and add filler metal as you did in Practice 31-8.

3. Add the filler metal along the top leading edge of the molten weld pool.

4. You may have to flash the torch off the molten weld pool occasionally to control the heat and avoid sagging.

5. Inspect the bead for uniformity and for visual defects.

6. Repeat this practice until you can consistently make welds that are visually defect free.

7. Turn off the gas cylinders and regulators and clean up your work area when you are finished welding.

INSTRUCTOR'S COMMENTS _____

_____

_____

■ PRACTICE 31-25

Name _____   Date _____

Base Metal Thickness _____   Filler Metal Diameter_____

Class _____   Instructor _____   Grade _____

OBJECTIVE: After completing this practice, you should be able to make a lap joint in the horizontal position.

EQUIPMENT AND MATERIALS NEEDED FOR THIS PRACTICE

1. Properly setup oxyacetylene welding equipment.

2. Proper safety protection (welding goggles, safety glasses, spark lighter, and pliers). Refer to Chapter 2 in the text for more specific safety information.

3. Two or more pieces of 16-gauge carbon steel sheet, approximately 1-1/2 in. (38 mm) wide by 6 in. (152 mm) long.

4. 1/16 in. (2 mm), 3/32 in. (2.4 mm), and 1/8 in. (3 mm) diameter RG45 filler rods. Determine by trial and error the size of filler rod with which you are most comfortable.

INSTRUCTIONS

1. Tack the two pieces together in a lap joint configuration and position the workpiece in a horizontal position.

2. Use the torch and add filler metal as you did in Practice 31-11.

3. Add the filler metal along the top leading edge of the molten weld pool.

4. You may have to flash the torch off the molten weld pool occasionally to control the heat and avoid sagging.

5. Inspect the bead for uniformity and for visual defects.

6. Repeat this practice until you can consistently make welds that are visually defect free.

7. Turn off the gas cylinders and regulators and clean up your work area when you are finished welding.

INSTRUCTOR'S COMMENTS _____

_____

_____

■ PRACTICE 31-26

Name _____ Date _____

Base Metal Thickness _____ Filler Metal Diameter_____

Class _____ Instructor _____ Grade _____

OBJECTIVE: After completing this practice, you should be able to make a tee joint in the horizontal position.

EQUIPMENT AND MATERIALS NEEDED FOR THIS PRACTICE

1. Properly setup oxyacetylene welding equipment.

2. Proper safety protection (welding goggles, safety glasses, spark lighter, and pliers). Refer to Chapter 2 in the text for more specific safety information.

3. Two or more pieces of 16-gauge carbon steel sheet, approximately 1-1/2 in. (38 mm) wide by 6 in. (152 mm) long.

4. 1/16 in. (2 mm), 3/32 in. (2.4 mm), and 1/8 in. (3 mm) diameter RG45 filler rods. Determine by trial and error the size of filler rod with which you are most comfortable.

INSTRUCTIONS

1. Tack the two pieces together in a tee joint configuration and position the workpiece in a horizontal position.

2. Use the torch and add filler metal as you did in Practice 31-12.

3. Add the filler metal along the top leading edge of the molten weld pool.

4. You may have to flash the torch off the molten weld pool occasionally to control the heat and avoid sagging.

5. Inspect the bead for uniformity and for visual defects.

6. Repeat this practice until you can consistently make welds that are visually defect free.

7. Turn off the gas cylinders and regulators and clean up your work area when you are finished welding.

INSTRUCTOR'S COMMENTS _____

_____

_____

■ PRACTICE 31-27

Name _____ Date _____

Base Metal Thickness _____ Filler Metal Diameter_____

Class _____ Instructor _____ Grade _____

OBJECTIVE: After completing this practice, you should be able to make a stringer bead in the overhead position.

EQUIPMENT AND MATERIALS NEEDED FOR THIS PRACTICE

1. Properly setup oxyacetylene welding equipment.

2. Proper safety protection (welding goggles, safety glasses, spark lighter, and pliers). Refer to Chapter 2 in the text for more specific safety information.

3. One or more pieces of 16-gauge carbon steel sheet, approximately 3 in. (76 mm) wide by 6 in. (152 mm) long.

4. 1/16 in. (2 mm), 3/32 in. (2.4 mm), and 1/8 in. (3 mm) diameter RG45 filler rods. Determine by trial and error the size of filler rod with which you are most comfortable.

INSTRUCTIONS

1. Position the workpiece in the overhead position.

2. Use the torch and add filler metal as you did in Practice 31-6.

3. It may be necessary to flash the torch away from the metal to avoid overheating and dripping. Move the hotter primary flame away from the puddle while still keeping the puddle within the secondary flame.

4. Establish a rhythm of moving the torch and adding filler metal so as to keep the bead uniform.

5. Inspect the completed bead for uniformity and freedom from defects.

6. Repeat this practice until you can consistently make welds that are visually defect free.

7. Turn off the gas cylinders and regulators and clean up your work area when you are finished welding.

INSTRUCTOR'S COMMENTS _____

_____

_____

■ PRACTICE 31-28

Name _____ Date _____

Base Metal Thickness _____ Filler Metal Diameter_____

Class _____ Instructor _____ Grade _____

OBJECTIVE: After completing this practice, you should be able to make a butt joint in the overhead position.

EQUIPMENT AND MATERIALS NEEDED FOR THIS PRACTICE

1. Properly setup oxyacetylene welding equipment.

2. Proper safety protection (welding goggles, safety glasses, spark lighter, and pliers). Refer to Chapter 2 in the text for more specific safety information.

3. Two or more pieces of 16-gauge carbon steel sheet, approximately 1-1/2 in. (38 mm) wide by 6 in. (152 mm) long.

4. 1/16 in. (2 mm), 3/32 in. (2.4 mm), and 1/8 in. (3 mm) diameter RG45 filler rods. Determine by trial and error the size of filler rod with which you are most comfortable.

INSTRUCTIONS

1. Tack the two pieces together in a butt joint configuration and position the workpiece in the overhead position.

2. Use the torch and add filler metal as you did in Practice 31-8.

3. Add the filler metal along the leading edge of the molten weld pool.

4. You may have to flash the torch off the molten weld pool occasionally to control the heat and avoid dripping.

5. Inspect the bead for uniformity and for visual defects.

6. Repeat this practice until you can consistently make welds that are visually defect free.

7. Turn off the gas cylinders and regulators and clean up your work area when you are finished welding.

INSTRUCTOR'S COMMENTS _____

_____

_____

■ PRACTICE 31-29

Name _____   Date _____

Base Metal Thickness _____   Filler Metal Diameter_____

Class _____   Instructor _____   Grade _____

OBJECTIVE: After completing this practice, you should be able to make a lap joint in the overhead position.

EQUIPMENT AND MATERIALS NEEDED FOR THIS PRACTICE

1. Properly setup oxyacetylene welding equipment.

2. Proper safety protection (welding goggles, safety glasses, spark lighter, and pliers). Refer to Chapter 2 in the text for more specific safety information.

3. Two or more pieces of 16-gauge carbon steel sheet, approximately 1-1/2 in. (38 mm) wide by 6 in. (152 mm) long.

4. 1/16 in. (2 mm), 3/32 in. (2.4 mm), and 1/8 in. (3 mm) diameter RG45 filler rods. Determine by trial and error the size of filler rod with which you are most comfortable.

INSTRUCTIONS

1. Tack the two pieces together in a lap joint configuration and position the workpiece in the overhead position.

2. Use the torch and add filler metal as you did in Practice 31-11.

3. Add the filler metal along the leading edge of the molten weld pool.

4. You may have to flash the torch off the molten weld pool occasionally to control the heat and avoid dripping.

5. Inspect the bead for uniformity and for visual defects.

6. Repeat this practice until you can consistently make welds that are visually defect free.

7. Turn off the gas cylinders and regulators and clean up your work area when you are finished welding.

INSTRUCTOR'S COMMENTS _____

_____

_____

■ PRACTICE 31-30

Name _____   Date _____

Base Metal Thickness _____   Filler Metal Diameter_____

Class _____   Instructor _____   Grade _____

OBJECTIVE: After completing this practice, you should be able to make a tee joint in the overhead position.

EQUIPMENT AND MATERIALS NEEDED FOR THIS PRACTICE

1. Properly setup oxyacetylene welding equipment.

2. Proper safety protection (welding goggles, safety glasses, spark lighter, and pliers). Refer to Chapter 2 in the text for more specific safety information.

3. Two or more pieces of 16-gauge carbon steel sheet, approximately 1-1/2 in. (38 mm) wide by 6 in. (152 mm) long.

4. 1/16 in. (2 mm), 3/32 in. (2.4 mm), and 1/8 in. (3 mm) diameter RG45 filler rods. Determine by trial and error the size of filler rod with which you are most comfortable.

INSTRUCTIONS

1. Tack the two pieces together in a tee joint configuration and position the workpiece in the overhead position.

2. Use the torch and add filler metal as you did in Practice 31-12.

3. Add the filler metal along the leading edge of the molten weld pool.

4. You may have to flash the torch off the molten weld pool occasionally to control the heat and avoid dripping.

5. Inspect the bead for uniformity and for visual defects.

6. Repeat this practice until you can consistently make welds that are visually defect free.

7. Turn off the gas cylinders and regulators and clean up your work area when you are finished welding.

INSTRUCTOR'S COMMENTS _____

_____

_____

■ PRACTICE 31-31

Name _____ Date _____

Base Metal Thickness _____ Filler Metal Diameter_____

Class _____ Instructor _____ Grade _____

OBJECTIVE: After completing this practice, you should be able to make a stringer bead on a piece of pipe in the 1G position.

EQUIPMENT AND MATERIALS NEEDED FOR THIS PRACTICE

1. Properly setup oxyacetylene welding equipment.

2. Proper safety protection (welding goggles, safety glasses, spark lighter, and pliers). Refer to Chapter 2 in the text for more specific safety information.

3. One or more pieces of schedule 40 mild steel pipe, approximately 2 in. (51 mm) in diameter by 6 in. (152 mm) long.

4. 1/16 in. (2 mm), 3/32 in. (2.4 mm), and 1/8 in. (3 mm) diameter RG45 filler rods. Determine by trial and error the size of filler rod with which you are most comfortable.

INSTRUCTIONS

1. Position the pipe in the 1G horizontal rolled position.

2. Start welding the bead at the 2 o'clock position weld up to the 12 o'clock position.

3. Stop and roll the pipe so that the crater is at the 2 o'clock position and weld again up to the 12 o'clock position.

4. Repeat this process until the weld bead extends all the way around the pipe.

5. Inspect the bead for straightness, uniform width and reinforcement, and for absence of visual defects.

6. Repeat this practice until you can consistently make welds that are defect free.

7. Turn off the gas cylinders and regulators and clean up your work area when you are finished welding.

INSTRUCTOR'S COMMENTS _____

_____

_____

■ PRACTICE 31-32

Name _____ Date _____

Base Metal Thickness _____ Filler Metal Diameter_____

Class _____ Instructor _____ Grade _____

OBJECTIVE: After completing this practice, you should be able to make a butt joint on pipe in the 1G position.

EQUIPMENT AND MATERIALS NEEDED FOR THIS PRACTICE

1. Properly setup oxyacetylene welding equipment.

2. Proper safety protection (welding goggles, safety glasses, spark lighter, and pliers). Refer to Chapter 2 in the text for more specific safety information.

3. Two or more pieces of schedule 40 mild steel pipe, approximately 2 in. (51 mm) in diameter by 3 in. (76 mm) long with the ends beveled as shown in textbook Figure 31-98.

4. 1/16 in. (2 mm), 3/32 in. (2.4 mm), and 1/8 in. (3 mm) diameter RG45 filler rods. Determine by trial and error the size of filler rod with which you are most comfortable.

INSTRUCTIONS

1. Tack the beveled ends together and position the weldment in the 1G horizontal rolled position.

2. Start welding the joint at the 2 o'clock position weld up to the 12 o'clock position.

3. Stop and roll the pipe so that the crater is at the 2 o'clock position and weld again up to the 12 o'clock position.

4. Repeat this process until the weld bead extends all the way around the pipe.

5. Inspect the bead for straightness, uniform width and reinforcement, and for absence of visual defects.

6. Repeat this practice until you can consistently make welds that are defect free.

7. Turn off the gas cylinders and regulators and clean up your work area when you are finished welding.

INSTRUCTOR'S COMMENTS _____

_____

_____

■ PRACTICE 31-33

Name _____   Date _____

Base Metal Thickness _____   Filler Metal Diameter_____

Class _____   Instructor _____   Grade _____

OBJECTIVE: After completing this practice, you should be able to make a stringer bead on a piece of pipe in the 5G position.

EQUIPMENT AND MATERIALS NEEDED FOR THIS PRACTICE

1. Properly setup oxyacetylene welding equipment.

2. Proper safety protection (welding goggles, safety glasses, spark lighter, and pliers). Refer to Chapter 2 in the text for more specific safety information.

3. One or more pieces of schedule 40 mild steel pipe, approximately 2 in. (51 mm) in diameter by 6 in. (152 mm) long.

4. 1/16 in. (2 mm), 3/32 in. (2.4 mm), and 1/8 in. (3 mm) diameter RG45 filler rods. Determine by trial and error the size of filler rod with which you are most comfortable.

INSTRUCTIONS

1. Position the pipe in the 5G horizontal fixed position.

2. Start welding the bead at the 6 o'clock position weld up to the 12 o'clock position.

3. Stop and begin welding again at the 6 o'clock position weld up to the 12 o'clock position around the other side of the pipe.

4. The weld bead should now extend all the way around the pipe.

5. Inspect the bead for straightness, uniform width and reinforcement, and for absence of visual defects.

6. Repeat this practice until you can consistently make welds that are defect free.

7. Turn off the gas cylinders and regulators and clean up your work area when you are finished welding.

INSTRUCTOR'S COMMENTS _____

_____

_____

■ PRACTICE 31-34

Name _____ Date _____

Base Metal Thickness _____ Filler Metal Diameter_____

Class _____ Instructor _____ Grade _____

OBJECTIVE: After completing this practice, you should be able to make a butt joint on pipe in the 5G position.

EQUIPMENT AND MATERIALS NEEDED FOR THIS PRACTICE

1. Properly setup oxyacetylene welding equipment.

2. Proper safety protection (welding goggles, safety glasses, spark lighter, and pliers). Refer to Chapter 2 in the text for more specific safety information.

3. Two or more pieces of schedule 40 mild steel pipe, approximately 2 in. (51 mm) in diameter by 3 in. (76 mm) long with the ends beveled as shown in textbook Figure 31-101.

4. 1/16 in. (2 mm), 3/32 in. (2.4 mm), and 1/8 in. (3 mm) diameter RG45 filler rods. Determine by trial and error the size of filler rod with which you are most comfortable.

INSTRUCTIONS

1. Tack the beveled ends together and position the weldment in the 5G horizontal fixed position.

2. Start welding the bead at the 6 o'clock position weld up to the 12 o'clock position.

3. Stop and begin welding again at the 6 o'clock position weld up to the 12 o'clock position around the other side of the pipe.

4. The weld bead should now extend all the way around the pipe.

5. Inspect the bead for straightness, uniform width and reinforcement, and for absence of visual defects.

6. Repeat this practice until you can consistently make welds that are defect free.

7. Turn off the gas cylinders and regulators and clean up your work area when you are finished welding.

INSTRUCTOR'S COMMENTS _____

_____

_____

■ PRACTICE 31-35

Name _____ Date _____

Base Metal Thickness _____ Filler Metal Diameter_____

Class _____ Instructor _____ Grade _____

OBJECTIVE: After completing this practice, you should be able to make a stringer bead on a piece of pipe in the 2G position.

EQUIPMENT AND MATERIALS NEEDED FOR THIS PRACTICE

1. Properly setup oxyacetylene welding equipment.

2. Proper safety protection (welding goggles, safety glasses, spark lighter, and pliers). Refer to Chapter 2 in the text for more specific safety information.

3. One or more pieces of schedule 40 mild steel pipe, approximately 2 in. (51 mm) in diameter by 6 in. (152 mm) long.

4. 1/16 in. (2 mm), 3/32 in. (2.4 mm), and 1/8 in. (3 mm) diameter RG45 filler rods. Determine by trial and error the size of filler rod with which you are most comfortable.

INSTRUCTIONS

1. Position the pipe in the 2G vertical position.

2. This bead is made similar to a horizontal stringer bead on a flat plate.

3. Start with a small bead and then increase the size. This will allow you to build a shelf to support the molten metal.

4. The weld bead should extend all the way around the pipe.

5. Inspect the bead for straightness, uniform width and reinforcement, and for absence of visual defects.

6. Repeat this practice until you can consistently make welds that are defect free.

7. Turn off the gas cylinders and regulators and clean up your work area when you are finished welding.

INSTRUCTOR'S COMMENTS _____

_____

_____

■ PRACTICE 31-36

Name _____   Date _____

Base Metal Thickness _____   Filler Metal Diameter _____

Class _____   Instructor _____   Grade _____

OBJECTIVE: After completing this practice, you should be able to make a butt joint on pipe in the 2G position.

EQUIPMENT AND MATERIALS NEEDED FOR THIS PRACTICE

1. Properly setup oxyacetylene welding equipment.

2. Proper safety protection (welding goggles, safety glasses, spark lighter, and pliers). Refer to Chapter 2 in the text for more specific safety information.

3. Two or more pieces of schedule 40 mild steel pipe, approximately 2 in. (51 mm) in diameter by 3 in. (76 mm) long with the ends beveled as shown in textbook Figure 31-103.

4. 1/16 in. (2 mm), 3/32 in. (2.4 mm), and 1/8 in. (3 mm) diameter RG45 filler rods. Determine by trial and error the size of filler rod with which you are most comfortable.

INSTRUCTIONS

1. Tack the beveled ends together and position the weldment in the 2G vertical position.

2. This bead is made similar to a horizontal butt joint on flat plates.

3. Start with a small bead and then increase the size. This will allow you to build a shelf to support the molten metal.

4. The weld bead should extend all the way around the pipe.

5. Inspect the bead for uniform width and reinforcement and for absence of visual defects.

6. Repeat this practice until you can consistently make welds that are defect free.

7. Turn off the gas cylinders and regulators and clean up your work area when you are finished welding.

INSTRUCTOR'S COMMENTS _____

_____

_____

■ PRACTICE 31-37

Name _____   Date _____

Base Metal Thickness _____   Filler Metal Diameter_____

Class _____   Instructor _____   Grade _____

OBJECTIVE: After completing this practice, you should be able to make a stringer bead on a piece of pipe in the 6G 45° fixed inclined position.

EQUIPMENT AND MATERIALS NEEDED FOR THIS PRACTICE

1. Properly setup oxyacetylene welding equipment.

2. Proper safety protection (welding goggles, safety glasses, spark lighter, and pliers). Refer to Chapter 2 in the text for more specific safety information.

3. One or more pieces of schedule 40 mild steel pipe, approximately 2 in. (51 mm) in diameter by 6 in. (152 mm) long.

4. 1/16 in. (2 mm), 3/32 in. (2.4 mm), and 1/8 in. (3 mm) diameter RG45 filler rods. Determine by trial and error the size of filler rod with which you are most comfortable.

INSTRUCTIONS

1. Position the pipe in the 6G 45° fixed inclined position.

2. Start welding the bead at the bottom and weld up to the top.

3. The bead shape will change as you move around the pipe.

4. Stop at the top and begin welding again at the bottom and weld up to the top around the other side of the pipe.

5. The weld bead should now extend all the way around the pipe.

6. Inspect the bead for straightness, uniform width and reinforcement, and for absence of visual defects.

7. Repeat this practice until you can consistently make welds that are defect free.

8. Turn off the gas cylinders and regulators and clean up your work area when you are finished welding.

INSTRUCTOR'S COMMENTS _____

_____

_____

■ PRACTICE 31-38

Name _____ Date _____

Base Metal Thickness _____ Filler Metal Diameter_____

Class _____ Instructor _____ Grade _____

OBJECTIVE: After completing this practice, you should be able to make a butt joint on pipe in the 6G 45° fixed inclined position.

EQUIPMENT AND MATERIALS NEEDED FOR THIS PRACTICE

1. Properly setup oxyacetylene welding equipment.

2. Proper safety protection (welding goggles, safety glasses, spark lighter, and pliers). Refer to Chapter 2 in the text for more specific safety information.

3. Two or more pieces of schedule 40 mild steel pipe, approximately 2 in. (51 mm) in diameter by 3 in. (76 mm) long with the ends beveled as shown in textbook Figure 31-106.

4. 1/16 in. (2 mm), 3/32 in. (2.4 mm), and 1/8 in. (3 mm) diameter RG45 filler rods. Determine by trial and error the size of filler rod with which you are most comfortable.

INSTRUCTIONS

1. Tack the beveled ends together and position the weldment in the 6G 45° fixed inclined position.

2. Start welding the joint at the bottom and weld up to the top.

3. The bead shape will change as you move around the pipe.

4. Stop at the top and begin welding again at the bottom and weld up to the top around the other side of the pipe.

5. The weld bead should now extend all the way around the pipe.

6. Inspect the bead for uniform width and reinforcement and for absence of visual defects.

7. Repeat this practice until you can consistently make welds that are defect free.

8. Turn off the gas cylinders and regulators and clean up your work area when you are finished welding.

INSTRUCTOR'S COMMENTS _____

_____

_____

CHAPTER 31: QUIZ 1

Name _____   Date _____

Class _____   Instructor _____   Grade _____

INSTRUCTIONS

Carefully read Chapter 31 in the textbook and answer each question.

MATCHING

In the space provided to the left of Column A, write the letter from Column B that best answers or completes the statement in Column A.

Column A	Column B
_____ 1. The general grouping of processes known as oxyfuel consists of a number of separate processes, all of which burn a fuel gas with _____.	a. mild steel
_____ 2. _____ is the most widely used fuel gas, but approximately 25 other gases are available.	b. iron
_____ 3. _____ is the easiest metal to gas weld.	c. size
_____ 4. One group of welding rods used for welding with an oxyfuel torch is designated with the prefix letter R, and another group of rods used for _____ is designated with the prefix letter B.	d. flux
_____ 5. The short ends of both welding and brazing rods can be fused together so that the amount of _____ can be minimized.	e. rod
_____ 6. Ferrous filler metals are welding rods that are mainly _____, but they may have other elements added to change their strength, corrosion resistance, weldability, or another physical property.	f. speed
_____ 7. Mild steel and low-alloy steel are the materials that are most frequently gas welded and are easily welded without a _____.	g. secondary
_____ 8. Cast iron filler rods for gas welding use the prefix R, which refers to the welding rod, in front of CI, which stands for _____.	h. acetylene
_____ 9. The oxygen and _____ flame is the only 100% nonpolluting fuel-gas flame.	i. mixing chamber
_____ 10. The torch tip _____ should be used to control the weld bead width, penetration, and speed.	j. hydrogen
_____ 11. Penetration is the _____ into the base metal that the weld fusion or melting extends from the surface, excluding any reinforcement.	k. leak-detecting solution

_____ 12. It is never safe to lower the _____ size if the tip's flame produces too much heat for your welding job.

l. brazing

_____ 13. The torch angle and the angle between the inner cone and the metal have a great effect on the _____ of melting and size of the molten weld pool.

m. vacuum

_____ 14. Welding _____ size and torch manipulation can be used to control the weld bead characteristics.

n. mixed

_____ 15. The molten weld pool must be protected by the _____ flame to prevent the atmosphere from contaminating the metal.

o. oxygen

_____ 16. The oxygen and fuel gas must be _____ completely before they leave the tip and create the flame.

p. connected

_____ 17. One method of mixing the gases uses equal or balanced pressures, and the gases are mixed in a _____.

q. flame

_____ 18. The injector works by passing the oxygen through a venturi, which creates a _____ to pull the fuel gas in and then mixes the gases together.

r. cast iron

_____ 19. The best protection for torch connections against damage and dirt is to leave the tip and hoses _____ when the torch is not in use.

s. depth

_____ 20. To find leaking valve seats, set the regulators to a working pressure; and with the torch valves off, spray the tip with a _____.

t. scrap filler metal

CHAPTER 31: QUIZ 2

Name _____   Date _____

Class _____   Instructor _____   Grade _____

INSTRUCTIONS

Carefully read Chapter 31 in the textbook and answer each question.

MATCHING

In the space provided to the left of Column A, write the letter from Column B that best answers or completes the statement in Column A.

	Column A	Column B
_____	1. Because no industrial standard tip size identification system exists, the student must become familiar with the size of the orifice (hole) in the tip and the _____ range for which it can be used.	a. ignition
_____	2. Metal-to-metal seal tips must be tightened with a _____; tips with an O-ring or a gasket may be tightened by hand.	b. same time
_____	3. Dirty tips can be cleaned using a set of _____.	c. out-of-position
_____	4. When cracking the valve, always stand to one side, point the valve away from anyone in the area, and be sure there are no sources of _____.	d. repaired
_____	5. The acetylene valve should never be opened more than one and one-half turns so that in an emergency it can be _____.	e. small
_____	6. Leaking cylinder valve stems should not be _____.	f. overflow
_____	7. The flat outside corner joint can be made with or without the addition of _____.	g. thickness
_____	8. On a flat butt joint if the sheets to be welded are of different sizes or thicknesses, then the torch should be pointed so that both pieces melt at the _____.	h. horizontal
_____	9. Heat is not distributed _____ in the flat lap joint, so the flame must be directed on the bottom sheet and away from the metal top sheet.	i. backfire or pop
_____	10. A problem that is unique to the tee joint is that a large percentage of the welding heat is reflected back on the torch, which can cause even a properly cleaned and adjusted torch to _____.	j. filler metal
_____	11. Whenever a weld is performed in a position other than flat, it is said to be _____ welding	k. keyhole

——— 12. When making a vertical weld, it is important to control the size of the molten weld pool, because if it increases beyond that which the shelf will support, the molten weld pool will _____ and drip down the weld.

l. difficult

——— 13. Horizontal welds, like vertical welds, must rely on some part of the weld bead to _____ the molten weld pool as the weld is made.

m. wrench

——— 14. When starting a horizontal bead, it is important to start with a _____ bead and build it to the desired size.

n. tubing

——— 15. The possibility of being burned increases greatly when welding in the _____ position.

o. molten weld pool

——— 16. Mild steel pipe and _____, both small diameter and thin wall, can be gas welded and is used to make structures, such as bicycle and motorcycle frames, gates, works of art, handrails, and light aircraft frames.

p. overhead

——— 17. The vertically fixed pipe requires a _____ weld.

q. tip cleaners

——— 18. The 45° fixed pipe position requires careful manipulation of the _____ to ensure a uniform and satisfactory weld.

r. support

——— 19. It is the combination of compound angles that makes the 6G position particularly _____.

s. uniformly

——— 20. If penetration is required when welding thin-wall tubing, then the weld will probably have a _____ to ensure 100% penetration.

t. turned off quickly

CHAPTER 31: QUIZ 3

Name _____   Date _____

Class _____   Instructor _____   Grade _____

INSTRUCTIONS

Carefully read Chapter 31 in the textbook and answer each question.

IDENTIFICATION

Identify the numbered items on the drawing by writing the letter next to the identifying term in the space provided.

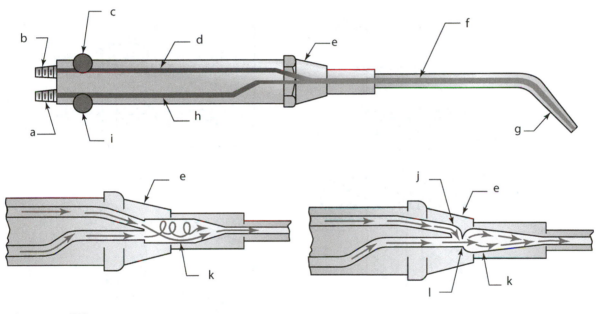

1. _____   TIP

2. _____   INJECTOR

3. _____   OXYGEN AND ACETYLENE MIXTURE

4. _____   ACETYLENE HOSE CONNECTION

5. _____   OXYGEN CONTROL VALVE

6. _____   MIXER

7. _____   OXYGEN

8. _____   ACETYLENE CONTROL VALVE

9. _____   MIXING CHAMBER

10. _____   OXYGEN HOSE CONNECTION

11. _____   ACETYLENE

12. _____   VENTURI

CHAPTER 31: QUIZ 4

Name _____   Date _____

Class _____   Instructor _____   Grade _____

INSTRUCTIONS
Carefully read Chapter 31 in the textbook and answer each question.

IDENTIFICATION
Identify the numbered items on the drawing by writing the letter next to the identifying term in the space provided.

1. _____   SAFETY RELEASE DISC

2. _____   CORRECT HEAT

3. _____   MOLTEN WELD METAL

4. _____   MAIN SEATING SEAL

5. _____   SHELF OF SOLIDIFIED WELD METAL

6. _____   BACK SEATING SEAL

7. _____   VALVE STEM

8. _____   WATCH THIS AREA

9. _____   TOO MUCH HEAT

CHAPTER 31: QUIZ 5

Name _____   Date _____

Class _____   Instructor _____   Grade _____

INSTRUCTIONS

Carefully read Chapter 31 in the textbook and answer each question.

IDENTIFICATION

Identify the numbered items on the drawing by writing the letter next to the identifying term in the space provided.

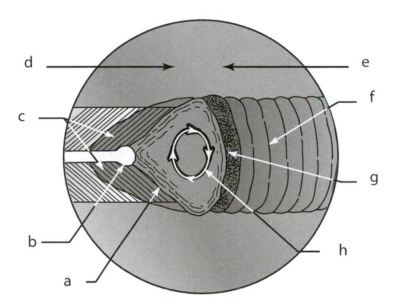

1. _____ DIRECTION OF WELDING

2. _____ PLASTIC WELD METAL

3. _____ PATH OF END OF WELDING ROD IN MOLTEN POOL

4. _____ ADVANCING EDGE OF POOL

5. _____ KEYHOLE

6. _____ AREAS OF VEE BEGINNING TO MELT

7. _____ SOLIDIFIED WELD METAL

8. _____ DIRECTION FLAM IS POINTING

Brazing, Braze Welding, and Soldering

Brazing Troubleshooting	
Problem	**Correction**
Brazing alloy will not melt and adhere to the base metal.	1. Braze on clean metal only. 2. Increase the heat of the base metal. 3. Use more flux. 4. Check brazing alloy to see if it is compatible with base metal. 5. Roughen work surface.
Brazing alloy flows away from joint.	1. Direct the heat into the joint. 2. Reduce heat directed at the joint. 3. Reposition the assembly so that gravity may help the flow of the brazing alloy into the joint. 4. Use more flux. 5. Make sure that the base metal is clean.
Capillary action will not occur.	1. Braze on clean metal only. 2. If the gap opening is not uniform, it may interrupt the capillary flow. 3. Assembly fit up may be too tight or too loose. 4. Too much heat can cause the flux to break down.

■ PRACTICE 32-1

Name _____ Date _____

Class _____ Instructor _____ Grade _____

OBJECTIVE: After completing this practice, the student should be able to braze stringer beads on 16-gauge carbon steel.

EQUIPMENT AND MATERIALS NEEDED FOR THIS PRACTICE

1. A properly lit and adjusted torch.

2. Proper safety protection (welding goggles, safety glasses, pliers, long-sleeved shirt, long pants, leather shoes or boots, and a spark lighter).

3. One piece of clean 16-gauge mild steel, 3 in. (76 mm) wide by 6 in. (152 mm) long.

4. Brazing flux and BRCuZn brazing rod.

INSTRUCTIONS

1. Light and adjust the torch to a neutral flame.

2. If a prefluxed rod is not used, you must flux the brazing rod by heating the end with the torch and dipping the hot end into the flux.

3. Place the sheet flat on a firebrick and hold the flame at one end.

4. Touch the flux-covered rod to the sheet and allow a small amount of brazing rod to melt onto the hot sheet.

5. Once the molten brazing metal wets the sheet, start moving the torch in a circular pattern while dipping the rod into the molten braze pool as you move along the sheet. If the size of the molten pool increases, you can control it by reducing the torch angle, raising the torch, traveling at a faster rate, or flashing the flame off the molten braze pool. Flashing the torch off a braze joint will not cause oxidation problems as it does when welding, because the molten metal is protected by a layer of flux.

6. As the braze bead progresses across the sheet, dip the end of the rod back in the flux, if a powdered flux is used, as often as needed to keep a small molten pool of flux ahead of the bead.

7. Repeat this practice until you can consistently produce brazed stringer beads that are straight with uniform width and buildup.

8. Turn off the gas cylinders and regulators and clean up your work area when you are finished brazing.

INSTRUCTOR'S COMMENTS _____

_____

_____

■ PRACTICE 32-2 AND PRACTICE 32-3

Name _____   Date _____

Class _____   Instructor _____   Grade _____

OBJECTIVE: After completing Practice 32-2, you should be able to braze butt joints in the flat position on 16-gauge carbon steel; and after completing Practice 32-3, you should be able to make the same butt joint with 100% penetration.

EQUIPMENT AND MATERIALS NEEDED FOR THIS PRACTICE

1. A properly lit and adjusted torch.

2. Proper safety protection (welding goggles, safety glasses, pliers, long-sleeved shirt, long pants, leather shoes or boots, and a spark lighter).

3. Two pieces of clean 16-gauge mild steel, 1-1/2 in. (38 mm) wide by 6 in. (152 mm) long.

4. Brazing flux and BRCuZn brazing rod.

INSTRUCTIONS

1. Place the metal flat on a firebrick, hold the plates tightly together, and make a tack braze at both ends of the joint. If the plates become distorted, they can be bent back into shape with a hammer before making another tack weld in the center.

2. Align the sheets so that you can comfortably make a braze bead along the joint.

3. Starting as you did in Practice 32-1, make a uniform braze along the joint.

4. Repeat this practice until a uniform braze can be made without defects.

5. Repeat this practice using more heat so that 100% penetration occurs.

6. Turn off the gas cylinders and regulators and clean up your work area when you are finished brazing.

INSTRUCTOR'S COMMENTS _____

_____

_____

■ PRACTICE 32-4

Name _____ Date _____

Class _____ Instructor _____ Grade _____

OBJECTIVE: After completing this practice, you should be able to braze tee joints with 100% joint penetration.

EQUIPMENT AND MATERIALS NEEDED FOR THIS PRACTICE

1. A properly lit and adjusted torch.

2. Proper safety protection (welding goggles, safety glasses, pliers, long-sleeved shirt, long pants, leather shoes or boots, and a spark lighter).

3. Two or more pieces of 16-gauge steel 1-1/2 in. (38 mm) wide by 6 in. (152 mm) long.

4. Brazing flux and BRCuZn brazing rod.

INSTRUCTIONS

1. Place the first piece of metal flat on a firebrick and hold or brace the second piece vertically on the first piece. The vertical piece should be within 5° of square to the bottom sheet.

2. Tack braze the pieces together in the tee joint configuration being sure that they are held tightly together.

3. Place the metal on a firebrick so the face of the braze will be flat, Figure 32-39.

4. Direct the flame on the sheets just ahead of the braze bead, being careful not to overheat the braze metal. If the bead has a notch, the root of the joint is still not hot enough to allow the braze metal to flow properly.

5. After the joint is completed and cooled, look at the back of the joint for a line of braze metal that flowed through.

6. Repeat this practice until the joint can be made without defects.

7. Turn off the gas cylinders and regulators and clean up your work area when you are finished brazing.

INSTRUCTOR'S COMMENTS _____

_____

_____

■ PRACTICE 32-5 AND PRACTICE 32-6

Name _____ Date _____

Class _____ Instructor _____ Grade _____

OBJECTIVE: After completing this practice, you should be able to make a brazed lap joint in the flat position, Practice 32-5; and with 100% penetration, Practice 32-6.

EQUIPMENT AND MATERIALS NEEDED FOR THIS PRACTICE

1. Properly setup, lit, and adjusted torch.

2. Proper safety protection (welding goggles, safety glasses, spark lighter, and pliers). Refer to Chapter 2 in the text for more specific safety information.

3. Two or more pieces of 16-gauge mild steel, 1-1/2 in. (38 mm) wide by 6 in. (152 mm) long.

4. Brazing flux and BRCuZn brazing rod.

INSTRUCTIONS

1. Braze tack the pieces together in a lap joint configuration.

2. Place the sheets in the flat position and align them so that you can comfortably make a braze bead along the joint.

3. Starting as you did in Practice 32-5, make a uniform braze bead along the joint.

4. Go slowly enough and apply enough heat so that you will achieve 100% penetration.

5. Inspect the bottom side of the bead to ensure that 100% penetration was achieved.

6. Repeat this practice until you can consistently make braze beads that are defect free.

7. Turn off the gas cylinders and regulators and clean up your work area when you are finished brazing.

INSTRUCTOR'S COMMENTS _____

_____

_____

■ PRACTICE 32-7

Name _____   Date _____

Class _____   Instructor _____   Grade _____

OBJECTIVE: After completing this practice, you should be able to make a brazed tee joint in the flat position using thin to thick metal.

EQUIPMENT AND MATERIALS NEEDED FOR THIS PRACTICE

1. Properly setup, lit, and adjusted torch.

2. Proper safety protection (welding goggles, safety glasses, spark lighter, and pliers). Refer to Chapter 2 in the text for more specific safety information.

3. One piece of 16-gauge mild steel plate, 1-1/2 in. (38 mm) wide by 6 in. (152 mm) long and one piece of 1/4 in. (6 mm) thick by 1-1/2 in. (38 mm) wide by 6 in. (152 mm) long.

4. Brazing flux and BRCuZn brazing rod.

INSTRUCTIONS

1. Braze tack the pieces together in a tee joint configuration with the thin plate in the vertical position.

2. The thin plate will heat up faster than the thick plate so direct the torch mostly on the thick plate. Both plates must be heated equally, Figure 33-42.

3. Direct the heat to the thicker plate while adding the brazing rod onto the thinner plate, Figure 33-43.

4. Make a braze along the joint that is uniform in appearance.

5. Inspect the bead for uniformity and for freedom from defects.

6. Repeat this practice until you can consistently make braze beads that are defect free.

7. Turn off the gas cylinders and regulators and clean up your work area when you are finished brazing.

INSTRUCTOR'S COMMENTS _____

_____

_____

■ PRACTICE 32-8

Name _____ Date _____

Class _____ Instructor _____ Grade _____

OBJECTIVE: After completing this practice, you should be able to make a brazed lap joint in the flat position using thin to thick metal.

EQUIPMENT AND MATERIALS NEEDED FOR THIS PRACTICE

1. Properly setup, lit, and adjusted torch.

2. Proper safety protection (welding goggles, safety glasses, spark lighter, and pliers). Refer to Chapter 2 in the text for more specific safety information.

3. One piece of 16-gauge mild steel plate, 1-1/2 in. (38 mm) wide by 6 in. (152 mm) long and one piece of 1/4 in. (6 mm) thick by 1-1/2 in. (38 mm) wide by 6 in. (152 mm) long.

4. Brazing flux and BRCuZn brazing rod.

INSTRUCTIONS

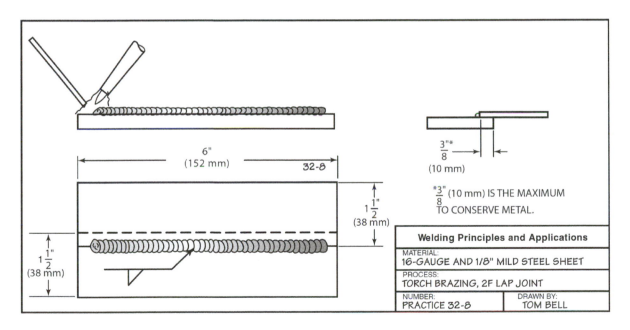

1. Braze tack the pieces tightly together in a lap joint configuration.

2. Place the tacked assembly on a firebrick in the flat position with the thin plate up.

3. The thin plate will heat up faster than the thick plate, so direct the torch mostly on the thick plate, Figure 32-45. Both plates must be heated equally.

4. Direct the heat to the thicker plate while adding the brazing rod.

5. Make a braze along the joint that is uniform in appearance.

6. Inspect the bead for uniformity and for freedom from defects.

7. Repeat this practice until you can consistently make braze beads that are defect free.

8. Turn off the gas cylinders and regulators and clean up your work area when you are finished brazing.

INSTRUCTOR'S COMMENTS _____

_____

_____

■ PRACTICE 32-9

Name _____   Date _____

Class _____   Instructor _____   Grade _____

OBJECTIVE: After completing this practice, you should be able to make a braze welded butt joint in the flat position using thick metal.

EQUIPMENT AND MATERIALS NEEDED FOR THIS PRACTICE

1. Properly setup, lit, and adjusted torch.

2. Proper safety protection (welding goggles, safety glasses, spark lighter, and pliers). Refer to Chapter 2 in the text for more specific safety information.

3. Use two pieces of metal that are 1-1/2 in. (38 mm) wide and 6 in. (152 mm) long. One piece is 3/8 in. (9.5 mm) or thicker and the other piece is 1/4 in. (6 mm) thick. Grind the edge round on the thicker plate as shown in Figure 32-47.

4. Brazing flux and BRCuZn brazing rod.

INSTRUCTIONS

1. Braze tack the pieces together in a butt joint configuration. See Figure 32-47 in the textbook.

2. Place the tacked assembly on a firebrick in the flat position.

3. The thinner plate will heat up faster than the thicker plate so direct the torch mostly on the thicker plate. Both plates must be heated equally.

4. The flame should be moved in a triangular motion so that the root is heated as well as the top of the bead, Figure 32-48.

5. Make a braze bead along the joint that is uniform in appearance.

6. Inspect the bead for uniformity and for freedom from brazing defects.

7. When the assembly is cool, do a bend test, as shown in Figure 32-49 in the textbook.

8. Repeat this practice until you can consistently make braze beads that are defect free.

9. Turn off the gas cylinders and regulators and clean up your work area when you are finished brazing.

INSTRUCTOR'S COMMENTS _____

_____

_____

■ PRACTICE 32-10

Name _____ Date _____

Class _____ Instructor _____ Grade _____

OBJECTIVE: After completing this practice, you should be able to make a braze welded tee joint in the flat position using thin to thick metal.

EQUIPMENT AND MATERIALS NEEDED FOR THIS PRACTICE

1. Properly setup, lit, and adjusted torch.

2. Proper safety protection (welding goggles, safety glasses, spark lighter, and pliers). Refer to Chapter 2 in the text for more specific safety information.

3. Use two pieces of metal that are 1-1/2 in. (38 mm) wide and 6 in. (152 mm) long. One piece is 3/8 in. (9.5 mm) or thicker and a other piece is 1/4 in. (6 mm) thick.

4. Brazing flux and BRCuZn brazing rod.

INSTRUCTIONS

1. Braze tack the pieces together in a tee joint configuration with the thin plate in the vertical position.

2. Place the tacked assembly on a firebrick in the flat position.

3. The thin plate will heat up faster than the thick plate so direct the torch mostly on the thick plate. Both plates must be heated equally.

4. Direct the heat mostly at the thick plate and add the braze rod on the thin plate.

5. Make a braze bead along the joint that is uniform in appearance.

6. Inspect the bead for uniformity and for freedom from defects.

7. When the assembly is cool, do a bend test, as shown in Figure 32-50 in the textbook.

8. Repeat this practice until you can consistently make braze beads that are defect free.

9. Turn off the gas cylinders and regulators and clean up your work area when you are finished brazing.

INSTRUCTOR'S COMMENTS _____

_____

_____

■ PRACTICE 32-11

Name _____   Date _____

Class _____   Instructor _____   Grade _____

OBJECTIVE: After completing this practice, you should be able to use braze welding to weld up a hole in the flat position.

EQUIPMENT AND MATERIALS NEEDED FOR THIS PRACTICE

1. Properly setup, lit, and adjusted torch.

2. Proper safety protection (welding goggles, safety glasses, spark lighter, and pliers). Refer to Chapter 2 in the text for more specific safety information.

3. One piece of 16-gauge mild steel plate, 3 in. (76 mm) wide by 3 in. (76 mm) long having a 1-in. (25-mm) diameter hole drilled through it.

4. Brazing flux and BRCuZn brazing rod.

INSTRUCTIONS

1. Place the piece in the flat position on two firebricks so that the hole is between them.

2. Start by running a stringer bead around the hole. See Figure 32-51 in the textbook.

3. When the bead is complete, turn the torch at a very steep angle and point it at the edge of the hole nearest the torch. Hold the end of the brazing rod in the flame so that both the bead around the hole and the rod meet at the same time. See Figure 32-52 in the textbook.

4. Put the rod in the molten bead and then flash the torch off to allow the molten metal to cool.

5. Repeat this process around the hole until it is completely filled.

6. When the braze weld is complete, it should be fairly flat with the surrounding metal.

7. Inspect the braze weld for uniformity and for freedom from defects.

8. Repeat this practice until you can consistently make braze beads to fill up holes that are defect free.

9. Turn off the gas cylinders and regulators and clean up your work area when you are finished brazing.

INSTRUCTOR'S COMMENTS _____

_____

_____

■ PRACTICE 32-12

Name _____   Date _____

Class _____   Instructor _____   Grade _____

OBJECTIVE: After completing this practice, you should be able to use braze welding to build up a flat surface.

EQUIPMENT AND MATERIALS NEEDED FOR THIS PRACTICE

1. Properly setup, lit, and adjusted torch.

2. Proper safety protection (welding goggles, safety glasses, spark lighter, and pliers). Refer to Chapter 2 in the text for more specific safety information.

3. One piece of 1/4-in. (6-mm) mild steel plate, 3 in. (76 mm) wide by 3 in. (76 mm) long.

4. Brazing flux and BRCuZn brazing rod.

INSTRUCTIONS

1. Place the piece in the flat position on a firebrick.

2. Start along one side of the plate and make a braze weld down that side.

3. When you get to the end, turn the plate 180° and braze back alongside the first braze covering about one-half of the first braze. See Figure 32-53 in the textbook.

4. Repeat this process until the side is completely covered with braze metal.

5. Turn the plate 90° and repeat the process being careful to get good fusion with the layer beneath. See Figure 32-54 in the textbook.

6. Be sure that the edges are built up enough so that they could be cut back square.

7. Repeat this process until the buildup is at least 1/4 in. (6 mm) thick.

8. Inspect the braze weld for uniformity and for freedom from defects.

9. Repeat this practice until you can use brazing to build up flat surfaces that are consistently defect free.

10. Turn off the gas cylinders and regulators and clean up your work area when you are finished brazing.

INSTRUCTOR'S COMMENTS _____

_____

_____

■ PRACTICE 32-13

Name _____   Date _____

Class _____   Instructor _____   Grade _____

OBJECTIVE: After completing this practice, you should be able to use braze welding to build up a rounded surface.

EQUIPMENT AND MATERIALS NEEDED FOR THIS PRACTICE

1. Properly setup, lit, and adjusted torch.

2. Proper safety protection (welding goggles, safety glasses, spark lighter, and pliers). Refer to Chapter 2 in the text for more specific safety information.

3. One piece of mild steel bolt or rod, 1/2 in. (13 mm) in diameter by 3 in. (76 mm) long.

4. Brazing flux and BRCuZn brazing rod.

INSTRUCTIONS

1. Place the mild steel bolt or rod in the flat position on a firebrick.

2. Start along one end and make a braze weld 1-1/2 in. (38 mm) long.

3. Turn the rod 180° and braze back alongside the first braze covering about one-half of the first braze. See Figure 32-55 in the textbook.

4. Rotate the rod and repeat the process being careful to get good fusion with the layer beneath.

5. Repeat this process until the rod is completely covered with braze metal and is 1 in. (25 mm) in diameter. See Figure 32-56 in the textbook.

6. Inspect the braze weld for uniformity and for freedom from defects.

7. Repeat this practice until you can use brazing to build up rounded surfaces that are consistently defect free.

8. Turn off the gas cylinders and regulators and clean up your work area when you are finished brazing.

INSTRUCTOR'S COMMENTS _____

_____

_____

■ PRACTICE 32-14

Name _____   Date _____

Class _____   Instructor _____   Grade _____

OBJECTIVE: After completing this practice, you should be able to silver braze copper pipe in the 2G position.

EQUIPMENT AND MATERIALS NEEDED FOR THIS PRACTICE

1. Properly setup, lit, and adjusted air MPS, air propane, or any air fuel gas torch. Be sure that the regulator pressure is set to the manufacturer's specifications for your fuel type and tip size.

2. Proper safety protection (welding goggles, safety glasses, spark lighter, and pliers). Refer to Chapter 2 in the text for more specific safety information.

3. Two or more short random length pieces of 1/2-in. to 1-in. (13-mm to 25-mm) copper pipe with matching copper pipe fittings.

4. Brazing flux and BCuP-2 to BCuP-5 brazing rod.

5. Steel wool, sanding cloth, and/or a wire brush.

INSTRUCTIONS

1. Clean the pipe O.D. and the fitting I.D. using steel wool, sanding cloth, or a wire brush. Then slide the fitting onto the pipe. Be sure that the pipe is seated at the bottom of the fitting.

2. Heat the brazing rod and make a bend about 3/4 in. (19 mm) from the end. Use this bend as a gauge so that you do not put too much brazing metal into the joint.

3. Place the pipe in the 2G position. Heat the pipe first but not too much Figure 32-57. When it is hot (but not glowing red), start heating the fitting. Keep the torch moving in order to uniformly heat the entire joint. The joint is at the correct temperature when the braze metal starts to wet the surface.

4. Move the flame to the back side of the pipe and feed the brazing rod into the joint. Move the torch and brazing rod all the way around the pipe and fitting so that the entire joint will be filled. There should be a slight fillet showing all the way around the joint. See Figure 32-58 in the textbook.

5. After cooling, hacksaw the joint apart for inspection. See Figures 32-59 to 32-63 in the textbook.

6. Repeat this practice until you can make defect free brazed pipe joints.

7. Turn off the gas cylinder and regulator and clean up your work area when you are finished brazing.

INSTRUCTOR'S COMMENTS _____

_____

_____

■ PRACTICE 32-15

Name _____   Date _____

Class _____   Instructor _____   Grade _____

OBJECTIVE: After completing this practice, you should be able to silver braze copper pipe in the 5G horizontally fixed position.

EQUIPMENT AND MATERIALS NEEDED FOR THIS PRACTICE

1. Properly setup, lit, and adjusted air MPS, air propane, or any air fuel gas torch. Be sure that the regulator pressure is set to the manufacturer's specifications for your fuel type and tip size.

2. Proper safety protection (welding goggles, safety glasses, spark lighter, and pliers). Refer to Chapter 2 in the text for more specific safety information.

3. Two or more short random length pieces of 1/2-in. to 1-in. (13-mm to 25-mm) copper pipe with matching copper pipe fittings.

4. Brazing flux and BCuP-2 to BCuP-5 brazing rod.

5. Steel wool, sanding cloth, and/or a wire brush.

INSTRUCTIONS

1. Clean the pipe O.D. and the fitting I.D. using steel wool, sanding cloth, or a wire brush. Then slide the fitting onto the pipe. Be sure that the pipe is seated at the bottom of the fitting.

2. Place the pipe in the 5G position. See Figure 32-64 in the textbook. Heat the pipe first but not too much. When it is hot (but not glowing red), start heating the fitting. Keep the torch moving in order to uniformly heat the entire joint. The joint is at the correct temperature when the braze metal starts to wet the surface.

3. Move the flame to the back side of the pipe and feed the brazing rod into the joint. Move the torch and brazing rod all the way around the pipe and fitting so that the entire joint will be filled. There should be a slight fillet showing all the way around the joint. See Figure 32-58 in the textbook.

4. After cooling, hacksaw the joint apart for inspection. See Figures 32-59 to 32-63 in the text-book.

5. Repeat this practice until you can make defect free brazed pipe joints.

6. Turn off the gas cylinder and regulator and clean up your work area when you are finished brazing.

INSTRUCTOR'S COMMENTS _____

_____

_____

■ PRACTICE 32-16

Name _____   Date _____

Class _____   Instructor _____   Grade _____

OBJECTIVE: After completing this practice, you should be able to silver braze copper pipe in the 2G vertical up position.

EQUIPMENT AND MATERIALS NEEDED FOR THIS PRACTICE

1. Properly setup, lit, and adjusted air MPS, air propane, or any air fuel gas torch. Be sure that the regulator pressure is set to the manufacturer's specifications for your fuel type and tip size.

2. Proper safety protection (welding goggles, safety glasses, spark lighter, and pliers). Refer to Chapter 2 in the text for more specific safety information.

3. Two or more short random length pieces of 1/2-in. to 1-in. (13-mm to 25-mm) copper pipe with matching copper pipe fittings.

4. Brazing flux and BCuP-2 to BCuP-5 brazing rod.

5. Steel wool, sanding cloth, and/or a wire brush.

INSTRUCTIONS

1. Clean the pipe O.D. and the fitting I.D. using steel wool, sanding cloth, or a wire brush. Then slide the fitting onto the pipe. Be sure that the pipe is seated at the bottom of the fitting.

2. Place the pipe in the 2G vertical position. See Figure 32-65 in the textbook. Heat the pipe first but not too much. When it is hot (but not glowing red), start heating the fitting. Keep the torch moving in order to uniformly heat the entire joint. The joint is at the correct temperature when the braze metal starts to wet the surface.

3. Move the flame to the top of the joint and feed the brazing rod into the joint. Move the torch and brazing rod all the way around the pipe and fitting so that the entire joint will be filled. There should be a slight fillet showing all the way around the joint. See Figure 32-58 in the textbook.

4. After cooling, hacksaw the joint apart for inspection. See Figures 32-59 to 32-63 in the textbook.

5. Repeat this practice until you can make defect free brazed pipe joints.

6. Turn off the gas cylinder and regulator and clean up your work area when you are finished brazing.

INSTRUCTOR'S COMMENTS _____

_____

_____

■ PRACTICE 32-17

Name _____ Date _____

Class _____ Instructor _____ Grade _____

OBJECTIVE: After completing this practice, you should be able to silver braze dissimilar metals.

EQUIPMENT AND MATERIALS NEEDED FOR THIS PRACTICE

1. Use the same equipment, materials, setup, and procedures as described in Practice 32-14.

2. All required PPE.

3. BAg or a similar brazing alloy.

4. Two pieces of metal, each a different type of metal, such as copper and bronze, brass and steel, stainless steel and copper, etc.

INSTRUCTIONS

1. If the brazing rod is not flux covered, heat the end of the rod and dip it in the flux.

2. Heat the parts to be joined; apply most of the heat on the larger part to help keep the heating uniform. Overheating can damage the parts.

3. When joining dissimilar metal parts, the fitup may not be ideal. To prevent the brazing metal from falling through any large gaps, frequently flash the flame off of the parts to allow the molten braze metal time to partially solidify.

4. When the complete joint is filled, do not move the parts until the braze alloy has had time to cool through its paste range.

5. Repeat this practice until you can make defect-free joints.

6. Turn off the cylinder, bleed the hoses, back out the regulator adjusting screws, and clean your work area when you are finished.

INSTRUCTOR'S COMMENTS _____

_____

_____

■ PRACTICE 32-18

Name _____   Date _____

Class _____   Instructor _____   Grade _____

OBJECTIVE: After completing this practice, you should be able to make a soldered tee joint in the flat position.

EQUIPMENT AND MATERIALS NEEDED FOR THIS PRACTICE

1. Properly setup, lit, and adjusted air MPS, air propane, or any air fuel gas torch. Be sure that the regulator pressure is set to the manufacturer's specifications for your fuel type and tip size.

2. Proper safety protection (welding goggles, safety glasses, spark lighter, and pliers). Refer to Chapter 2 in the text for more specific safety information.

3. Two or more pieces of 18- to 24-gauge mild steel sheet, 1-1/2 in. (38 mm) wide by 6 in. (152 mm) long.

4. Flux-cored tin-lead or tin-antimony wire solder or solid wire solder and a container of liquid flux.

5. Steel wool, sanding cloth, and/or a wire brush.

INSTRUCTIONS

1. Using the steel wool, sanding cloth, and/or a wire brush, clean the metal surface where the solder joint will be placed.

2. Hold one piece vertically on the other piece and spot solder both ends. If flux cored solder is not being used, paint the liquid flux on at this time.

3. Place the piece in the flat position and point the torch flame in the same direction that you will be soldering. Flash the torch on and off of the joint while adding solder to the joint to control the heat and keep the molten pool small.

4. Continue flashing the torch on and off and adding solder until you reach the end of the joint. When the joint is completed, the solder bead should be uniform.

5. Inspect the bead for uniformity and visual defects.

6. Repeat this practice until you can make defect free soldered joints.

7. Turn off the gas cylinder and regulator and clean up your work area when you are finished brazing.

INSTRUCTOR'S COMMENTS _____

_____

_____

■ PRACTICE 32-19

Name _____   Date _____

Class _____   Instructor _____   Grade _____

OBJECTIVE: After completing this practice, you should be able to make a soldered lap joint in the flat position.

EQUIPMENT AND MATERIALS NEEDED FOR THIS PRACTICE

1. Properly setup, lit, and adjusted air MPS, air propane, or any air fuel gas torch. Be sure that the regulator pressure is set to the manufacturer's specifications for your fuel type and tip size.

2. Proper safety protection (welding goggles, safety glasses, spark lighter, and pliers). Refer to Chapter 2 in the text for more specific safety information.

3. Two or more pieces of 18- to 24-gauge mild steel sheet, 1-1/2 in. (38 mm) wide by 6 in. (152 mm) long.

4. Flux cored tin-lead or tin-antimony wire solder or solid wire solder and a container of liquid flux.

5. Steel wool, sanding cloth, and/or a wire brush.

INSTRUCTIONS

1. Using the steel wool, sanding cloth, and/or a wire brush, clean the metal surface where the solder joint will be placed.

2. Place one piece tightly on the other piece in a lap joint configuration and spot solder (tack) both ends, Figure 32-68. If flux cored solder is not being used, paint the liquid flux on at this time.

3. Place the piece in the flat position and point the torch flame in the same direction that you will be soldering. Flash the torch on and off of the joint while adding solder to the joint to control the heat and keep the molten pool small.

4. Continue flashing the torch on and off and adding solder until you reach the end of the joint. When the joint is completed, the solder bead should be uniform.

5. Inspect the bead for uniformity and visual defects.

6. Repeat this practice until you can make defect free soldered joints.

7. Turn off the gas cylinder and regulator and clean up your work area when you are finished brazing.

INSTRUCTOR'S COMMENTS _____

_____

_____

■ PRACTICE 32-20

Name _____   Date _____

Class _____   Instructor _____   Grade _____

OBJECTIVE: After completing this practice, you should be able to solder copper pipe in the 2G vertical down position.

EQUIPMENT AND MATERIALS NEEDED FOR THIS PRACTICE

1. Properly setup, lit, and adjusted air MPS, air propane, or any air fuel gas torch. Be sure that the regulator pressure is set to the manufacturer's specifications for your fuel type and tip size.

2. Proper safety protection (welding goggles, safety glasses, spark lighter, and pliers). Refer to Chapter 2 in the text for more specific safety information.

3. Two or more short random length pieces of 1/2-in. to 1-in. (13-mm to 25-mm) copper pipe with matching copper pipe fittings.

4. Solid tin-lead or tin-antimony wire solder and a container of liquid flux.

5. Steel wool, sanding cloth, and/or a wire brush.

INSTRUCTIONS

1. Clean the pipe O.D. and the fitting I.D. using steel wool, sanding cloth, or a wire brush. Then apply the liquid flux to the pipe and the fitting.

2. Slide the fitting onto the pipe. Be sure that the pipe is seated at the bottom of the fitting. Twist the pipe in the fitting to be sure that the flux is being applied completely around the inside of the joint.

3. Make a bend about 3/4 in. (19 mm) from the end of the wire solder. Use this bend as a gauge so that you do not put too much solder into the joint.

4. Place the pipe in the 2G vertical down position, Figure 32-69. Heat the pipe and fitting uniformly. The joint is at the correct temperature when the solder starts to wet the surface.

5. When the solder starts to wet, move the flame away from the pipe and feed the solder into the joint, wiping it all the way around the joint. See Figure 32-70 in the textbook.

6. After cooling, hacksaw and pry the joint apart for inspection. See Figure 32-71 in the textbook.

7. Repeat this practice until you can make defect free soldered pipe joints.

8. Turn off the gas cylinder and regulator and clean up your work area when you are finished brazing.

INSTRUCTOR'S COMMENTS _____

_____

_____

■ PRACTICE 32-21

Name _____ Date _____

Class _____ Instructor _____ Grade _____

OBJECTIVE: After completing this practice, you should be able to solder copper pipe in the 1G horizontal rolled position.

EQUIPMENT AND MATERIALS NEEDED FOR THIS PRACTICE

1. Properly setup, lit, and adjusted air MPS, air propane, or any air fuel gas torch. Be sure that the regulator pressure is set to the manufacturer's specifications for your fuel type and tip size.

2. Proper safety protection (welding goggles, safety glasses, spark lighter, and pliers). Refer to Chapter 2 in the text for more specific safety information.

3. Two or more short random length pieces of 1/2-in. to 1-in. (13-mm to 25-mm) copper pipe with matching copper pipe fittings.

4. Solid tin-lead or tin-antimony wire solder and a container of liquid flux.

5. Steel wool, sanding cloth, and/or a wire brush.

INSTRUCTIONS

1. Clean the pipe O.D. and the fitting I.D. using steel wool, sanding cloth, or a wire brush. Then apply the liquid flux to the pipe and the fitting.

2. Slide the fitting onto the pipe. Be sure that the pipe is seated at the bottom of the fitting. Twist the pipe in the fitting to be sure that the flux is being applied completely around the inside of the joint.

3. Make a bend about 3/4 in. (19 mm) from the end of the wire solder. Use this bend as a gauge so that you do not put too much solder into the joint.

4. Place the pipe in the 1G horizontal rolled position. Heat the pipe and fitting uniformly. The joint is at the correct temperature when the solder starts to wet the surface.

5. When the solder starts to wet, move the flame away from the pipe and feed the solder into the joint, wiping it from the 10 o'clock to the 2 o'clock position. Roll the pipe 90° and repeat. See Figure 32-72 in the textbook.

6. After cooling, hacksaw and pry the joint apart for inspection. See Figure 32-71 in the textbook.

7. Repeat this practice until you can make defect free soldered pipe joints.

8 Turn off the gas cylinder and regulator and clean up your work area when you are finished brazing.

INSTRUCTOR'S COMMENTS _____

_____

_____

■ PRACTICE 32-22

Name _____   Date _____

Class _____   Instructor _____   Grade _____

OBJECTIVE: After completing this practice, you should be able to solder copper pipe in the 4G vertical up position.

EQUIPMENT AND MATERIALS NEEDED FOR THIS PRACTICE

1. Properly setup, lit, and adjusted air MPS, air propane, or any air fuel gas torch. Be sure that the regulator pressure is set to the manufacturer's specifications for your fuel type and tip size.

2. Proper safety protection (welding goggles/safety glasses, spark lighter, and pliers). Refer to Chapter 2 in the text for more specific safety information.

3. Two or more short random length pieces of 1/2-in. to 1-in. (13-mm to 25-mm) copper pipe with matching copper pipe fittings.

4. Solid tin-lead or tin-antimony wire solder and a container of liquid flux.

5. Steel wool, sanding cloth, and/or a wire brush.

INSTRUCTIONS

1. Clean the pipe O.D. and the fitting I.D. using steel wool, sanding cloth, or a wire brush. Then apply the liquid flux to the pipe and the fitting.

2. Slide the fitting onto the pipe. Be sure that the pipe is seated at the bottom of the fitting. Twist the pipe in the fitting to be sure that the flux is being applied completely around the inside of the joint.

3. Make a bend about 3/4 in. (19 mm) from the end of the wire solder. Use this bend as a gauge so that you do not put too much solder into the joint.

4. Place the pipe in the 4G vertical up position. Heat the pipe and fitting uniformly. The joint is at the correct temperature when the solder starts to wet the surface.

5. When the solder starts to wet, move the flame away from the pipe and feed the solder into the joint, wiping it all the way around the joint.

6. After cooling, hacksaw and pry the joint apart for inspection. See Figure 32-71 in the textbook.

7. Repeat this practice until you can make defect free soldered pipe joints.

8. Turn off the gas cylinder and regulator and clean up your work area when you are finished brazing.

INSTRUCTOR'S COMMENTS _____

_____

_____

■ PRACTICE 32-23

Name _____   Date _____

Class _____   Instructor _____   Grade _____

OBJECTIVE: After completing this practice, you should be able to solder aluminum to copper.

EQUIPMENT AND MATERIALS NEEDED FOR THIS PRACTICE

1. Properly setup, lit, and adjusted air MPS, air propane, or any air fuel gas torch. Be sure that the regulator pressure is set to the manufacturer's specifications for your fuel type and tip size.

2. Proper safety protection (welding goggles/safety glasses, spark lighter, and pliers). Refer to Chapter 2 in the text for more specific safety information.

3. One or more pieces of 1/8-in. (3-mm) by 1-1/2-in. (38-mm) square aluminum plate and a copper penny.

4. Solid tin-lead or tin-antimony wire solder and a container of liquid flux.

5. Steel wool, sanding cloth, and/or a wire brush.

INSTRUCTIONS

1. Clean the surface of the aluminum plate and the penny until all coatings and/or oxides have been completely removed.

2. Tin the aluminum by heating it slightly and then dripping molten solder on it. Do this by heating the solder in the flame. Use no flux. Rub the molten solder around on the aluminum using a piece of steel wool. Continue to heat the plate and rub the solder with the steel wool until the surface is tinned. See Figure 32-73 in the textbook.

3. Repeat step 2 with the copper penny. Flux should be used on the penny.

4. Place the tinned surface of the penny in contact with the tinned surface of the aluminum. See Figure 32-74 in the textbook.

5. Heat the two until the solder melts and flows out from between the penny and the plate.

6. After cooling, check the bond by trying to break the joint.

7. Repeat this practice until you can make defect free aluminum to copper soldered joints.

8. Turn off the gas cylinder and regulator and clean up your work area when you are finished brazing.

INSTRUCTOR'S COMMENTS _____

_____

_____

CHAPTER 32: QUIZ 1

Name _____ Date _____

Class _____ Instructor _____ Grade _____

INSTRUCTIONS

Carefully read Chapter 32 in the textbook and answer each question.

MATCHING

In the space provided to the left of Column A, write the letter from Column B that best answers or completes the statement in Column A.

Column A	Column B
_____ 1. Brazing and soldering are both classified by the American Welding Society as _____, which means that the filler metal is melted and the base material or materials are not melted.	a. low-temperature
_____ 2. Brazing occurs at a temperature _____ 840°F (450°C), and the parts being joined must be fitted so that the joint spacing is very small, approximately 0.025 in. (0.6 mm) to 0.002 in. (0.06 mm).	b. overlapping
_____ 3. Braze welding can be done with the same brazing alloys as brazing, so the only difference is it does not need _____ to pull filler metal into the joint.	c. fluxes
_____ 4. Soldering takes place at temperatures _____ 840°F (450°C).	d. oxides
_____ 5. An advantage of brazing and soldering is that because the base metal does not have to melt, a _____ heat source can be used.	e. paste range
_____ 6. A brazed joint can be made that has a _____ strength four- to five-times higher than the filler metal itself.	f. wetting
_____ 7. For a solder or braze joint, the shear strength depends on the amount of _____ area of the base parts.	g. molten metal bath
_____ 8. For most soldered or brazed joints, fatigue resistance is usually fairly _____.	h. below
_____ 9. The compatibility of the base materials to the _____ will determine the corrosion resistance.	i. tensile
_____ 10. Fluxes used in brazing and soldering must remove any _____ that form as a result of heating the parts.	j. induction
_____ 11. _____ are available in many forms, such as solids, powders, pastes, liquids, sheets, rings, and washers, and are also available mixed with the filler metal, inside the filler metal, or on the outside of the filler metal.	k. above

—— 12. Brazing and soldering fluxes will remove light surface oxides, promote _____, and aid in capillary action.

—— 13. Brazing and soldering methods are grouped according to the method with which _____ is applied: torch, furnace, induction, dipped, or resistance.

—— 14. An advantage of torch brazing is that a torch is very _____.

—— 15. In the _____ and soldering method, the parts are heated to their brazing or soldering temperature by passing them through or putting them into a furnace.

—— 16. The _____ method of heating uses a high-frequency electrical current to establish a corresponding current on the surface of the part.

—— 17. Two types of dip brazing or soldering are used: molten flux bath and _____.

—— 18. The _____ method of heating uses an electric current that is passed through the part, and the resistance of the part to the current flow results in the heat needed to produce the bond.

—— 19. The ultrasonic method uses high-frequency _____ to produce the bond or to aid with heat in the bonding.

—— 20. A _____ is the temperature range in which a metal is partly solid and partly liquid as it is heated or cooled.

l. portable

m. sound waves

n. filler metal

o. resistance

p. capillary action

q. furnace brazing

r. liquid-solid phase bonding processes

s. heat

t. low

CHAPTER 32: QUIZ 2

Name _____   Date _____

Class _____   Instructor _____   Grade _____

INSTRUCTIONS
Carefully read Chapter 32 in the textbook and answer each question.

MATCHING
In the space provided to the left of Column A, write the letter from Column B that best answers or completes the statement in Column A.

Column A	Column B
_____ 1. Soldering alloys are usually identified by their major _____ elements.	a. expensive
_____ 2. _____ solders must not be used where lead could become a health hazard in things such as food and water.	b. elevated
_____ 3. Tin-antimony is the most common solder used in _____ because it is lead-free.	c. sil-phos
_____ 4. Cadmium-silver solder alloys have excellent wetting, flow, and strength characteristics, but they are _____.	d. alloying
_____ 5. The American Welding Society's classification system for brazing alloys uses the letter B to indicate that the alloy is to be used for _____.	e. brazing rods
_____ 6. Copper-zinc alloys are the most popular brazing alloys and are available as regular and _____ alloys.	f. hard spots
_____ 7. The copper-zinc filler rods are often grouped together and known as _____.	g. silver-based
_____ 8. Copper-phosphorus and copper-silver-phosphorus are sometimes referred to as phos-copper or _____.	h. powdered
_____ 9. Silver-copper alloys can be used to join almost any metal, ferrous or nonferrous, except aluminum, magnesium, zinc, and a few other _____ metals.	i. tin-lead
_____ 10. Nickel alloys are used for joining materials that need high strength and corrosion resistance at an _____ temperature.	j. corrosive
_____ 11. Nickel and nickel alloys are increasingly used as a substitute for _____ alloys.	k. low-fuming
_____ 12. Aluminum-silicon brazing filler metals can be used to join most _____ sheet and cast alloys.	l. overheat

——— 13. Silver and gold are both used in small quantities when joining metals that will be used under _____ conditions, when high joint ductility is needed, or when low electrical resistance is important.

m. plumbing

——— 14. The _____ between the parts being joined greatly affects the tensile strength of the finished part.

n. brittle

——— 15. Using prefluxed rods is easier for students than using _____ flux, which has to have the hot tip of the brazing rod dipped to apply.

o. chemical or solvent

——— 16. Surface cleaning may be as simple as wiping it off with a shop rag or as extensive as _____ cleaning.

p. moving

——— 17. Unlike welding or hardsurfacing, braze buildup has no _____ that make remachining difficult.

q. brazing

——— 18. To prevent overheating with an oxyacetylene flame, keep the torch _____ and hold the flame so that the inner cone is approximately 1 in. (25 mm) from the surface.

r. spacing

——— 19. Most BCuP alloys can be used to join many dissimilar metals, but you should avoid using them on ferrous metals such as steels and stainless steels, because the phosphorus can form a _____ joint.

s. aluminum

——— 20. Both tin-lead and tin-antimony solders have a low melting temperature, and if an oxyacetylene torch is used, it is very easy to _____ the solder.

t. low-melting

CHAPTER 32: QUIZ 3

Name _____   Date _____

Class _____   Instructor _____   Grade _____

INSTRUCTIONS

Carefully read Chapter 32 in the textbook and answer each question.

IDENTIFICATION

Identify the numbered items on the drawing by writing the letter next to the identifying term in the space provided.

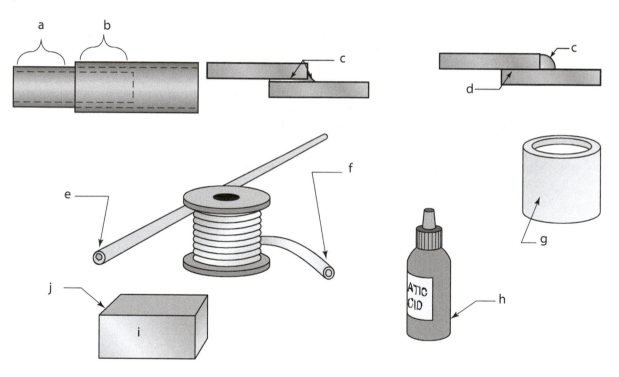

1. _____   FLUX-COVERED BRAZING ROD

2. _____   BRAZE METAL

3. _____   FLUX-CORED SOLDERING WIRE

4. _____   HEAT HERE 1st

5. _____   SOLID FLUX

6. _____   LIQUID FLUX

7. _____   HEAT HERE 2nd

8. _____   POWDERED FLUX

9. _____   NO BRAZE METAL

10. _____   SAL AMMONIAC

CHAPTER 32: QUIZ 4

Name _____ Date _____

Class _____ Instructor _____ Grade _____

INSTRUCTIONS

Carefully read Chapter 32 in the textbook and answer each question.

IDENTIFICATION

Identify the numbered items on the drawing by writing the letter next to the identifying term in the space provided.

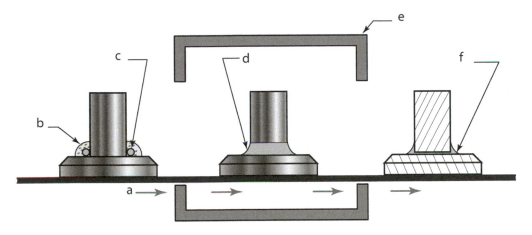

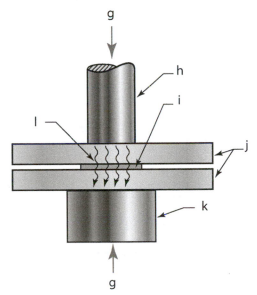

1. _____ FILLER MELTS AND FLOWS

2. _____ FORCE

3. _____ SOUND WAVES

4. _____ ANVIL

5. _____ FURNACE

6. _____ CONVEYOR

7. _____ FILLER ALLOY

8. _____ PARTS BEING JOINED

9. _____ TRANSDUCER

10. _____ BRAZED JOINT

11. _____ FILLER METAL

12. _____ FLUX

NOTES